SYNUCLEIN AND THE COELACANTH

SYNUCLEIN AND THE COELACANTH

The Molecular and Evolutionary Origins of Parkinson's Disease

JAMES M. GRUSCHUS
Molecular Biophysicist, Washington, DC, United States

ELSEVIER
ACADEMIC PRESS
An imprint of Elsevier

Academic Press is an imprint of Elsevier
125 London Wall, London EC2Y 5AS, United Kingdom
525 B Street, Suite 1650, San Diego, CA 92101, United States
50 Hampshire Street, 5th Floor, Cambridge, MA 02139, United States
The Boulevard, Langford Lane, Kidlington, Oxford OX5 1GB, United Kingdom

Illustrations Courtesy of Maria Mills

Notices
Knowledge and best practice in this field are constantly changing. As new research and experience broaden our understanding, changes in research methods, professional practices, or medical treatment may become necessary.

Practitioners and researchers must always rely on their own experience and knowledge in evaluating and using any information, methods, compounds, or experiments described herein. In using such information or methods they should be mindful of their own safety and the safety of others, including parties for whom they have a professional responsibility.

To the fullest extent of the law, neither the Publisher nor the authors, contributors, or editors, assume any liability for any injury and/or damage to persons or property as a matter of products liability, negligence or otherwise, or from any use or operation of any methods, products, instructions, or ideas contained in the material herein.

Library of Congress Cataloging-in-Publication Data
A catalog record for this book is available from the Library of Congress

British Library Cataloguing-in-Publication Data
A catalogue record for this book is available from the British Library

ISBN: 978-0-323-85707-9

For information on all Academic Press publications visit our website at https://www.elsevier.com/books-and-journals

Publisher: Nikki Levy
Acquisitions Editor: Melanie Tucker
Editorial Project Manager: Kristi Anderson
Production Project Manager: Kiruthika Govindaraju
Cover Designer: Alan Studholme

Typeset by TNQ Technologies

To the memories of my grandmother Gladys Gruschus, my father Jack Gruschus, Sr., and my aunt Suzie Gouett.

Contents

Preface

"The cause of Parkinson's disease is unknown" (*Wikipedia, The Free Encyclopedia*, 9 May 2020). This sentiment pervades the current scientific literature on Parkinson's disease. My hope in writing this book is not just to explain the science of Parkinson's disease in layman's terms, but also to show how the disparate fields of Parkinson's research, from molecular biophysics to neuroanatomy, are all pieces of a larger puzzle that, when put together, explain the origin and cause of the disease.

Why do only humans develop Parkinson's disease? Other neurodegenerative diseases have animal parallels, Alzheimer's in chimpanzees, multiple sclerosis in macaques, and Lou Gehrig's disease in dogs, but nothing like Parkinson's has been seen in any species but humans. A protein shared with our closest ocean-dwelling, cold-blooded relative lies at the heart of this mystery. *Synuclein and the Coelacanth* delves into the causes of Parkinson's disease and how the evolution of the human brain has left us uniquely vulnerable. Genetic risk factors, environmental toxins, and neuroanatomy are woven together in a journey that goes from subatomic physics to socioeconomics. Surprising connections among neurodegenerative disease, intelligence, and innate immunity are explored. Follow the storied history of neurosurgery, its darker episodes, and its bold successes, including deep brain stimulation. Finally, new therapeutic agents are being developed that hope to go beyond just treating the symptoms of Parkinson's and actually halt the disease. The journey ends where it started, with the mystery of the disease's unique origins revealed and new avenues for science to explore.

This book is written for academics and nonacademics alike, for medical professionals and those who just want to learn more about Parkinson's disease. So sit back, get comfortable, and enjoy the journey!

Acknowledgments

Big thanks to my cousin Jonathan Hulse for interviewing my aunt and transcribing the dialog, and to Ehud Goldin and especially Mike Allen for their help with the manuscript.

CHAPTER 1

Out of the Depths – The synuclein proteins, their evolution, and the genetic code

Still.
Drifting.
Flashes of bioluminescence.
A presence approaching. Squid? Lunge, snap!
Too far, it got away.
Drifting.
Still.

The coelacanth is a large fish, with females typically reaching 5 feet or more. They feed at night, drift feeders, floating on the current, waiting for prey to swim close. During the day they sleep in underwater caves. Long, somewhat oval in shape, they are covered with large, dark blue-gray scales sparsely speckled with lighter colored ones. They have curious tails, with an extension down the middle that ends in an additional minitail. In 1938, when the ichthyologist J.L.B. Smith was shown a mysterious fish caught off the South African coast, the distinctive tail was one of the first things he recognized. He identified the coelacanth not by its similarity to other living fish, but by its similarities to fossils of species thought extinct for over 60 million years.

Synuclein and the Coelacanth
ISBN 978-0-323-85707-9
https://doi.org/10.1016/B978-0-323-85707-9.00005-8

The coelacanth is a living fossil with a unique status. All other living fossils, gingko trees, horseshoe crabs, and so on, were known from living species first and fossils second. There are two living species, the Western Indian Ocean coelacanth, *Latimeria chalumnae*, and the Indonesian coelacanth, *Latimeria menadoensis*. The *Latimeria* part of the name honors Marjorie Courtenay-Latimer, the museum curator who saved the first coelacanth specimen for examination by J.L.B. Smith, whereas *chalumnae* and *menadoensis* are nods to the locations where the first specimens of each species were found, near the Chalumna River of South Africa and the Indonesian island Manado. They typically live near underwater volcanic slopes, which might explain their absence from the recent fossil record. I imagine lava is not conducive to fossil formation.

By the way, the captain of the fishing boat that caught the coelacanth in 1938 was Hendrik Goosen. As far as I know, the only lasting acknowledgment of his contribution to science is his name on a commemorative brass plate at a dock in East London, South Africa, which, adding insult to injury, is named Latimer's Landing. At least, they could have named the dock after the fisherman.

Coelacanths have a second unique distinction. Among the cold-blooded denizens of the oceans, they are our closest relatives. Four of their fins are especially thick, two below in front and two below in the back. Together with the freshwater lungfishes, they are the lobe-finned fishes, the closest relatives of land vertebrates, the tetrapods, which include all amphibians, reptiles, and birds, as well as mammals. Because of this close relationship, and likely their public relations—friendly living fossil status, the coelacanth was chosen for full DNA sequencing of its genome.

The results of the genome sequencing were reported in 2013 in a paper authored by a multinational team of nearly 100 researchers [1]. They confirmed that coelacanths have in fact evolved more slowly than tetrapods. It has been suggested that this slow evolution is the result of the lack of evolutionary pressures in their environment, but I suspect it has more to do with their long life span, estimated at greater than 60 years, and the correspondingly slow reproductive rate. Their genome is around three billion bases long, in which a base corresponds to a single bit of genetic information, and is comparable in size to human and many other tetrapod genomes. The West African lungfish, by contrast, has an unusually large genome, greater than 50 billion bases. Because of this, only a small number of genes from the lungfish were chosen for sequencing. Comparison of lungfish and coelacanth genes with the corresponding ones in tetrapods suggest that lungfish are actually slightly more closely related to us.

The report on the sequencing of the coelacanth genome also sought to identify genes that they have in common with other bony fishes but which have been lost in tetrapods, positing that the lost genes must have been replaced by other existing or novel genes helpful for adapting to land. Of genes involved with body and organ development, the region losing the most genes was the brain. However, it was one set of brain proteins that was *not* lost that first piqued my own research interest in the coelacanth: the synucleins.

Synucleins are small proteins found primarily in neurons. They are thought to be part of the cellular machinery involved in the release of neurotransmitters, the chemical messengers at synapses that allow neurons to communicate with each other. Several other cell types have synucleins, including red and white blood cells and platelets, chromaffin cells, which release adrenaline into the bloodstream, and the beta cells of pancreatic islets, which release insulin, to name a few. The tumor cells of some cancers, such as breast cancer and melanoma, possess synucleins, whereas most often the corresponding noncancerous tissues have little or none. Synucleins are specific to vertebrates; no invertebrates have them, not even sea squirts and lancelets, the closest relatives of vertebrates.

Typically, species have three versions of synuclein proteins, which in coelacanths and humans are named alpha-, beta-, and gamma-synuclein. In 1988, the first synuclein was discovered in the course of research on the electricity-producing organ of the Pacific electric ray *Torpedo californicus* [2]. The protein was observed in high amounts in synapses and cell nuclei of neurons; hence the name, *sy* for synapse and *nuclein* for nucleus. Ironically, later studies failed to find high levels of synuclein in neuronal cell nuclei, and it has been suggested that the original discovery might have used antibodies specific to both synuclein and an unknown nuclear protein [3]. But it was too late; the name "synuclein" stuck.

In 1993, a synuclein similar to that in the electric ray was found in humans, discovered using a protein fragment found in senile plaque from brains of deceased patients who had had Alzheimer's disease [4]. This region of synuclein is called the nonamyloid-beta component region, because Alzheimer amyloid plaques are primarily made from another protein fragment called amyloid-beta. Then, in 1997, researchers at the National Human Genome Research Institute, National Institutes of Health (NIH) found a mutation in human synuclein that led to familial early-onset Parkinson's disease [5]. The human synuclein was alpha-synuclein and is now known to have a central role in the development of Parkinson's disease.

It was not just that coelacanths retained the same synucleins as humans that initially caught my attention, it was that the proteins were so similar, especially the alpha and beta versions. Proteins consist of chains of amino acids, and the sequence of amino acids for coelacanth alpha-synuclein is 83% identical to the human sequence, 85% identical for beta-synuclein, but only 53% for gamma-synuclein. In contrast, the alpha- and beta-synuclein sequences of our more distantly related watery cousins, the ray-finned fish, are less than 60% identical. What is behind the surprising similarity of human and coelacanth alpha- and beta-synuclein?

To put this into perspective, consider another vertebrate-specific protein, hemoglobin, the oxygen carrying protein in our blood. This is clearly an important protein, and yet the coelacanth and human amino acid sequences are only 51% identical. Comparing other corresponding pairs of proteins in humans and coelacanths, I found similar variations, from around 40% to 100% identity. What causes some coelacanth proteins, such as alpha- and beta-synuclein, to be more similar to their human counterparts than others?

Protein sequences are encoded by the sequence of bases in their corresponding DNA genes. In each generation, there is a chance that mutations (that is, alterations, deletions or additions of one or more bases) can occur in the DNA of egg or sperm cells. These mutations then have a chance to be passed on to the next generation. If the mutation occurs in an important location, changing an important amino acid in a protein, for instance, progeny with this mutation could be less fit or have some reproductive disadvantage. On the other hand, DNA mutations that cause inconsequential changes in the amino acid sequence do not affect reproductive fitness. It is even possible for the mutation to confer some sort of advantage. Thus, over time, proteins can accumulate harmless mutations, a process called genetic drift, as well as favorable mutations, which is called positive selection. The unlucky progeny carrying unfavorable mutations will be less likely to pass them on, which is called negative selection, and this weeding out process is what leads to certain protein sequences being more conserved. All of these processes are just manifestations of survival of the fittest, but that is not the key point. What I really want to explore is why some genes are less tolerant to mutations than others.

Several kinds of mutations are harmful to proteins. Understanding why requires learning some basics of protein molecular structure. The sequence of amino acids dictates, for the most part, the structure of the protein molecule, or the shape it folds into. There are 20 amino acids, and each is like a particular kind of bead, and a protein is a string of those beads. For instance, certain amino acids interact favorably with water whereas other amino acids do not. Those that like water are called hydrophilic and those that do not, hydrophobic. When the chain of amino acids is in a watery environment inside the cell, the trick is to keep the bits that like water at the surface of the protein, and the unfavorable bits tucked away inside.

These hydrophobic and hydrophilic bits are called side chains because they branch from the main molecular chain, called the peptide backbone, which connects the sequence of amino acids of the protein. Each amino acid has a side chain that determines which environment it prefers. If a mutation occurs that replaces a hydrophobic side chain with a hydrophilic one at the center of the protein, it can destabilize the structure. In addition, the function of a protein molecule typically depends on its assuming the correct shape. For enzymes, proteins that catalyze chemical reactions, maintaining the correct shape of the pocket where the chemical binds is critical. For two proteins that interact, the correct interface between the proteins is most important, like fitting together a pair of three-dimensional Tetris pieces. Any mutation that changes the shape so that the two proteins do not fit together, or the chemical no longer fits, is unfavorable.

I am a molecular biophysicist. In particular, I determine the structure of proteins and their complexes. For several years, I helped with beta testing software, written by Xiongwu Wu in the laboratory of Bernie Brooks at the NIH. The software predicts how two proteins interact to form a complex. Complementary shape is clearly the most important factor, but other aspects of the protein surfaces make important contributions, as well. Often, proteins have not just hydrophobic amino acids in the protein center but also a few at the surface. When two proteins interact, the hydrophobic bits of one protein surface tend to pair with hydrophobic amino acids on the surface of the interacting protein. Certain hydrophilic amino acids have electric charge, some positive, others negative. Opposites attract, and matching positive surface regions with negative ones on the interacting proteins is also important.

So how does all this help explain why some protein sequences, such as alpha- and beta-synuclein, are more conserved, accumulating fewer mutations over time? The answer seems to be how many interaction partners the protein has. In cells, we have the amusing situation in which the more promiscuous the protein is in its interactions, the more conservative its sequence is. To illustrate, consider the calcium signaling protein calmodulin. Cells use calcium signaling for numerous processes, such as muscle contraction and the firing of neurons. In the presence of calcium, calmodulin alters its shape, exposing more hydrophobic side chains, priming it for interaction with the dozens, perhaps hundreds, of its protein partners, including alpha-synuclein. Its coelacanth and human sequences are 100% identical. Every amino acid is important for its structure, its calcium binding, and/or its protein interactions. In systems biology jargon, such proteins are "hub" proteins, important nodes in the network of protein interactions.

In contrast to calmodulin, hemoglobin is not a hub protein. Apart from its oxygen and carbon dioxide transport duties, it apparently has little interaction with other proteins. Alpha- and beta-synuclein lie in between; they seem to have important interaction partners, not quite as many as calmodulin, but enough to result in higher conservation of their sequences. Proteins can also have amino acid sequence variations within a species. These nonharmful variations are called polymorphisms, and most protein sequences have them, although for the same reasons described earlier, hub proteins have fewer polymorphisms. In addition, for many proteins, such as the one involved in cystic fibrosis, having one mutated copy of the gene does not cause the disease, because the other, nonmutated copy provides enough of the working protein to keep cells healthy. For some of these mutations, such as the hemoglobin sickle cell mutation, having just one mutant copy from a parent can even confer a protective effect, against malaria in this case, although having two is still harmful. So far, no nonharmful polymorphisms in the protein sequences of alpha- and beta-synuclein have been found; instead, all known mutant sequences discovered so far are rare and appear to be potentially harmful, even when just one gene copy is mutated.

There is another potential dimension to protein conservation: the DNA sequence of the gene that encodes the protein and the corresponding messenger RNA (mRNA) that carries this information to be translated by the cell's protein synthesizing machinery. DNA is made from four types of bases, designated by the letters A, C, G, and T, which perhaps you have seen in diagrams of the DNA double-helix structure, described in 1953 by James Watson and Francis Crick using x-ray data from Maurice Wilkins and Rosalind Franklin [6]. The sequence of DNA bases of the gene encodes its corresponding protein, but there are additional bases before, after, and in between the coding parts that also help direct whether and when the gene should be read, or transcribed, into mRNA.

mRNA is also made up of four bases, similar to the corresponding four DNA bases, but in a single chain, rather than a double helix. Its structure is more flexible and can fold in interesting ways, sometimes interacting with other RNA molecules that also have a role in gene regulation. The structure that mRNA assumes, as well as its base sequence, helps direct cellular machinery that cuts out the noncoding sequences, splicing the coding parts into one continuous mRNA strand that then becomes translated into the protein sequence by another cellular complex of RNA and proteins called a ribosome. You might expect that the spliced mRNA sequence must be just as conserved as the protein it encodes, but this is not true.

Proteins are chains in which each link is one of the 20 types of amino acids, but mRNA is a chain made of only four kinds of bases. How can sequences using just four bases of mRNA encode a chain made of 20 different kinds of amino acids? The trick is to use triplets of bases; one amino acid is encoded by three bases in the mRNA sequence. This is the genetic code, and I had the honor of working with one of its discoverers, Nobel Laureate Marshall Nirenberg. In the 1990s, when I was in the laboratory of James Ferretti at the NIH, my colleague Désirée Tsao and I determined the structure of a protein–DNA complex that Nirenberg was studying, involved in fruit fly neural development.

When we would go to his laboratory and present our results to Nirenberg, I remember him always saying, "Well, I think that's just terrific!" He was a pleasure to work with. In 1961, over 30 years before I met him, Nirenberg, along with postdoctoral fellow J. Heinrich Matthaei, discovered that when a long chain of RNA made with the base uracil (which corresponds to the T, or thymine, base in DNA) was added to a cell free protein expression system, a protein chain consisting of just the amino acid phenylalanine was synthesized [7]. At the time, the idea that triplets of bases coded for the 20 amino acids was already established; they just did not know which triplets corresponded to which amino acids. Their discovery that the triplet TTT coded the amino acid phenylalanine was the very first piece of the genetic code.

There are 64 ways to arrange four kinds of bases into groups of three: four in the first position times four in the second times four in the third ($4 \times 4 \times 4 = 64$). Three of these base triplets do not code for amino acids; instead, they signal the end of the sequence. The other 61 triplets code for the 20 amino acids. Two amino acids have just one corresponding triplet: for example tryptophan, whose DNA triplet is TGG. Others have up to six triplets, such as serine, which is coded by AGT, AGC, TCT, TCC, TCA, and TCG. Codons that encode the same amino acid are called synonymous. Because of this redundancy, with several mRNA triplets encoding the same amino acid, the DNA and mRNA sequences do not necessarily have as much sequence identity between species as their corresponding proteins. If you took an amino acid sequence and you randomly chose for each amino acid a corresponding triplet, and then repeated the process, again randomly choosing a corresponding triplet for each amino acid, and then compared the two resulting mRNA sequences, only about three of four bases would be the same; that is, the two mRNA sequences would be about 75% identical on average.

So how do the coelacanth and human mRNA synuclein sequences compare? Recall that the amino acid sequences of alpha-synuclein are 83%

identical for the two species. If we randomly chose corresponding mRNA triplets, for the identical amino acids we would expect roughly 75% mRNA agreement. For the nonidentical amino acids, there would be only about 25% agreement, because there are four bases from which to randomly choose. Using these percentages, the mRNA identity between human and coelacanth alpha-synuclein should be around 67%.[1]

The actual coelacanth and human alpha-synuclein mRNA sequences are 79% identical, far higher than would be predicted randomly choosing mRNA triplets, and 79% mRNA identity means the coding regions of the DNA gene have 79% identity as well. For comparison, consider calmodulin with its 100% identical protein sequences for human and coelacanth. The corresponding mRNA sequences are 87% identical, more than the 75% obtained from randomly choosing synonymous codons, but considerably less than 100%. For alpha-synuclein, the drop in sequence identity from protein to mRNA is just four percentage points, from 83% to 79%. I admit there are nuances in the mathematics that I have glossed over. Still, the percentage of identity in the alpha-synuclein mRNA and DNA sequences suggests that at least some of the amino acid conservation might actually be due to conservation at the DNA—mRNA level; that is, there could also be important cellular interactions with the alpha-synuclein DNA and/or mRNA sequences that are intolerant of mutations.

Important protein and DNA—mRNA interactions can explain why alpha- and beta-synuclein have more conserved sequences, but why were synucleins retained in tetrapod brains when many other brain development proteins were lost or replaced? An important clue comes from the synucleins in other fish. The other fish consist of the ray-finned fishes, which include all living bony fishes except for coelacanths and lungfish; cartilaginous fishes, which include sharks, skates, and rays; and the jawless fishes, which include lampreys and hagfish. The jawless fishes split from the fish evolutionary tree first, followed by the cartilaginous fishes, and then the bony fishes split into ray-finned and lobe-finned fishes.

In the lamprey, there are three synuclein proteins: one closer in sequence to human alpha- and beta-synucleins (59% and 57% identical, respectively), one similar to gamma-synuclein (67%), and one other with less than 50% identity to all three human synucleins. Three cartilaginous fish genomes have been sequenced: the Australian ghost shark, the whale shark, and the Greenland shark. They have alpha-, beta-, and

[1] That is, 75% of 83% plus 25% of 17%, equaling 67%.

gamma-synuclein, like tetrapods and coelacanths; for instance, the ghost shark has 72%, 69%, and 59% identity with human alpha-, beta-, and gamma-synuclein proteins.[2] About 20 ray-finned fish genomes have been sequenced. Curiously, Although they all have beta-synuclein, many lack alpha-synuclein, and they have two forms of gamma-synuclein. Their synucleins have diverged the furthest from human synucleins. The Japanese rice fish, medaka, is one that has all four synucleins, with sequence identities of 58% and 58% to human alpha- and beta-synucleins, respectively, and 50% and 46% for its two gamma-synucleins to human gamma-synuclein.

Based on these sequence comparisons, it appears that after the ray-finned and lobe-finned fishes split, the ray-finned fish synucleins started evolving in a different direction. This suggests that to understand why human synucleins are so similar to coelacanth synucleins, one might ask what brain or neurological traits differ between ray-finned and lobe-finned fishes, but that tetrapods and lobe-finned fishes share. One such trait is obvious. To examine it, let us revisit our coelacanth.

The light above brightens.
Dawn is approaching.
Time to return to the cave.

In videos of coelacanths, there is something clearly different in how they swim compared with other fish. Their curious tail twists clockwise and counterclockwise. Their thick lobe fins wave up and down, the left front and right back fins together, and then right front and left back together. The motion of the fins looks like tetrapod motion on land.

How did coelacanths evolve this way of swimming? Most likely, the common ancestor we share with coelacanths did not live where coelacanths live today, but in shallow seas. Imagine such a sea, thick with fronds and stalks of waving seaweed. Having strong fins to muscle through dense vegetation could certainly be an advantage. Perhaps the opposing up and down motion of each pair of fins proved a more efficient way to move through the plant life. Lungfish live in freshwater versions of such environments. Having evolved lobe fins, our ancestors then evolved to have

[2] Ironically, the species in which the first synuclein protein was discovered, the Pacific electric ray, a cartilaginous fish, has not had its genome sequenced. That first synuclein was a gamma-synuclein, but whether the Pacific electric ray also has alpha- and beta-synuclein is not yet known.

their thick fins push them onto land, but the ancestral coelacanth apparently went the opposite direction, back to the depths of the open sea.

Of course, traits other than locomotion might explain the conservation of the alpha- and beta-synuclein sequences. For example, in his book *Your Inner Fish*, Neil Shubin explains how hiccups are a side effect of tetrapods evolving to breathe air. Nerves in the neck, once involved in breathing using gills, had to be extended down to the diaphragm to work the lungs, and in doing so, they became more exposed to the vagaries of gastronomical distress. Coelacanths have vestigial lungs, once used for breathing, but now oil-filled, perhaps serving as a kind of swim bladder. Maybe the synucleins somehow helped in the redirection and/or function of the nerves that tetrapods needed to be able to breathe air.

It is also possible that there is no outwardly apparent trait connected with synuclein conservation. Perhaps lobe-finned fish evolved a novel neurochemistry, with alpha- and beta-synuclein as part of their neurochemical tool kit, which gave ancestral tetrapods an advantage in evolving to live on land. What of gamma-synuclein? It could be important, too, but maybe, like hemoglobin, its function requires less interaction with other proteins. Whatever the explanation, whether directly aiding survival or otherwise improving reproductive success, the conservation of alpha- and beta-synuclein means they must have been important in the evolutionary path from coelacanth to human.

References

[1] Amemiya CT, et al. The African coelacanth genome provides insights into tetrapod evolution. Nature April 18, 2013;496:311–6.

[2] Maroteaux L, Campanelli JT, Scheller RH. Synuclein: a neuron-specific protein localized to the nucleus and presynaptic nerve terminal. Journal of Neuroscience August 1, 1988;8(8):2804–15.

[3] Huang Z, et al. Determining nuclear localization of alpha-synuclein in mouse brains. Neuroscience December 29, 2011;199:318–32.

[4] Uéda K, et al. Molecular cloning of cDNA encoding an unrecognized component of amyloid in Alzheimer disease. Proceeding of the National Academy of Sciences of the United States of America December 1, 1993;90(23):11282–6.

[5] Polymeropoulos MH, et al. Mutation in the α-synuclein gene identified in families with Parkinson's disease. Science January 27, 1997;276:2045–7.

[6] Watson JD, Crick FH. Molecular structure of nucleic acids; a structure for deoxyribose nucleic acid. Nature April 25, 1953;171(4356):737–8.

[7] Nirenberg MW, Matthaei JH. The dependence of cell-free protein synthesis in *E. coli* upon naturally occurring or synthetic polyribonucleotides. Proceedings of the National Academy of Sciences of the United States of America October 15, 1961;47:1588–602.

CHAPTER 2

Bad News, Good News — The demographics and genetic and environmental risk factors for Parkinson's disease

My hypochondriacal imagination presents the following scene to my mind's eye:

I'm sitting alone, waiting in the examination room. Was that a twitch in my hand? It hasn't had any tremors this morning, but I feel one could come any second. There is a knock at the door, and the doctor enters. "Hello. I'm Dr. Lewey. How are you feeling today?" After a few more questions, he examines me, having me flip my hands back and forth, touch my nose. Then he has me rapidly keep pinching my forefingers and thumbs. He moves my arms around, telling me to try

Synuclein and the Coelacanth
ISBN 978-0-323-85707-9
https://doi.org/10.1016/B978-0-323-85707-9.00012-5

to relax, which I'm having a hard time doing. He has me walk heel to toe (did I just wobble?), then takes me out to the hallway and has me walk down to the end and back. Back in the examination room, he looks over a folder of what, I guess, contains my medical history. "Well, Mr. Gruschus, it appears that …"

About 60,000 people in the United States receive a diagnosis of Parkinson's disease each year. The symptoms, both movement- and nonmovement-related, are progressively debilitating and sometimes mind-extinguishing. A diagnosis is definitely bad news, but not all the news is bad. With careful drug regimens and surgical options, the debilitating symptoms can be managed for years, even decades. Parkinson's disease usually progresses slowly, as long as dementia is not present, and these days, it is thought of as a disease you die with, not of. The risk for dying from pneumonia is somewhat increased [1], most likely owing to motor symptoms affecting swallowing, but most Parkinson's patients live full lives, dying of unrelated, natural causes. In contrast, dementias such as Alzheimer's disease, the most common neurodegenerative disease, kill the mind and then body, with no medical treatments that significantly alter the course of the disease. To put things into perspective, one has a 10 times higher risk of getting a non—Parkinson's related dementia than developing Parkinson's disease, so if I were doomed to have a neurodegenerative disease, having Parkinson's would be the lucky draw. A very unlucky few get multiple neurodegenerative diseases.

Most diagnoses of Parkinson's disease require at least two of the following four movement-related symptoms be present over a period of time: resting tremor; slowness of movement, called bradykinesia; rigidity in the arms, legs, or trunk, often accompanied by muscle and joint pain; and postural instability, that is, difficulty correcting perturbations to balance. Other distinctive symptoms can also be present, including a stooped posture, shuffling gait, dizziness when standing up from a lying position, speaking softly, and handwriting that becomes progressively smaller. There can be non—movement related symptoms as well: fatigue, depression, cognitive impairment, visual hallucinations, constipation, poor sense of smell, and sleep disturbances. The last three non—movement related symptoms often precede the movement-related symptoms by 10 years or more. Exactly which movement and non—movement related symptoms patients have and when they appear is highly variable, and because of this some have called Parkinson's a boutique disease. Symptoms can vary from day to day and come and go hour to hour.

Neurologists cannot diagnose Parkinson's disease in patients with 100% certainty, at least not while the patients are still living. The diagnosis can be confirmed only when autopsy shows signs of alpha synuclein—related pathology in specific regions of the brain, which will be described in detail in subsequent chapters. Diseases with synuclein-related pathology are called synucleinopathies, of which Parkinson's disease is considered the most common. Another common synucleinopathy is dementia with Lewy bodies (DLB), in which cognitive impairment and the movement symptoms associated with Parkinson's disease appear together within a year of each other.

Pathologically, DLB is similar to Parkinson's, so much so that many consider them to be varying expressions of the same disease. DLB has an incidence about four to five times lower than classic Parkinson's disease. Also, in roughly 50% of classic Parkinson's patients, dementia will appear more than a year after the movement-related symptoms; such cases are called Parkinson's disease with dementia (PDD). Upon autopsy, many DLB and PDD brains show the senile plaques associated with Alzheimer's disease alongside alpha synuclein—related pathology. Other, much rarer synucleinopathies include multiple system atrophy (MSA), pure autonomic failure (PAF), and some forms of neurodegeneration with brain iron accumulation (NBIA). Many of the symptoms for MSA and NBIA overlap with Parkinson's disease, although they differ in brain pathology, and in MSA the progression of the disease is usually faster. The risk for developing MSA is roughly one in 10,000 [2], and for PAF and the synucleinopathic form of NBIA, about one in a million [3].

Other brain pathologies can cause the same symptoms as Parkinson's disease; such cases are called parkinsonism. Autopsy studies of the brains of patients have shown that in the past, about one-fourth of those with a Parkinson's diagnosis actually had non synuclein—related pathology. Some had vascular issues, such as ministrokes, that damaged the same regions typically damaged in Parkinson's. Others involved pathology of another protein, tau; such cases are called tauopathies. From 1916 to 25, there was an outbreak of a brain fever, *encephalitis lethargica*, possibly viral, in Europe and North America that left many people with Parkinsonian symptoms, although since then, parkinsonism resulting from viral infection has been rare. Other causes include neurotoxin exposure, tumors, and brain trauma. I suspect that with the better brain imaging technology available today, the percentage of misdiagnosed Parkinson's disease has been reduced to a significant degree.

How many people in the world have Parkinson's disease? Estimates vary; most place the number at around 10 million cases. For the United States, estimates range from 600,000 to one million. One study done in 2010 by Allison Wright Willis and coauthors at the Washington University School of Medicine, St. Louis, found that 1.6% of just under 30 million

Medicare patients over age 65 years had a diagnosis of Parkinson's disease [4].[1] As for the lifetime risk of developing Parkinson's disease, the most common figure one comes across is one in 100, a 1% chance. For the United States this figure is probably slightly low and lies somewhere between 1.2% and 1.6%.[2] As the population continues to age, the lifetime risk of developing Parkinson's will likely inch upward.

If we took a random sample of 100 patients in the United States with a diagnosis of Parkinson's disease, what would be their most prominent demographic feature? The greatest risk factor for Parkinson's is age. One has about a 10 times greater chance of being given a diagnosis of Parkinson's disease at age 75 than at 55. For example, using data from a study of Parkinson's patients at an HMO in northern California, only 4% were below the age of 50, 10% were in the age bracket of 50–59, 26% were in the age bracket of 60–69, 44% were in the age bracket of 70–79, and 16% were aged 80 or greater [5]. The average age at diagnosis varies from study to study, from 60 to 68 years.

Another prominent feature of our 100 US patients would be that about 60 would be male. Most studies around the world, although not all, show this male bias. Perhaps sex hormones have a role in disease risk. Some have speculated that there might be a genetic risk factor on one of the sex chromosomes, such as the *MAO-B* gene on the X chromosome [6], which might be more beneficial to women or more harmful to men.

Curiously, in studies outside the United States, the male–female ratio for Parkinson's disease is lower, even in Europe. For example, in a study of over three million UK citizens posted at www.parkinsons.org.uk, only 56% of Parkinson's patients were male, and several other European countries had even lower male percentages. Japan actually has more females with

[1] This percentage translates into about 770,000 people over age 65 living with Parkinson's in the United States, assuming the same 1.6% figure and taking the total US population as 323 million with 15% of the population over age 65. Because this number excludes those with a diagnosis of Parkinson's who are aged less than 65 years, the total number should be slightly higher.

[2] The average US life expectancy is approaching 79 years, and about 75% of Americans will live past 65 years; thus, one would expect a lifetime risk closer to the 1.6% prevalence seen in the Medicare study. That is, the lifetime risk would be predicted to be about 75% of 1.6%, which equals 1.2%, plus some additional percentages. One addition to the lifetime risk comes from people over age 65 who did not have Parkinson's disease at the time of the Medicare study, but who will develop the disease in the future. Another addition is from those who had Parkinson's disease, but who did not make it to age 65. Thus, the total lifetime risk is expected to be greater than a 1.2% chance.

Parkinson's disease than males [7]. This suggests there might be some environmental factor in the United States that increases the risk for American males. Perhaps American males are more exposed to deleterious chemical agents than in Europe and elsewhere. Certain pesticides and herbicides are known to increase Parkinson's disease risk, such as rotenone and paraquat, and studies have shown that agricultural workers have about double the risk of developing Parkinson's disease compared with the general population [8], although other researchers have failed to see a significant correlation [9,10].

The US Medicare study tested "... the hypothesis that Parkinson disease is associated with rural living." Despite the putative higher incidence of Parkinson's disease in agricultural workers, they found that the incidence of Parkinson's was actually *lower* overall in rural counties. On the other hand, studies in other countries have seen a higher incidence (including agricultural workers) in rural areas [11,12]. In any case, agricultural workers make up less than 2% of the US population, so it seems unlikely that pesticides and herbicides alone could account for the higher male incidence in the United States.

The Medicare study found that Parkinson's disease had the highest prevalence in counties that stretched from the Midwest to New England, encompassing the Rust Belt states. They suggested that the higher incidence might be the result of some sort of industrial chemical exposure, such as solvents or toxic metals such as manganese, mercury, or lead. Whether males have an even higher prevalence than females in this Parkinson disease belt was not explicitly stated in the study, however.

Another possibility is that disease risk is inversely related to vitamin D levels; perhaps people in the regions of highest increased risk tend to get less sunlight than other regions. Vitamin D deficiency is more common in Parkinson's patients than in those without the disease, but whether the onset of deficiency predates or is a consequence of the disease is not yet known [13]. In addition, if more sunlight were protective, why would agricultural workers have higher risk?

What else might explain the male bias in the US Medicare study? Several studies have shown that brain trauma roughly doubles one's risk of developing Parkinson's disease [14,15]. Perhaps American males have more head trauma than their European counterparts. On the other hand, whereas Americans have their football, Europeans have theirs, too, in which the more heading of the ball, the better, so which is worse? Studies have connected repeated, nonconcussion-causing blows to the head with another neurodegenerative disease, chronic traumatic encephalopathy, which is a kind of tauopathy. In contrast, for Parkinson's disease, studies showed that increased risk is correlated with concussions, in particular those involving temporary loss of consciousness, in which longer loss of consciousness is correlated with higher risk. I used to play rugby, and although I only had one loss-of-consciousness concussion in my years of playing, my hypochondriacal imagination keeps asking, "Was once enough?"

The potential connection between Parkinson's disease and workers in one male-dominated occupation, firefighters, has received a surprising amount of attention on the Internet. Fires can produce thousands, perhaps even millions of toxic chemicals in their smoke. Carbon monoxide poisoning, especially with loss of consciousness, can cause temporary parkinsonism and possibly increases risk for developing Parkinson's disease later in life [16]. The claims of a scientific connection between firefighting and Parkinson's appear to stem from a single, non—peer reviewed presentation at a conference in 1990 [17]. Oddly, I was unable to find any peer-reviewed papers in the nearly 30 years since then confirming the epidemiological connection. Personally, I would not be surprised if Parkinson's disease is more prevalent among firefighters, but more careful investigation is needed before any firm conclusion can be drawn.

Another demographic aspect of our 100 people with Parkinson's that deviates from the general US population is the percentage of whites compared with blacks and Asians. For example, in the Medicare study, blacks and Asians made up 8.0% and 1.6%, respectively, of the total number of beneficiaries aged 65 years and greater, but only 5.2% and 1.1% of those with Parkinson's disease. Adjusting for tobacco use made no difference. Part of the discrepancy for blacks can be explained by their shorter life expectancy compared with whites, 75 versus 79 years. Still, even when adjusted for this, whites had a significantly higher incidence of Parkinson's.

As for Asians, in the United States they live 8 years longer than whites on average, so the lower prevalence of Parkinson's for Asians is opposite what one would predict. The Medicare study found that prevalence among Hispanics was slightly lower than among non-Hispanic whites, although another study found the opposite [5]. In the Medicare study, Hispanics comprised only 2% of the total beneficiaries, which is puzzling because Hispanics currently comprise more than 15% of the total US population. However, beneficiaries were included in the study only if they were aged 65 years or older, so the lower percentage reflects, at least in part, the lower number of those who self-identified as Hispanic in previous decades.

When seeking causes for the differing Parkinson's disease prevalence among different ethnicities in the United States, one must consider both genetic and environmental factors. Two major factors, age and tobacco use, have already been ruled out. Socioeconomic and dietary factors might have some role. Whether alcohol consumption influences risk is uncertain. Some studies, including one with the eye-catching title "Voluptuary habits and clinical subtypes of Parkinson's disease," show a protective effect [18], whereas in other studies no obvious correlation was seen [19]. One study found that long-term use of methamphetamine (crystal meth) or other amphetamine-related drugs is associated with triple the risk for developing the disease, whereas for other drugs, such as cocaine, no increased risk has been seen [20]. Curiously, another study found that patients who had been prescribed amphetamine (Adderall) or methylphenidate (Ritalin) at any time were *less* likely to develop Parkinson's disease [21]. There also might be genetic differences between ethnicities that influence Parkinson's prevalence.

Ashkenazi Jews have a higher prevalence of familial Parkinson's disease than the rest of the United States population, as do certain ethnic groups in other parts of the world. Most familial cases are due to mutations in either the *LRRK2* gene, pronounced "lark 2," or the *GBA1* gene, both of which will be discussed later here and in subsequent chapters. Parkinson's disease cases with clear genetic causes have been a terrible curse to certain families but a boon to science, providing invaluable clues regarding the biological basis of the disease. Indeed, the central importance of alpha-synuclein in the disease was established by both pathological examination and genetic studies showing the correlation between alpha-synuclein mutations and familial Parkinson's disease.

Interestingly, many studies have seen that smoking correlates with lower (yes, lower) incidence of Parkinson's, first reported in 1968 by M. Dean Nefzger and coauthors at the National Research Council in Washington, DC [22]. The protective correlation still seems to hold even when factors such as age are considered. Although tobacco use does not explain racial differences in prevalence, it is still an environmental factor that could influence the risk for Parkinson's. Some have hypothesized that nicotine might have some protective effect. On the other hand, others have wondered whether there might be a more indirect explanation, that the same genetic factor that predisposes one to nicotine addiction also makes one less prone to developing Parkinson's disease.

Several studies have seen that caffeine also has an apparent protective effect; those who consume more have a lower incidence than those who do not, although not so dramatic an effect as is seen with tobacco use [23]. Many researchers have hypothesized that caffeine, like nicotine, might directly protect neurons from Parkinson's-related degeneration. These are reasonable hypotheses; both caffeine and nicotine directly interact with receptor proteins at neuronal synapses, affecting neurotransmitter levels. However, I still cannot help but wonder whether smoking and/or caffeine consumption truly have a protective effect or instead reflect some sort of genetic factor that leads one to crave nicotine and/or caffeine, a factor that also somehow protects one from Parkinson's disease.

Let us recap the prominent features of our 100 Americans with Parkinson's disease. They are predominantly older, and 86 are aged greater than 59 years. Based on the study of Medicare beneficiaries in the United States, 60 of them are male, 90 are non-Hispanic white, five are black, two are Hispanic (which is probably an underestimation), one is Asian, and two are of another race.

In terms of environmental risk factors, perhaps four of the 100 are agricultural workers, because they make up almost 2% of the population, but they have double the risk for Parkinson's disease, at least according to some studies. In other words, two of the agricultural workers would have Parkinson's just because of the normal one in 100 chance of developing the disease and two more would have had Parkinson's owing to their environment, perhaps from exposure to pesticides and herbicides. Given the lower than average Parkinson's prevalence in rural US counties, it would seem that incidental, low-level pesticide and herbicide exposure is not a significant risk factor. Another population that might also be at risk similar to agricultural workers is nonagricultural pest and weed control personnel, although they make up a tiny fraction of the total population.

In the studies on brain trauma, about 12% of study subjects reported some sort of trauma involving at least momentary loss of consciousness in their past, and the percentage with Parkinson's was roughly double that seen for the controls. This means around 24 of the 100 people with Parkinson's disease have had some sort brain trauma. Again, for half of these, we assume Parkinson's disease was triggered, or perhaps existing preclinical Parkinson's was somehow accelerated, by the brain trauma, and for the other half we assume the normal one in 100 chance everyone has of developing the disease. One of the more disquieting observations was that the earlier in age the brain trauma occurred, the greater the likelihood was of Parkinson's later in life. So much for the invincibility of the young.

How many more of our 100 people come by the disease from environmental factors: in particular, the factors responsible for Midwest—Northeast Parkinson disease belt? This is a difficult question that the Medicare study chose not to address, probably wisely. Even the concept of what constitutes cause is problematic. We can make a crude estimate, however, considering cause as any factor that correlates with increased disease risk in the geographical region. In the figure showing prevalence by county, those in the Midwest—Northwest belt had prevalence rates at least double the national average. First, let us assume that the area covered by the belt accounts for about 20% of the US population, just over 60 million, which is what one gets adding the populations of Illinois, Indiana, Ohio, Pennsylvania, and New York.[3]

Next, let us say the disease prevalence in the belt is double the national average, just slightly lower than in the Medicaid study. You might think that double the prevalence would translate to 40% of our 100 people with Parkinson's: that is, double 20%. However, because of a mathematical headache called normalization, the actual number is smaller, about 33%, with half of the 33% having the disease owing to the unidentified environmental factors in the Parkinson belt and the other half having the disease as a result of normal chance. Normalization ensures all percentages add to 100%. Another mathematical wrinkle is that the normal chance of developing Parkinson's should be adjusted downward slightly to *not* include those who develop the disease from the Parkinson belt factor(s). However, given the approximate nature of this crude estimate, such methodological exactitude is probably overkill.

All told, 30 people, more or less, in our group of 100 appear to have Parkinson's caused, or at least influenced, by environmental factors: two from pesticide or herbicide exposure; 12 (actually closer to 11, taking into account normalization) from brain trauma; and perhaps 17 from the mysterious Rust Belt factor(s). Of course, likely other environmental factors have yet to be identified, and some people could be at risk from multiple factors, Northeastern hard-hitting athletes who farm on the side, for instance. In addition, genetic factors are involved. Indeed, around 10% of Parkinson's patients have a clear family history of the disease.

[3] Probably Michigan and New Jersey should also be included, but then again, the populations in the rural regions of the listed states should probably be excluded. In any case, 60 million is only meant to give a rough estimate.

In the rest of the chapter, we will explore the contribution genetics has in Parkinson's disease. But before that, I want to discuss one more possible source of the disease. In addition to being caused by genetic and environmental factors, Parkinson's disease might also happen just by chance. In fact, an underlying assumption of disease risk factors is that they modify a base chance of getting the disease. Like cancer, it could be that sometimes you get Parkinson's just from bad luck. With cancer, this can result from a spontaneous error in DNA repair or replication. For Parkinson's disease, it might be due to a spontaneous interaction of alpha-synuclein molecules that chance to adopt their pathological form, a butterfly effect that, once started, spreads the disease from the initial site to the rest of the brain.

In 1986, neurologist Larry Golbe was attending the funeral of one of his Parkinson's disease patients who had drowned in a swimming pool. The patient's brother, also at the funeral, sought out Golbe, worried that he, too, had the disease. Golbe examined the brother and confirmed his fears. Intrigued, Golbe interviewed the family and found four more cases. In his interviews, he learned that the family came from the Campania region of Italy, from a village called Contursi, although he was not to know how important this fact was until later. Several months had passed, when he happened to examine another Parkinson's patient of Italian descent and learned her family was from the same village.

Thus began the saga of the Contursi kindred, ultimately leading to the discovery 10 years later, by a team of scientists at Rutgers University and the National Institutes of Health (NIH), of the first gene mutation responsible for familial Parkinson's disease [24]. The story is told in detail in Jon Palfreman's book *Brain Storms—The Race to Unlock the Mysteries of Parkinson's Disease*. The gene responsible was *SNCA*, the gene that encodes alpha-synuclein. It is customary to write gene names with italics. The mutation caused a single change in the sequence of the protein, changing the 53rd amino acid from alanine to threonine. Those who inherited a copy of the mutated gene from either of their parents almost always developed the disease, a pattern of disease inheritance called autosomal dominant. On average, half of the offspring of the affected parent inherited the mutation. Autosomal means the gene is not on a sex chromosome.

In the next few years, more kindred studies revealed additional *SNCA* mutations, as well as families that had one or two extra copies of *SNCA*, brought about by a genetic glitch called gene duplication or triplication, respectively. The family with gene triplication had an especially aggressive form of the disease [25]. Mutations in other genes were discovered as well. One of these was the gene locus *PARK8* that encodes LRRK2, also known as dardarin, named for the Basque word for trembling [26]. The gene is now more commonly notated *LRRK2*, after the protein. Disease-causing mutations of *LRRK2* are far more common than *SNCA* mutations, but they are still relatively rare in the general population, with only one to two carriers per 10,000 people. The percentage of carriers can be much higher in certain populations, notably in Basques, Berbers, and Ashkenazi Jews.

Like *SNCA*, Parkinson's disease caused by mutations in *LRRK2* is inherited in an autosomal dominant fashion, although some of the mutations do not always cause the disease. For example, the most common mutation, called G2019S, which results in a change from the amino acid glycine to serine in position 2019 of the protein sequence, causes Parkinson's disease in only about a third of carriers, although the percentage varies considerably depending on ethnicity [27]. When a mutation affects only a fraction of carriers, the mutation is said to have reduced penetrance, where the penetrance equals the percentage of carriers who develop the disease. Thus, the G2019S mutation has a penetrance of about 33%.

Another mutation that was found during these exciting early years was in the gene *PRKN*, which encodes the protein parkin [28]. As you might have guessed, the gene and protein had not yet been named until the Parkinson's kindred study that fingered them. The first patients in this study had a rare, juvenile form of parkinsonism. Unlike the disease-causing alpha-synuclein and *LRRK2* mutations, these early-onset patients had inherited two copies of mutated *PRKN* genes, one from each parent. This pattern of inheritance is called autosomal recessive, in which on average only a fourth of the offspring in affected families develops the disease. Other recessive gene mutations include the *PARK7* gene, encoding the DJ-1 protein [29] and the *PTEN-induced putative kinase 1* (*PINK1*) gene, encoding the protein PINK1 [30].

The gene mutations in *SNCA*, *LRRK2*, *PRKN*, *PARK7*, and *PINK1* are called Mendelian mutations. The term "Mendelian" comes from the name of the monk and geneticist Gregor Mendel, who first discovered, using pea plants in a monastery garden, that genes are inherited one from each parent, leading to dominant and recessive traits. With Mendelian mutations, if you inherit one mutant gene for dominant inheritance, or two mutant genes for recessive inheritance, you develop the disease, although not 100% of the time in the case of reduced penetrance.

A few additional, very rare Mendelian mutations were found in these early years, but disconcertingly, taking all these newly discovered mutations into account, one could explain only a small fraction of familial Parkinson's disease cases, less than 20%. Remember, cases of familial Parkinson's disease make up only about 10% of total cases of Parkinson's. In other words, less than 20% of 10%, just under 2%, of all cases of Parkinson's were clearly associated with a known Mendelian mutation. Perhaps some of the familial cases actually resulted from environmental causes, environmental factors that were shared in extended families.

Putting aside the question of how an extended family can share an environmental risk factor that presumably does not affect unrelated neighbors, more evidence was presented in the late 1990s that seemed to indicate environment had an overwhelming role. The evidence came from twin studies. In such studies, one looks for concordance: that is, cases in which both twins have the trait, in this case, Parkinson's disease. The studies include both identical and fraternal twins. For identical twins, both should have the same trait. For fraternal twins, half should share the same trait for dominantly inherited traits, one-fourth for recessively inherited traits. The first studies found no significant evidence that genetics had any role at all; the chance that identical twins both had Parkinson's was practically no different from that for fraternal twins [31].

The earlier twin studies tacitly assumed that the trait (i.e., Parkinson's disease) would manifest at roughly the same age. However, Parkinson's is a highly variable disease, including the age of onset. The final twin study was done in 1999 by a collaboration including researchers at the Parkinson's Institute in California [32]. They used a database of 19,842 white male twins from the National Academy of Sciences/National Research Council World War II Veteran Twins Registry, by far the largest number of twins studied to that point. This time, the study examined whether the results changed if they broadened the criteria to include cases in which a twin might be exhibiting possible early signs of Parkinson's disease. Still, the results were little different from the earlier studies, and showed Mendelian inheritance only for the four pairs of identical twins who had received Parkinson's disease diagnoses before age 50.

After the flurry of rare Mendelian gene mutation discoveries in the late 1990s, the search for genetic causes of Parkinson's disease stalled for several years. Then, in 2004, two papers came out connecting Parkinson's disease to a rare disorder, Gaucher disease [33,34]. The Parkinson's inheritance pattern was not Mendelian, but still, in extended families of patients with

Gaucher disease, the researchers found that Parkinson's disease occurred in more than the expected one in 100 people. Gaucher disease is a recessive disorder in which one inherits two defective copies of the gene *GBA1*. *GBA1* encodes the enzyme glucocerebrosidase, which breaks down certain lipids, which are fat molecules. With two defective *GBA1* copies, the lipids accumulate in special compartments inside cells called lysosomes, and thus the disease caused by inheriting two defective copies of the gene is called a lysosomal storage disease. Several other lysosomal storage diseases result from defects in other enzymes that break down similar lipids.

Over the next 5 years, numerous studies came out comparing the frequency of *GBA1* mutations in Parkinson's disease patients versus healthy controls. In 2009, these results were combined and summarized in a paper authored by NIH researcher Ellen Sidransky and over 60 coauthors [35]. The combined results showed that Parkinson's patients were five times more likely to harbor a *GBA1* mutation than were controls. The effect was dominant; that is, having just one *GBA1* mutation was sufficient to increase risk. *GBA1* mutations were the first of a new type of genetic cause of Parkinson's disease, so-called genetic risk factors.

Genetic risk factors alter the chance that one will inherit a trait, in this case, Parkinson's. In the case of *GBA1*, a mutation increases the chance of developing the disease by roughly a factor of five, from a one in 100 chance without a mutation to one in 20 with a mutation. *GBA1* mutations are fairly common, just under 1% for the general population, and much more common in certain populations, Ashkenazi Jews in particular. Given how common *GBA1* mutations are, why did the twin studies fail to show a significant genetic contribution?

The answer to the twin paradox is in the math. If two identical twins have a *GBA1* mutation, each has a one in 20 chance of developing Parkinson's disease, so the chance that at least one gets the disease is double that, one in 10. What is the chance that the other twin will also get the disease? If one ignores any other genetic risk factors, the chance they both have the disease falls to one in 20 squared, that is, only one in 400 sets of identical twins with *GBA1* mutations will both get the disease. This should be considered a lower bound, because identical twins carry identical DNA; thus, they both must carry any additional genetic risk factors. Even so, the risks would have to be very much higher to make any significant impact in the twin studies. The bottom line is that twin studies are a good test for diseases caused by Mendelian gene mutations, but not for genetic risk factors.

In the 2000s, at the same time researchers were discovering that *GBA1* mutations were risk factors for Parkinson's disease, a major technological advance in DNA analysis was revolutionizing genetics research. The advance was the DNA microarray, or DNA chip. Using these chips, researchers could quantitate things such as messenger RNA expression levels in tissues and cell cultures, and most importantly for genetics research, they could determine the DNA polymorphisms present in a person's genome. Recall that a polymorphism is a variation in the DNA sequence of a gene that has no obvious deleterious effects for the protein encoded by it: in other words, not a Mendelian mutation. The technology advanced quickly from being able to test for hundreds to thousands, and then to hundreds of thousands of polymorphisms on a single chip. These were called single nucleotide polymorphism (SNP, pronounced "snip") chips.

Research into genetic risk factors blossomed with use of these chips, yielding a statistical bonanza of genetic information on studied populations. Using an approach called a genome-wide association study (GWAS), risk factors could be identified by comparing the frequency of each polymorphism for people with a particular trait, such as red hair or Parkinson's disease, with the frequency of the same polymorphism for those who did not have the trait. How does this work exactly? A polymorphism can occur at a DNA base position within or outside the coding regions of a gene, or even in between genes. Let us say a particular polymorphism involves a change from the nucleotide, or base, A (adenine), which happens to be the more commonly found base in that particular position, to the base G (guanine). Let us say that people with red hair have the base G with a frequency of 0.20, that is, 20%, in that position, whereas non—red hair people have G with a frequency of 0.17.

The statistically important question to ask is how likely the difference in polymorphism frequency for redheads is just due to chance. This situation is related to the question of how one can tell whether a die is loaded; that is, whether it has a greater tendency to give a particular number on a roll. If the die is not loaded, the average frequency for any particular number for a six-sided die will by 1:6, or approximately 0.17. Let us say you suspect the number six comes up more frequently than 1:6 of the time. How many times would have to roll the die to prove it is loaded? This depends on couple of things, such as how confident you want to be of the result. In most science fields, the bare minimum standard of confidence is 95%. In other words, you would want your result to be correct at least 95% of the time.

Another question is how much more frequently the loaded number is expected to come up. With regard to the earlier polymorphism example, how many times would we have to roll the die (i.e., how many people would have to be tested) to be confident that the redhead versus non-redhead frequency of the DNA base G, 0.20 versus 0.17, is not just a statistical fluke?[4] For the die problem, it turns out that we would have to roll it nearly 600 times to be 95% confident that the result of 0.20 for rolling a six was not a fluke. We would have to test the same number, around 600 redheads, to get the same level of confidence, and because they account for only about 4% of the population, this would translate to over 15,000 random subjects.

Luckily, random subjects are not a requirement; one can specifically recruit redheads, or people with Parkinson's, or people with whatever trait one wishes to study. In addition, one also needs to recruit a sufficient number of controls, people who do not have the trait, enough to make sure the polymorphism frequencies are known with sufficient accuracy. The math is practically the same, involving fancy functions such as the binomial distribution and its continuous extension, the normal distribution, which in layman's terms is basically a bell curve. The main measure reported in GWAS is the odds ratio, the polymorphism frequency for those with the trait divided by the frequency for the controls. The more the odds ratio differs from 1, and the more subjects in the study, the greater the confidence that the result is real.

In practice, a result with 95% confidence in a GWAS does not even warrant notice; confidences of greater than 99.999999% are more the norm. The confidence number used in these studies is called a p-value, given by 1 minus the confidence percentage expressed as a decimal value (the percentage divided by 100), and it represents the chance that the result is a statistical fluke. A confidence of 95% translates to a p-value of 0.05%, and 99.999999% translates to a p-value of 0.00000001.

Why all the extra zeroes? It is because of something called the Bonferroni correction, named for an Italian mathematician, although its modern use in GWAS is due to the statistician Olive Jean Dunn. To explain, say I am testing a dataset (i.e., the SNP frequencies in a set of genomes for those

[4] Technically speaking, it matters whether we are asking the question of "redhead versus non-redhead" or "redhead versus all": that is, the average frequency of G in nonredheads compared with the average frequency of G for everyone. However, for uncommon traits, both situations are practically the same. For example, because redheads account for only about 4% of the population, the average G frequencies for nonredheads and for everyone will be practically the same (0.170 versus 0.172).

with the trait of interest). If I only require 95% confidence, a p-value of 0.05, in the result for any particular SNP, the chance that at least one of them will have the desired p, just by chance, increases the more SNPs I test. To correct for this, I need to divide the desired p-value, 0.05, by the number of SNPs being tested, typically in the hundreds of thousands.

A typical SNP microarray these days can detect over half a million different SNPs. This is impressive; unfortunately, in the human genome, there are about 15 million common SNPs with frequencies greater than 1%. Chances are that the true SNP resulting in a risk factor will not be on the chip. However, all is not lost. In a GWAS analysis, SNPs close to the true SNP can also show up with relatively low p-value. This is because within a population, the closer the measured SNP is to the true SNP, the more likely their nucleotide variations are correlated.

To explain this correlation, recall that in normal cells, each DNA chromosome has two copies, but each parent passes on just one of each chromosome to the offspring, and these passed-on chromosomes have undergone a process called recombination. In recombination, each chromosome passed on to the offspring is made of a mosaic of chunks coming from the parent's corresponding two chromosome copies. Each of the chromosomes passed to the offspring is whole, containing the complete set of chromosomal genes, but each one is a mixture, one of the father's two chromosomes and one of the mother's two chromosomes. The chunks are large, containing about half a million common SNPs each on average. Because of the largeness of the chunks, the chance that neighboring SNPs get passed down together each generation is high.

Over time, SNPs passed down together for generations eventually separate, because the boundary between the chunks in recombination can vary. Once in a rare while, the boundary will fall between the true SNP and its correlated neighbor. The closer they are, the lower the chance that the recombination boundary will happen to fall between them. The consequence of this in a GWAS is that there will be clusters of low *P* value SNPs near the true SNPs. These clusters are called loci; one cluster is called a locus. In GWAS papers, the clusters are displayed using Manhattan plots, in which the horizontal axis represents the position in the genome and the vertical axis is highest for the SNPs with the lowest p-value.[5] The loci show up as peaks, as "skyscrapers in the genetic skyline."

[5] The value plotted on the vertical axis is the absolute value of the log of the *P* value.

Advances in SNP chip technology have helped fuel the recent DNA ancestry and genealogy craze. In this case, the traits tested for are ethnic backgrounds, and rather than focus on one SNP at a time, they compare all of the customer's SNPs with the average SNP frequencies measured for each ethnic population. Then, using fancy mathematics, they predict which combination of ethnic backgrounds predict the customer's SNPs with highest probability, although these predictions can be dismayingly different, depending on which company, using its own particular genomic database and proprietary prediction algorithm, is chosen. Nevertheless, and more important for science, many of the companies have encouraged their customers to allow their DNA results to be used by researchers. Combined with more traditional databases from institutes and foundations around the world, the result has been a gold mine for medical research, in particular for Parkinson's disease.

The GWAS results for Parkinson's disease were stunning. Whereas just 15 years earlier, researchers had questioned whether there was any significant genetic contribution for the bulk of Parkinson's cases, the evidence for specific risk factors has become overwhelming. What was the standout result, the risk factor skyscraper that towered above the rest? Alpha-synuclein [36].

"Whoa!" I hope you are saying, "You said in the last chapter that alpha-synuclein had no polymorphisms!" Actually, what I wrote was, "So far, no nonharmful polymorphisms in the *protein sequences* of alpha and beta-synuclein have been found" (italics added). However, there are polymorphisms between and outside the coding regions of the *SNCA* gene. The GWAS loci associated with *SNCA* occur in regions preceding the first coding region of the gene, as well as the noncoding regions within the gene, called introns, and even past the last coding region of the gene, called the 3′-untranslated region. Such regions often contain regulatory elements, sequences of DNA that can bind proteins called enhancers and repressors that can ramp up or tamp down gene expression. As with the familial Parkinson's cases with *SNCA* duplication and triplication, the GWAS result strengthens the argument that altering alpha-synuclein levels, without necessarily changing the protein sequence, is sufficient to cause the disease.

GBA1 is also a risk factor locus, but the SNP with the most significant p-value is not one of the mutations known to cause Gaucher disease. Instead, it is a more common polymorphism occurring with a frequency of about 1.5% in the general population. Unlike the *SNCA* locus, the *GBA1* SNP is in the coding region, changing the amino acid in the 326th position from glutamate, with a negative side chain, to lysine, with a positive side chain. This is an intriguing result, one that we will revisit in a later chapter. *LRRK2* is also a risk locus, although rarer Mendelian genes such as *PRKN* did not make the cut.

Perhaps the most surprising risk factor locus includes the *MAPT* gene, which encodes the tau protein, the primary component of the neurofibrillary tangles most commonly associated with Alzheimer's disease. However, caution is warranted in interpreting the *MAPT* result, because tau mutations cause a number of other diseases, many of which can include parkinsonism symptoms, such as progressive supranuclear palsy (PSP) and frontotemporal-dementia-and-parkinsonism-linked-to-chromosome17 (FTDP-17). It could be that some of the Parkinson's disease patients in the GWAS analysis actually had a misdiagnosed, tauopathy-related disease.

All told, the various GWAS analyses have found around 30 risk factors for Parkinson's disease. Not all of the risk factors increase the chance of developing the disease; some are protective. For instance, in the 2017 report by Diana Chang et al., the less common nucleotide associated with the *MAPT* locus (in particular, the SNP with the lowest p-value, at the top of the *MAPT* skyscraper) is found in 22% of the over 300,000 control subjects without Parkinson's, but it occurs with a frequency of only 19% in the 6746 Parkinson's patients. Because the prevalence of tauopathy-related parkinsonism is considerably lower than Parkinson's disease, about six in 100,000 for PSP [37] and perhaps one in 1,000,000 for FTDP-17, this suggests that the lowest p-value *MAPT* mutation truly is protective for Parkinson's disease.

The less common nucleotide associated with the highest confidence *SNCA* SNP does increase risk, however, going from a frequency of 35% in the control group to 41% for Parkinson's patients. This 6% difference means the *SNCA* risk factor is perhaps responsible for even more cases of Parkinson's disease in the general population than those attributable to *GBA1* mutations, which occur with a frequency just under 1% in controls and 5% in Parkinson's patients. In addition, the highest confidence *SNCA* polymorphism might be just the tip of the iceberg, because studies have suggested that the *SNCA* gene might harbor multiple risk factor SNPs [38].

Although the total contribution of genetic factors leading to Parkinson's disease can be estimated using GWAS results, there is a better way to estimate this total, known as genome-wide complex trait analysis (GCTA). One problem with GWAS results is that it provides no information about how the various risks factors might interact. Some risk factors might simply add independently to risk, some may supersede or cancel others, whereas some sets of risk factors might contribute synergistically when present together. In contrast to GWAS, in GCTA, the individual risk factors are not the focus; instead, one calculates how similar the entire genomes of patients are to each other on average, compared with controls.

GCTA studies have produced estimates of the genetic contribution to Parkinson's disease from 27% to 40% [39,40], combining results from American and European cohorts. Like GWAS, GCTA studies use SNP chip data. Because currently SNP chips can measure only a small fraction of the common polymorphisms present in the human genome, let alone the rare ones, GCTA results are necessarily a lower-bound estimate of the full genetic contribution. Curiously, the GCTA study by Margaux F. Keller and Michael A. Nalls at NIH, along with coauthors including the International Parkinson's Disease Consortium and the Wellcome Trust Case Control Consortium 2, found less genetic contribution for early-onset cases of Parkinson's disease. Recall that in the twin studies, they found the opposite, 100% concordance for identical twins with early-onset Parkinson's, but very much lower concordance, similar to that for fraternal twins, for late-onset disease. This illustrates one of the weaknesses of GCTA studies: namely, that they miss genetic contributions from rare mutations, such as those that lead to early-onset Parkinson's disease.

Researchers of lysosomal storage diseases have seen hints that mutations in other lysosomal proteins, in addition to *GBA1* mutations, might be associated with Parkinson's disease. These mutations are much rarer than *GBA1* mutations, far too rare to show up in GWAS analyses. To address the question of whether other lysosomal protein mutations might contribute to Parkinson's disease risk, a new statistical technique has been developed, similar to the GWAS and GCTA approaches, but which uses full DNA sequencing in addition to SNP chips.

In a study employing this new statistical technique by Laurie A. Robak at Baylor University, along with coauthors including, once again, the International Parkinson's Disease Genomics Consortium, full sequencing results of the protein coding regions of all known lysosomal proteins known to cause lysosomal storage diseases were included for over 600 people [41]. With full sequencing, all SNPs, rare or common, are detected. They found that those with Parkinson's disease had a higher chance of having at least one rare, putatively damaging mutation in the set of lysosomal proteins studied. Their results are intriguing, but still preliminary, in my opinion. For instance, the study failed to include a set of biologically related proteins unrelated to Parkinson's disease as a control. Nevertheless, this kind of analysis holds great promise, and I look forward to future work in this area.

In our group of 100 Parkinson's patients, how many have the disease owing to genetic factors? Of the Mendelian mutations known to cause Parkinson's, perhaps two have *LRRK2* mutations. Chances are that none of the 100 cases are attributable to the other Mendelian mutations; they are too rare. Five of them have mutations in the *GBA1* gene: four of them have the disease owing to the influence of the *GBA1* mutations, whereas the fifth has the disease just because of the normal one in 100 chance. At least six of them have the disease because of *SNCA* loci risk factors. The other GWAS risk factors identified in the study by Chang et al., counting those for which the less common nucleotide increases risk, contribute about 11 more cases. This gives a total of 23 patients with genetic causes, less than the lower bound of 27 cases implied by GCTA studies, which suggests that more genetic risk factors have yet to be identified.

So far, in my simplistic accounting, about 30 of our 100 Parkinson's patients have the disease as a result of environmental causes, and at least 23 are afflicted owing to genetic factors, although in both cases, the numbers are likely underestimates. The full picture promises to be even more complicated, however, because it appears that certain genetic factors can also influence environmental risk factors. For instance, polymorphisms in the *CYP2D6* gene, which is involved in the metabolism of many molecules, including potential toxins in the environment, can potentially modulate the risk for pesticides and herbicides associated with Parkinson's disease [42]. Polymorphisms in two genes associated with caffeine, *CYP1A2*, involved in its metabolism and *ADORA2A*, a neurotransmitter receptor to which caffeine binds in the brain, might influence caffeine's putative protective role [43].

After all the scientific sleuthing is said and done, it could be that for most cases, both environmental and genetic factors come together to trigger, accelerate, or otherwise promote Parkinson's disease. Rare indeed will be the Parkinson's patient for whom no risk factors can be identified; the combined chance of having every common nucleotide in all of the SNPs of all known loci correlated with increased disease risk is quite low. These factors are hints as to what causes Parkinson's disease, and most of this book explores these hints, to help understand the how and the why of the disease. But, before that, I would like to introduce you to my family.

References

[1] Gorell JM, Johnson CC, Rybicki BA. Parkinson's disease and its comorbid disorders: an analysis of Michigan mortality data, 1970 to 1990. Neurology October 1994;44(10):1865–8.

[2] Bjornsdottir A, et al. Incidence and prevalence of multiple system atrophy: a nationwide study in Iceland. Journal of Neurology, Neurosurgery and Psychiatry February 2013;84(2):136–40.

[3] Kalman B, et al. An international registry for neurodegeneration with brain iron accumulation. Orphanet Journal of Rare Diseasesvolume 2012;7:1.

[4] Willis AW, et al. Geographic and ethnic variation in Parkinson disease: a population-based study of US Medicare beneficiaries. Neuroepidemiology April 2010;34:143–51.

[5] Van Den Eeden SK, et al. Incidence of Parkinson's disease: variation by age, gender, and race/ethnicity. American Journal of Epidemiology June 1, 2003;157(11):1015–22.

[6] Mellick GD, et al. Variations in the monoamine oxidase B (MAOB) gene are associated with Parkinson's disease. Movement Disorders March 1999;14(2):219–24.

[7] Kimura H, et al. Female preponderance of Parkinson's disease in Japan. Neuroepidemiology November 2002;21(6):292–6.

[8] Frigerio R, et al. Chemical exposures and Parkinson's disease: a population-based case-control study. Movement Disorders October 2006;21(10):1688–92.

[9] Feldman AL, et al. Occupational exposure in parkinsonian disorders: a 43-year prospective cohort study in men. Parkinsonism & Related Disorders November 2011;17(9):677–82.

[10] Firestone JA, et al. Occupational factors and risk of Parkinson's disease: a population-based case-control study. American Journal of Industrial Medicine December 18, 2009;53(3):217–23.

[11] Barbeau A, et al. Ecogenetics of Parkinson's disease: prevalence and environmental aspects in rural areas. Canadian Journal of Neurologic Science February 1987;14(1):36–41.

[12] Sade MY, et al. Parkinson's disease prevalence and proximity to agricultural cultivated fields. Parkinson's Disease August 18, 2015:576564.

[13] Wang L, et al. Vitamin D from different sources is inversely associated with Parkinson disease. Movement Disorders April 2015;30(4):560–6.

[14] Gao J, et al. Head injury, potential interaction with genes, and risk for Parkinson's disease. Parkinsonism & Related Disorders March 2015;21(3):292–6.

[15] Crane PK, et al. Association between traumatic brain injury and late life neurodegenerative conditions and neuropathological findings. JAMA Neurology September 1, 2016;73(9):1062–9.

[16] Lai CY, et al. Increased risk of Parkinson disease in patients with carbon monoxide intoxication: a population-based cohort study. Medicine (Baltimore) May 2015;94(19):e869.

[17] Minerbo GM, Jankovic J. Prevalence of Parkinson's disease among firefighters. Neurology 1990;40(Suppl. 1):348. 42nd Annual Meeting of the AAN, Miami, Abstracts.

[18] Nicoletti A, et al. Voluptuary habits and clinical subtypes of Parkinson's disease: the FRAGAMP case-control study. Movement Disorders October 30, 2010;25(14):2387–94.

[19] Bettiol SS, et al. Alcohol consumption and Parkinson's disease risk: a review of recent findings. Journal of Parkinson's Disease 2015;5(3):425–42.

[20] Curtin K, et al. Methamphetamine/amphetamine abuse and risk of Parkinson's disease in Utah: a population-based assessment. Drug and Alcohol Dependence January 1, 2015;146:30–8.

[21] Cepeda MS, et al. Comprehensive real-world assessment of marketed medications to guide Parkinson's drug discovery. Clinical Drug Investigation 2019;39(11):1067–75.

[22] Nefzger MD, Quadfasel FA, Karl VC. A retrospective study of smoking in Parkinson's disease. American Journal of Epidemiology September 1968;88(2):149–58.
[23] Munoz DG, Fujioka S. Caffeine and Parkinson disease. A possible diagnostic and pathogenic breakthrough. Neurology January 30, 2018;90(5):205–6.
[24] Polymeropoulos MH, et al. Mutation in the α-synuclein gene identified in families with Parkinson's disease. Science January 27, 1997;276:2045–7.
[25] Singleton AB, et al. alpha-Synuclein locus triplication causes Parkinson's disease. Science October 31, 2003;302(5646):841.
[26] Paisán-Ruíz C, et al. Cloning of the gene containing mutations that cause PARK8-linked Parkinson's disease. Neuron November 18, 2004;44(4):595–600.
[27] Lee AJ, et al. Penetrance estimate of LRRK2 p.G2019S mutation in individuals of non-Ashkenazi Jewish ancestry. Movement Disorders October 2017;32(10):1432–8.
[28] Kitada T, et al. Mutations in the parkin gene cause autosomal recessive juvenile parkinsonism. Nature April 1998;392(6676):605–8.
[29] Abou-Sleiman PM, et al. The role of pathogenic DJ-1 mutations in Parkinson's disease. Annals of Neurology September 2003;54(3):283–6.
[30] Valente EM, et al. Hereditary early-onset Parkinson's disease caused by mutations in PINK1. Science May 21, 2004;304(5674):1158–60.
[31] Johnson WG, Hodge SE, Duvoisin R. Twin studies and the genetics of Parkinson's disease—a reappraisal. Movement Disorders 1990;5(3):187–94.
[32] Tanner CM, et al. Parkinson disease in twins: an etiologic study. Journal of the American Medical Association January 27, 1999;281(4):341–6.
[33] Aharon-Peretz J, Rosenbaum H, Gershoni-Baruch R. Mutations in the glucocerebrosidase gene and Parkinson's disease in Ashkenazi Jews. New England Journal of Medicine November 4, 2004;351(19):1972–7.
[34] Goker-Alpan O, et al. Parkinsonism among Gaucher disease carriers. Journal of Medical Genetics December 2004;41(12):937–40.
[35] Sidransky E, et al. Multicenter analysis of glucocerebrosidase mutations in Parkinson's disease. New England Journal of Medicine October 22, 2009;361(17):1651–61.
[36] Chang D, et al. A meta-analysis of genome-wide association studies identifies 17 new Parkinson's disease risk loci. Nature Genetics October 2017;49(10):1511–6.
[37] Schrag A, Ben-Shlomo Y, Quinn NP. Prevalence of progressive supranuclear palsy and multiple system atrophy: a cross-sectional study. Lancet November 20, 1999;354 (9192):1771–5.
[38] Mueller JC, et al. Multiple regions of alpha-synuclein are associated with Parkinson's disease. Annals of Neurology April 2005;57(4):535–41.
[39] Keller MF, et al. Using genome-wide complex trait analysis to quantify 'missing heritability' in Parkinson's disease. Human Molecular Genetics November 15, 2012;21(22):4996–5009.
[40] Hamza TH, Payami H. The heritability of risk and age at onset of Parkinson's disease after accounting for known genetic risk factors. Journal of Human Genetics October 1, 2010;55(4):241–3.
[41] Robak LA, et al. Excessive burden of lysosomal storage disorder gene variants in Parkinson's disease. Brain December 1, 2017;140(12):3191–203.
[42] Singh NK, et al. Gene-gene and gene-environment interaction on the risk of Parkinson's disease. Current Aging Science 2014;7(2):101–9.
[43] Chuang YH, et al. Gene-environment interaction in Parkinson's disease: coffee, ADORA2A, and CYP1A2. Neuroepidemiology 2016;47(3–4):192–200.

CHAPTER 3

The Family Tremor — REM sleep behaviour disorder and essential tremor

"Well, Mr. Gruschus, it appears that ..."

I do not have Parkinson's disease. If anything, I might have the initial symptoms of another condition, essential tremor. What is essential tremor? Why, it is the most common movement disorder! The lifetime risk of getting it is nearly five times greater than getting Parkinson's. Many of you might be asking, "How can that be? I've never even heard of essential tremor." If you are how I used to be, you may have seen people with essential tremor and assumed they had Parkinson's disease. I know better now.

Like so many Parkinson's cases, essential tremor involves involuntary, rhythmic shaking, most commonly in the hands. Sometimes the tremor arises in other places, such as in the muscles of the neck, causing uncontrollable no-no or yes-yes shaking of the head, atypical of Parkinson's disease. Katharine Hepburn had essential tremor. In contrast to Parkinson's, the tremor does not occur when at rest. Instead, it is an action tremor, one that manifests when trying to do something, such as holding a glass, or speaking. The symptoms usually become worse with age. Eventually, signing one's name, or even saying it, can become challenging.

Synuclein and the Coelacanth
ISBN 978-0-323-85707-9
https://doi.org/10.1016/B978-0-323-85707-9.00008-3

Despite my hypochondriacal imagination, I have not actually seen a neurologist about whether I have essential tremor. Why do I think I might develop it? It runs in my family. My father and his mother had the condition, and my aunt still does. More than half of people with essential tremor have a family history of the condition, although the exact percentage varies from study to study. Twin studies found the disease to be highly heritable, although the amount of heritability varies widely depending on which study you read, with concordances from 45% to 93% for identical twins, and from 27% to 29% for nonidentical twins [1,2]. This suggests that, unlike Parkinson's, essential tremor is primarily caused by dominant, Mendelian mutations, although because the identical twin concordance is less than 100%, some mutations could have reduced penetrance, and environmental factors could also contribute.

My father and grandmother both had action tremor in their hands, whereas for my aunt, the neck region was affected first, causing voice and head tremors. The age at onset of the tremor was similar for all three, my father in his late seventies and my aunt and grandmother in their early eighties, although for all three I seem to recall telltale symptoms even earlier, and certainly the onset was gradual. The gradual onset is typical, although they were quite a bit older at onset than average. The average age at onset of essential tremor is in the early forties, even earlier for those with a strong family history of the condition; juvenile cases are common [3].

I have something in common in with my grandmother: we both chose science as our career path. Gladys May Johnson, born in 1899 in Kentucky, was part of the early vanguard of 20th-century women entering science. She got her bachelor's degree in chemistry and a master's degree in psychology in the 1920s at Washington University in St. Louis, Missouri. After graduation, she got a position with the Ralston Purina Company in the personnel management division, what we now call human resources, designing and implementing employee personality tests. By 1930, she was already married and had resigned her job to raise her daughter; 4 years later, my father arrived. The marriage did not last, and to support her children, she became a high school math teacher, eventually moving her family to Silver City, New Mexico. She had several students who went on to be famous, including astronaut and US Senator Harrison Schmitt.

While I was in high school, my family moved to Albuquerque, New Mexico, where my grandmother had also moved in her retirement. I would visit her every Friday, bringing a big bag of tortilla chips to eat while we first watched *Dallas* on TV, then *Falcon Crest* right after. This is when I first saw her tremor; she was in her early eighties. Without a trace of self-consciousness, she told me she had essential tremor. If anything, she seemed a trifle smug about it. She assured me it was benign and that there was nothing to worry about. Cleary, she was not worried. As a child, she had contracted polio and had been bedridden for years. Eventually, first using crutches and then just leg braces, she regained the ability to walk. By adulthood, all that was left was a lingering weakness in one ankle. I suppose to her, essential tremor was a relatively minor nuisance compared with what she had already gone through.

To my discredit, I promptly forgot about my grandmother having essential tremor. I did not forget that she had a tremor; I just forgot she had told me what it was called. Flash forward 30 years. I am visiting my father Jack Sr., in Albuquerque. Like his mother and me, his first inclination had been to pursue a career in science, either archeology or geology, but instead, he chose accounting to provide a stable income for his new family. He liked his work, but after decades of orchestrating the financial minutiae of bids for government contracts at a large defense contractor, he missed the human factor. After retirement, he continued working part of the year as a tax accountant, enjoying the face-to-face interactions with his income tax customers. As a bonus, he would brag that he used the extra money he earned to go on chartered deep sea fishing trips with his buddies during the off-season. To be honest, though, he had planned and saved so judiciously for his retirement that I suspect he could easily have afforded the trips anyway.

Near the end of my visit, my father's companion Tina told me some disturbing news. My father had been moving during his sleep, seemingly acting out his dreams, sometimes kicking or flailing his arms. Even when not flailing about in his sleep, his legs would often move rhythmically. She had been struck several times and had to start sleeping in a different bed. By this time, I had begun to study alpha-synuclein, and the symptoms she was describing sounded like rapid eye movement (REM) sleep behavior disorder (RBD), a possible early sign of Parkinson's disease. I had a sinking feeling. My father was in his late seventies then, and his tremor had also just become noticeable. I heard the worry in Tina's voice and assured her that whatever was going on, it would progress very slowly, and not to worry.

In truth, back at the time I was reassuring Tina, I knew much more about the proteins involved in Parkinson's disease (and still do) than the actual disease itself. I resolved to learn more. Unfortunately, rather than first learning more about the tremor itself, I focused on the early signs of Parkinson's. I learned that the periodic leg movement he had during sleep often accompanies REM sleep behavior disorder. Did my father have other symptoms seen in the preclinical stage of Parkinson's disease? He had a poor sense of smell, but he had always had that, and in fact, so do I. When someone would ask, "Do you smell that?" we both would answer, "Smell what?" Was he constipated or have other gastrointestinal issues? I did not know and was uncomfortable calling up Tina to ask, let alone asking my father. She did mention one curious symptom, a sharp cramping pain in his upper esophagus region, perhaps a cricopharyngeal spasm, often triggered by eating chicken. He would alleviate it by drinking a hot beverage, or sometimes by making himself vomit, but I am unaware whether this type of cramp falls within the spectrum of Parkinson's disease symptoms.

Flash forward another 5 years. My partner and I have just returned from a wonderful visit to New Mexico, the Land of Enchantment. We visited the Anasazi ruins of Chaco Canyon and toured one of the oldest continually occupied communities in the United States, Acoma Pueblo. We searched for (and found!) marine fossils on a mountaintop over 10,000 feet above sea level, and we ate as much spicy New Mexico cuisine as we could squeeze into our weeklong visit, with one special exception. The night before we left, my father prepared fish he had caught on a recent trip, dorado (mahi-mahi) and yellowfin tuna, baked to perfection in a yogurt dill sauce. It was a family feast, with both my father's and Tina's family there. I had one serving of fish, waited until the others seemed sated, and then had more. I noticed my father's tremor. It was worse, but he obviously preferred to ignore it. A day after we returned home to Washington, DC, I received a call from my younger brother. My father had died of a heart attack.

I remember saying something to Tina 5 years before about not worrying too much about the sleep and tremor issues, something to the effect that I would not be surprised if in the end, these problems turned out to be eclipsed by unrelated medical issues. I wish it had taken longer for this to come true. My mother also died of a heart attack. For both my parents, there was no slow decline or drawn-out suffering, not a bad way to go, really. Of course, my hypochondriacal imagination has already presumed this will be my end, too. But as my partner Rene constantly reminds me, I could be killed crossing the street tomorrow, and he means this not as a philosophical statement, but to scold me for how blithely I cross streets.

Later the same year that my father died, Trump was elected president. The cuts Trump was proposing to the National Institutes of Health (NIH) budget were so severe that I started planning for being out of job. It was in those months of uncertainty that I first had the idea of writing this book.

Thanks, Trump! I would relocate to Albuquerque, because of its lower cost of living and to be nearer my relatives. Perhaps I would spend a few months writing the book before getting a new job, maybe teaching at a local college, or even a programming job. As the months wore on, it became clear that Congress had no intention of heeding the president's proposed NIH budget, so I put my planning aside. In early fall, I mentioned what I had been planning to a friend, Jeb Stenhouse. Jeb, a policy wonk in the Environmental Protection Agency, had been making his own plans for jumping ship. "Why not write the book anyway?" he asked. Good point. I started researching the literature in earnest, and 2 months later, I started writing.

As I kept fact-checking and reviewing the literature, now and then I would come across a mention of essential tremor and would think, "I should look that up," but then I would be distracted by some new neural study or novel protein interaction. The weeks wore on, and with several chapters already begun, I once again ran across essential tremor mentioned in yet another paper, and I finally did a Web search. I read one website, then a paper, and then another, and the realization slowly dawned: my father's tremor was not the resting tremor of Parkinson's disease; it was the action tremor of essential tremor!

With this realization, did I recall what my grandmother had told me 35 years earlier? Nope. My mind was too busy asking, "But what about my father's REM sleep behavior disorder? Is that somehow related to essential tremor?" Unclear. Two studies observed a strong connection [4,5], whereas another saw no connection at all [6]. What about a connection between Parkinson's disease and essential tremor? Yep. A small group of researchers had been studying the connection between essential tremor and Parkinson's for over a decade, and there were even hints that synuclein-related pathology was involved in essential tremor, too [7].

I am not sure whether it was the excitement of coming across this unexpected Parkinson's disease essential tremor connection, but it took a while, maybe a night or three, before the memory of my grandmother telling me she had essential tremor finally bubbled back up through my subconscious. Makes me wonder what other pertinent memories have been squirreled away too deep—or worse, fallen through the holes. Life lesson learned: I need to take more notes! Actually, I just email myself things now; it's so much easier to search through.

People with essential tremor have an increased risk for developing Parkinson's disease. How much of a risk increase depends, once again, on which study you read, but the risk seems to be at least four times greater [7]. Consistent with this, essential tremor is more common among relatives of people with Parkinson's than in the general population. One feature possibly connecting essential tremor with Parkinson's is asymmetry of the symptoms [8]. For people with essential tremor, most have symmetric symptoms; that is, both hands shake equally. A minority have essential tremor more on one side than the other. Interestingly, for those who go on to develop Parkinson's disease, the same side to show essential tremor first is usually the first to show parkinsonian symptoms. My family's essential tremor is the asymmetric kind. For my father and aunt, the tremor was more prominent on the left side, at least at first. Likewise, my aunt remembers her mother's tremor being worse in the left hand.

One reason why studies of essential tremor vary so widely is the small numbers of people involved in the studies. For instance, the study that saw the strongest connection between potential symptoms of REM sleep behavior disorder and essential tremor involved only 46 people with the condition [4]. Because essential tremor is a milder condition than Parkinson's disease, it is unsurprising that essential tremor has received less study. Another confounding factor is that some studies excluded people with essential tremor who went on to develop Parkinson's disease, whereas other studies did not. This is the case with the two most controversial studies in the field.

The two studies that exemplify the controversy involve autopsies of the brains of patients with essential tremor: one study by a group at Columbia University in New York City, led by Elan D. Louis [9], and the other by scientists in the Arizona Study of Aging and Neurodegenerative Disorders, led by Holly A. Shill [10]. Both studies are ongoing, with new reports coming out every few years as more deceased patients are autopsied. The two studies actually agree for the most part in their major findings, that patients with essential tremor typically show pathology in one of two brain regions, in either the cerebellum or the locus coeruleus. The cerebellum helps coordinate muscular movement, speech, and balance, and when it is damaged, it can interfere with these functions as well as cause tremor. The locus coeruleus, which is Latin for "blue place," named for the bluish pigmentation of its cells, is located in the upper part of the brain stem, just next to the cerebellum, and is a relay center between other parts of the brain and the cerebellum; it is important for attention and arousal.

The controversy centers on alpha-synuclein pathology in the locus coeruleus. The Columbia group saw moderate to extensive synuclein pathology in the locus coeruleus of several patients with essential tremor, often with synuclein pathology in other regions of the brain as well, though less pronounced [11]. Although the Arizona group also observed essential tremor cases with locus coeruleus pathology, they saw mostly mild synuclein pathology, and only in a small fraction, 3 of 21 patients, or 13%, a percentage similar to what they saw in controls (10%) [12]. For most patients, the Arizona group observed no synuclein pathology.

One reason for the discrepancy could be that the Columbia group initially included patients who had both essential tremor and Parkinson's disease along with patients who had essential tremor alone. The Arizona study excluded patients who had both conditions. The Arizona group's approach makes sense. In Parkinson's disease, the locus coeruleus is one place where alpha-synuclein pathology is often seen. If one wants to make the case that synuclein pathology in the locus coeruleus is correlated with essential tremor, Parkinson's cases should be excluded because any synuclein pathology observed could be due to Parkinson's disease rather than essential tremor. This turns out to be trickier than it sounds, however, because many people with essential tremor also develop resting tremor along with action tremor, but often without the other cardinal symptoms of Parkinson's disease: bradykinesia, postural instability, and rigidity. Parkinson's patients can develop action tremor, too. With so few brains available to autopsy, there is pressure to include as many essential tremor cases as possible, even those who might be showing symptoms of other disorders.

As the Columbia group added more cases to their ongoing study, now excluding patients who had both Parkinson's and essential tremor, the percentage of essential tremor patients with significant alpha-synuclein pathology in the locus coeruleus fell from 33% of six patients in 2006 [13] (excluding the two with both essential tremor and Parkinson's disease) to 19% of 31 patients in 2007 [11], and then to 15% of 40 patients in 2009 [9]. As the numbers from the Columbia group grew closer over time to those of the Arizona group, it looked increasingly likely that the initial results from the Columbia group were a statistical fluke.

"Not so fast!" cried other experts in essential tremor. Patients with essential tremor have a much greater chance of developing Parkinson's disease, and if these studies excluded patients who also had Parkinson's disease, that could bias the results toward lower percentages. This is especially true if only those with locus coeruleus pathology are the ones more prone to getting Parkinson's disease. "And what of those patients with a pattern of moderate to extensive alpha-synuclein pathology in the locus coeruleus, compared with less pronounced pathology in the other usual regions seen for Parkinson's?" to paraphrase the Columbia group. They had a point; whereas locus coeruleus pathology is common in Parkinson's, it is almost always the case with Parkinson's patients that other regions have equal or more extensive synuclein pathology upon autopsy [14].

An additional confounding factor is that both Columbia and Arizona studies included patients and controls with mild cognitive impairment, often a prelude to dementia. In a 1990 study by Robert H. Perry et al. at the University of Newcastle on Tyne, UK [15], the researchers found upon autopsy that patients with psychiatric conditions, mostly mild cognitive impairment, but otherwise neurologically healthy, had a greater chance of having alpha-synuclein pathology: 9% versus those without such conditions, at 2%. Thus, the inclusion of patients and controls with mild cognitive impairment could inflate the percentages of synuclein pathology in both patients and controls, further obscuring any actual causative connection between essential tremor and alpha-synuclein.

The University of Newcastle study is notable for another result, too. Two of 131 controls with no psychiatric or neurological conditions had substantial neuron loss in the locus coeruleus compared with the other controls. This means that one can have substantial neuron loss in the locus coeruleus and not have essential tremor. Indeed, over a lifetime, one typically loses about 50% of neurons in the locus coeruleus [15]. Perhaps these controls were in the preclinical phase of essential tremor, and it would have emerged had they lived longer. But what if the neuron loss in the locus coeruleus is not the cause of the condition, but instead a spillover effect of pathology occurring in an adjacent part of the brain?

A 2017 study by scientists in Germany, Luxemburg, and The Netherlands focused on characterizing alpha-synuclein pathology in the cerebellum of Parkinson's patients [16]. Previous studies had occasionally seen synuclein pathology in the cerebellum, but much less than in other brain regions, so the cerebellar pathology received little attention. However, because the tremor and balance symptoms seen in Parkinson's disease are similar to those seen in disorders involving the cerebellum, this group of scientists thought that a more thorough study of synuclein pathology in the cerebellum was warranted. Interestingly, one of the regions most affected by synuclein pathology was the pars cerebellaris loci coerulei, which lies in the cerebellum immediately adjacent the locus coeruleus, connected to it by "cellular bridges of variable width" with similarly pigmented cell bodies, as described by Heiko Braak in a paper from 1975 [17]. It makes me wonder whether synuclein pathology in this cerebellar region might actually be the cause of essential tremor, at least in some cases.

Many questions remain regarding the alleged alpha-synuclein essential tremor connection. For those with a strong family history of essential tremor, is there more or less alpha-synuclein pathology compared with others without an obvious family history? For those with asymmetric symptoms, do they have more or less synuclein pathology compared with those with symmetric essential tremor, and if there is synuclein pathology, does it show the same asymmetry? The most immediate path to the answers to these questions is more brains, a lot more. To clarify, more autopsies are needed of people who had essential tremor, especially siblings (even better, twins) to address these questions definitively. Of course, future, noninvasive imaging technologies pinpointing alpha-synuclein pathology in the living brain could render these autopsies unnecessary, but that day has not yet arrived. When I die, I want my brain to be donated to a brain repository of people who had essential tremor, even if I do not develop the condition (they need controls, too!). I hope I can convince my brothers to do the same, and maybe some of you readers might consider it, too.

With so many people with essential tremor having family histories of the disease, one would think finding the associated Mendelian mutations would be straightforward. Family kindred studies have indeed turned up several candidate genes, but follow-up studies have mostly failed to find any of the same mutations in others with essential tremor, and when they did find them, the mutations turned out to be rare. Genome-wide association studies have also been performed, and a few polymorphisms have been identified that might be risk factors, but risk factors alone cannot explain the high heritability of the condition [18]. The autism spectrum disorders share the same predicament, because they are highly heritable, but with no obvious Mendelian mutations for the vast majority of those affected.

Others have hypothesized that inheritance of essential tremor might involve multiple gene variants, a combination of rare mutations, perhaps along with common polymorphisms, as well as environmental factors, the so-called complex/polygenic model [19]. For example, say two dominant mutations inherited together cause the condition. Then, the children of the affected parent have a 25% chance of getting both mutations and inheriting the condition, compared with a 50% chance if inheriting just one mutation were necessary. The essential tremor concordance rate for nonidentical twins is 27%—29%, consistent with a polygenic mode of inheritance. Kindred studies typically look for single mutations, present in affected members and absent in unaffected ones. These analyses could also have looked for pairs of mutations or more complicated combinations of genetic factors, but as far as I know, no such polygenic analyses have been conducted for essential tremor to date.

Epigenetics might have a role in this putative complex/polygenic mode of inheritance [19]. During embryonic development and later in life, genes can be chemically modified, usually to silence their transcription. This process, epigenetics, is so named because it can happen after and in addition to the genetic information that an embryo has inherited. The X chromosome is a good example; women have two copies, and men, one. In females, transcribing both copies of some X chromosome genes can produce too much of the corresponding proteins to a negative effect. To counter this, one copy, from either the mother or father, becomes chemically modified and packed differently to prevent or reduce its being transcribed. It is not just the X chromosome that gets modified; many autosomal genes also undergo epigenetic gene modification. That epigenetics might have a role in inheriting essential tremor is an intriguing idea, especially because environment can influence which genes become silenced. However, in a way, invoking epigenetics to help explain heritability is just passing the genetic buck, because ultimately, genes are responsible for the proteins responsible for the chemical modifications to genes; that is, genes code the proteins that carry out epigenetics.

Yet another reason why genetic causes may have eluded discovery is that essential tremor might not be just one condition, but rather several conditions with different underlying pathologies that produce similar symptoms [20]. The Columbia and Arizona studies already reported two types of pathology: in the cerebellum and the locus coeruleus. The presence or absence of alpha-synuclein pathology could indicate different genetic causes, too. Studies combining pathology and genetics have the potential to address these questions (more brains needed!).

On a recent visit to New Mexico, I was talking with my aunt Suzie, my father's sister, at a family get-together. It was not until afterward that I realized she had no head or voice tremor. Years before, when I last saw her, I remember both being very noticeable. Was she taking medication for the tremor? I wondered. Maybe one of the L-dopa drugs? I resolved to find out. This was still before I had been reacquainted with essential tremor. I got my cousin's son involved, Jonathan Hulse, who was in the premed program at the University of New Mexico (now graduated and in the UNM School of Medicine), and asked him to interview his grandmother for me. Here is some of the interview:

"He told me that his book is about [...] Parkinson's disease," Jonathan explains.

"Ooooh! Why?" Suzie asks.

"Ummm ... I think that's what he deals with at his work."

"You're kidding? Does he have the tremor?"

"He does, or he thinks that he's starting to get it."

"Yeah, he's about the right age to start."

"Yeah, he notices his hands shake when his blood sugar is low or if he's had too much caffeine."

"Go ahead and sit down, if you like," Suzie tells Jonathan.

"So his book, though, is mostly just about the molecular biology and the genetics of Parkinson's disease."

"Interesting. Now I'm interested."

"So it's, like, about the things that are going on in the brain and stuff like that, that cause Parkinson's disease."

"Uh-hm, uh-hm. And you know what helps is marijuana, a little bit."

"I think I've heard something about that."

"Yeah. Your dad told me all about that!" Suzie says with a laugh.

"Has it helped your tremor at all?"

"Oh, I think it does, but it doesn't cure it. It comes back."

"So it just helps with the symptoms a little bit," says Jonathan.

"It helps for the time being!" answers Suzie with another laugh.

First, let me point out that medical marijuana is completely legal in New Mexico. In fact, New Mexico was the first state to legally recognize the medicinal properties of marijuana back in 1978, although medical marijuana was not fully legalized there until 2007. Second, how I would have loved to have seen my father's reaction to my aunt's revelation! He was a straitlaced type, and I imagine he would have been mortified, at least mildly, to have learned his big sister had become a pot smoker. Here is more of the interview:

"… You're going to be a good doctor, by the way," Suzie tells her grandson.

"Thank you. So, have you, you've met with a doctor and talked about [your tremor] before, right?" Jonathan asks.

"I have, but my doctor's also into health, so he didn't get real excited," Suzie answers with a laugh.[1]

"What do you mean?"

"I asked him if there was a doctor here in Albuquerque that was knowledgeable that I could go see. And he did find someone for me, but she was a little flaky. And so … let's see … how can I say this?" She pauses for a second. "I've heard that Botox was good, and I can't remember for what, I think it was shaking, so I thought. And she wouldn't do it. She was kind of like, 'Oh, I don't think we need to do that right now,' or something. She was kind of hard to deal with. And so, I just go to a doctor that does Botox, and I have him put some in my lines here," she gestures at her forehead, "and it does help me. So, I'm using him for something that it isn't supposed to be helpful for, I guess. Does that make sense?"

"Sure," Jonathan answers.

"It helps my tremors."

"So, this doctor that you get the Botox from … ?"

"He's not a doctor, I don't think. He's just, uh, is a, uh … you call him doctor, but I don't know if it's a true doctor."

"Is he, like, a doctor of osteopathic medicine?"

"Something like that."

"So, the doctor that he referred you to, that you said you didn't like, was she, like, a regular doctor?"

"Yeah, but she was into nervous tremors, I guess."

"Did she give you a diagnosis?"

"No."

[1] By "into health," I believe Suzie is saying her doctor is into alternative medicine.

First, how irritating that her "regular" doctor did not offer or discuss a possible diagnosis. Second, apparently it was marijuana, Botox, or some combination of both that had completely suppressed my aunt Suzie's tremor when I had last talked to her. There are anecdotal reports about marijuana helping to suppress essential tremor, but no formal studies. A small Brazilian study reported that cannabidiol was effective in treating REM sleep behavior [21], but no follow-up studies have been performed. I imagine it is hard enough to obtain funding to study medical marijuana for more serious afflictions, let alone a relatively mild condition such as essential tremor. From what I have read, for Botox to help head and voice tremors, it has to be injected into neck muscles, not the forehead, so maybe my aunt was getting both locations treated, one for the tremor and the other to reduce wrinkles.

One substance with a strong connection to essential tremor is alcohol. For about half of people with essential tremor, alcohol helps (yes, helps!) reduce symptoms [22]. Just one drink a day is often sufficient. This is not the case for the resting tremor of Parkinson's; alcohol usually makes no difference. My father and my grandmother rarely drank, so I do not know how the tremor responded to alcohol. I am the only alcohol enthusiast in my immediate family; a good thing because my partner is studying to become a sommelier!

I learned one more tidbit about my aunt's condition at another recent family dinner. Apparently, she has lost her sense of smell, from litter box odor to food burning in the kitchen. My father's sense of smell was never *that* bad. So, both my father and my aunt had shown an early possible sign of Parkinson's disease: REM sleep behavior disorder for my father and anosmia, the loss of the sense of smell, for my aunt. What about my grandmother? She died of pneumonia at age 93. She already had advanced dementia by that time; the last time we saw each other, her face lit up, but she barely spoke. Did she have dementia with Lewy bodies, Alzheimer's disease, or some other condition? We will never know; no brain autopsy was performed, but she did not seem to have had any of the hallucinations or delusions that sometimes accompany dementia with Lewy bodies. Odds are that it was Alzheimer's; by age 93, almost half of people have the condition.

As mentioned in the interview, I have had isolated episodes of what might be action tremor, possibly enhanced by a combination of low blood sugar and too much caffeine. Did the worst episodes occur when I had drunk no alcohol on the previous days? Probably, but I cannot recall for sure. One especially embarrassing incident happened at work, while I was stretching my arm forward to put a test tube into a very expensive spectrometer at NIH. Just as I reached the hole at the top of the instrument where the sample is inserted, my hand twitched, breaking the delicate glass of the special tube inside the hole, spilling broken shards and precious protein solution down into the heart of the instrument.

For a moment, I stood in disbelief. As the reality sank in, I tried to devise a way to rescue the situation, maybe with a powerful Shop-Vac, or at the very least, I could try to get the glass shards out the spectrometer with masking tape balled up at the end of a wire. I actually tried the masking tape idea. It failed. I would have to confess my disaster. I spent the next day with the facilities manager removing the core of the instrument, removing the glass bits and cleaning away the protein residue with a series of different solvents. To their credit, neither my boss nor the facilities manager said anything negative; accidents happen.

Something else happened that makes me think even more that I inherited my father's condition. I was dreaming about the intersection of two dirt roads, my body lying on the reddish, dusty road surface. An elephant was approaching (maybe I was dreaming I was in Africa?) and was about to step on me. I had to roll out of the way! I rolled right out of bed.

At the start of the chapter, I wrote that essential tremor is the most common movement disorder. It affects more than seven million in the United States [23] and appears to be comparably common worldwide [24]. Part of the reason why the number of people is higher than for Parkinson's is that essential tremor starts at an earlier age on average. Even if just a fraction of essential tremor cases are associated with alpha-synuclein pathology, it could be that essential tremor, not Parkinson's disease, is the most common synucleinopathy. What does this mean for Parkinson's disease? It means that studying essential tremor could also help reveal more about alpha synuclein—associated diseases and, hopefully, lead to new ways of helping those affected. More brains needed!

References

[1] Tanner CM, et al. Essential tremor in twins: an assessment of genetic vs environmental determinants of etiology. Neurology October 23, 2001;57(8):1389—91.

[2] Clark LN, Louis ED. Challenges in essential tremor genetics. Revue Neurologique June 1, 2016;171(6—7):466—74.

[3] Louis ED, Dogu O. Does age of onset in essential tremor have a bimodal distribution? Data from a tertiary referral setting and a population-based study. Neuroepidemiology February 2007;29(3—4):208—12.

[4] Lacerte A, et al. Increased prevalence of non-motor symptoms in essential tremor. Tremor and Other Hyperkinetic Movements (NY) September 2, 2014;4:162.
[5] Barbosa R, et al. Probable REM-sleep behavior disorder and dysautonomic symptoms in essential tremor. Tremor and Other Hyperkinetic Movements (NY) December 29, 2017;7:522.
[6] Adler CH, et al. Probable RBD is increased in Parkinson's disease but not in essential tremor or restless legs syndrome. Parkinsonism & Related Disorders July 2011;17(6):456–8.
[7] Tarakad A, Jankovic J. Essential tremor and Parkinson's disease: exploring the relationship. Tremor and Other Hyperkinetic Movements 2019;8:589.
[8] Ghika A, et al. Motor and non-motor features: differences between patients with isolated essential tremor and patients with both essential tremor and Parkinson's disease. Tremor and Other Hyperkinetic Movements (NY) August 14, 2015;5:335.
[9] Louis ED, et al. Torpedoes in Parkinson's disease, Alzheimer's disease, essential tremor, and control brains. Movement Disorders August 15, 2009;24(11):1600–5.
[10] Shill HA, Adler CH, Beach TG. Lewy bodies in essential tremor are no different than in controls. Parkinsonism & Related Disorders February 2016;23:106–7.
[11] Louis ED, et al. Neuropathological changes in essential tremor: 33 cases compared with 21 controls. Brain December 2007;130(12):297–307.
[12] Shill HA, et al. Pathologic findings in prospectively ascertained essential tremor subjects. Neurology April 15, 2008;70(16):1452–5. pt. 2.
[13] Louis ED, et al. Neuropathologic findings in essential tremor. Neurology June 13, 2006;66(11):1756–9.
[14] Mills KA, et al. Gait function and locus coeruleus Lewy body pathology in 51 Parkinson's disease patients. Parkinsonism & Related Disorders December 2016;33:102–6.
[15] Perry RH, Irving D, Tomlinson BE. Lewy body prevalence in the aging brain: relationship to neuropsychiatric disorders, Alzheimer-type pathology and catecholaminergic nuclei. Journal of the Neurological Sciences December 1990;100(1–2):223–33.
[16] Seidel K, et al. Involvement of the cerebellum in Parkinson disease and dementia with Lewy bodies. Annals of Neurology June 2017;81(6):898–903.
[17] Braak H. On the pars cerebellaris loci coerulei within the cerebellum of man. Cell and Tissue Research July 8, 1975;160(2):279–82.
[18] Müller SH, et al. Genome-wide association study in essential tremor identifies three new loci. Brain: A Journal of Neurology 2016;139(12):3163–9.
[19] Testa CM. Key issues in essential tremor genetics research: where are we now and how can we move forward? Tremor and Other Hyperkinetic Movements (NY) 2013;3. pp. tre-03-105-1843-1.
[20] Espay AJ, et al. Essential pitfalls in "essential" tremor. Movement Disorders March 2017;32(3):325–31.
[21] Chagas MH, et al. Cannabidiol can improve complex sleep-related behaviours associated with rapid eye movement sleep behaviour disorder in Parkinson's disease patients: a case series. Journal of Clinical Pharmacy and Therapeutics October 2014;39(5):564–6.
[22] Growdon JH, Shahani BT, Young RR. The effect of alcohol on essential tremor. Neurology March 1975;25(3):259.
[23] Louis ED, Ottman R. How many people in the USA have essential tremor? Deriving a population estimate based on epidemiological data. Tremor and Other Hyperkinetic Movements 2014;4:259.
[24] Benito-León J, et al. Prevalence of essential tremor in three elderly populations of central Spain. Movement Disorders April 2003;18(4):389–94.

CHAPTER 4

The Knockout – The history of the knockout mouse and the surprising results for the synuclein proteins

Animals have played a vital role in the shaping of human history. Which one has been the most influential? Historians, economists, and anthropologists might argue for the horse, the cow, or maybe the dog, but when it comes to modern medicine, most scientists would probably choose the common house mouse, *Mus musculus*. Starting in the early 20th century, experiments with laboratory mice have helped to reveal the causes of cancer and unravel the mysteries of the immune system. More Nobel Prizes in medicine have involved research on mice than any other laboratory animal. This revolution in modern medical research started when Harvard geneticist William E. Castle bought some mice in 1902, just after the rediscovery of Gregor Mendel's work on the genetics of peas.

Synuclein and the Coelacanth
ISBN 978-0-323-85707-9
https://doi.org/10.1016/B978-0-323-85707-9.00010-1

Castle bought the mice from Abbie E.C. Lathrop, a former schoolteacher, who, owing to failing health, quit teaching and moved to a farm in Granby, Massachusetts [1]. There, after a failed attempt at poultry farming, she turned to breeding fancy mice and rats with the help of her friends Ada Warner Gray and Edith Chapin. Granby is over 80 miles from Boston. Why did Castle buy mice from Granby rather than somewhere nearer Boston? Perhaps it was simply a matter of marketing, just advertising in the right journal, or maybe Lathrop had already developed a reputation for the best mice in New England.

Lathrop kept detailed journals of her breeding projects, which made her mice more attractive to researchers. She even collaborated with experimental pathologist Leo Loeb at the University of Pennsylvania, coauthoring several scientific papers on cancerous skin lesions that had appeared in her mouse strains. Her failing health did not improve, and she died in 1918 of pernicious anemia at age 40, before her final paper was published. Abbie E.C. Lathrop is buried in West Cemetery, Granby, with her friends Ada and Edith, who later were buried nearby.

In 1922, a protégé of Castle, geneticist Clarence Cook "C.C." Little, brought mice from Harvard with him when he founded a new summer laboratory in Bar Harbor, Maine [2]. The laboratory was named the Roscoe B. Jackson Memorial Laboratory after an automobile magnate donor, and it soon became the primary source of laboratory mice in the United States. One of Little's inbred mouse strains, bred from Lathrop's mice, was C57BL/6J, known as Black 6, and today it is the world's most widely used strain for making genetically modified mice. Its genome was published in 2002, a year after the human genome, making it the second vertebrate to have its genome sequenced.

Little's contributions in the development of laboratory mice have been invaluable to medical science, but in other ways his legacy is mixed [2]. He was strong supporter of the rights of women and minorities, and was president from 1936-1939 of the American Birth Control League (ABCL), which went on to become Planned Parenthood. However, Little was also an outspoken proponent of eugenics and euthanasia, and had served as president of the American Eugenics Society in 1929. In the 1930s and 1940s, he could have redeemed himself in history's eye, because he had a leading role in establishing the American Cancer Society, where he asserted that inhaling the fine particles in tobacco smoke was unhealthy. However, redemption was lost in the 1950s, when he moved on to serve as the scientific advisory board director of the Tobacco Industry Research Committee, a trade association funded by the tobacco industry that lobbied Congress and funded pro-tobacco smoking research. In 1957, Little recanted his previous stance on the unhealthiness of tobacco smoke [3], stating that research "produced no evidence that cigarette smoking or other tobacco use contributes to the origin of lung cancer."

In 1974, the first transgenic mouse (that is, the first mouse with foreign DNA incorporated into its genome) was created by Rudolf Jaenisch, now at MIT, and Beatrice Mintz at the Fox Chase Cancer Center [4]. The DNA, from a simian virus, was injected into surgically removed, fertilized mouse embryos and reimplanted in foster mother mice, resulting in mice born with the viral DNA throughout their bodies. It took them 2 more years of research to prove that the viral DNA had really been incorporated into the mouse genome. Meanwhile, Martin J. Evans, working with Richard Gardner at Oxford, had shown that injecting embryonal carcinoma cells into fertilized eggs and implanting them into foster mice produced chimeric mice [5], whose tissues contained a mixture of the carcinoma and normal cells, although the carcinoma cells were not passed on to offspring. Chimeric mice had been created earlier by Mintz and coworkers in the 1960s by combining embryos from different mouse strains. The difference with Evans' work was that the embryonal carcinoma cells he injected could be grown in cell culture, which opened the door to genetic manipulation of the cells outside the embryo. In scientific jargon, such manipulation is called *in vitro*, Latin for "in glass," as opposed to *in vivo*, "in a living thing."

What the researchers needed were embryonic cells that could be genetically manipulated in cell culture, but without the genetic abnormalities of the carcinoma cells that prevented them from being passed on to offspring. Evans teamed up with embryologist Matt Kaufman to do just that, and by 1981 they had succeeded [6], as did their colleague Gail Martin later that year [7]. Their cultured embryonic stem cells could be injected into embryos to create chimeric mice, and by 1984 they had proven that the mice were able to pass the genetic changes on to their offspring [8]. Evans and Kaufman wrote in their 1981 paper that their embryonic stem cells could be used "as a vehicle for the transfer into the mouse genome of mutant [genes]." The next question was how to use the vehicle: that is, how to manipulate the embryonic stem cells genetically. Viruses are fine for randomly inserting foreign DNA, but what if one wanted to insert DNA at a particular location in the genome?

The answer was homologous recombination. We have already encountered recombination in Chapter 2, where the single-copy chromosomes in germ cells, eggs in the mother or sperm cells in the father, are formed by mixing corresponding chunks of the DNA from each respective parent's dual-copy chromosomes. In bacteria, it was already known that small bits of DNA called plasmids could transfer foreign DNA into bacterial genomic DNA. Higher organisms appeared to use similar cellular machinery to carry out homologous recombination in the formation of germ cells.

Perhaps injected foreign DNA with homology to particular genes in higher organisms such as humans could be incorporated at the site of the gene using homologous recombination, similar to what had been observed in bacteria. Most scientists were skeptical; the size of the human genome, and those of other vertebrates, is 1000 times larger than the typical bacterial genome. The chances that injected DNA could find the correct gene site seemed miniscule, like finding a needle in a haystack. Nevertheless, by the early 1980s, several teams of scientists were attempting this seemingly impossible task.

At the forefront of these scientists was the group led by Mario R. Capecchi at the Salk Institute in La Jolla, California. In 1982, they found that when DNA for the viral enzyme thymidine kinase (tk) was injected into cultured mammalian cells, multiple copies of the injected DNA were often inserted into the cell's genomic DNA in series, so-called "head-to-tail concatemers" [9]. They were able to show that the multiple copies were probably generated by homologous recombination. Apparently, once a copy of the injected DNA was inserted into the cell's DNA, the inserted copy recruited more copies to be inserted at the same site, using the cellular machinery for homologous recombination.

In the paper describing their results, Capecchi and his team speculated that "It will be interesting to determine whether we can exploit this machinery to 'target' a gene by homologous recombination to a specific chromosomal location." By 1985, another team of scientists, led by Oliver Smithies at the University of Wisconsin, had achieved this very goal [10]. Targetting the human beta-globin gene, they used homologous recombination to insert a DNA sequence containing a neomycin (*neo*) resistance gene flanked by human beta-globin gene DNA sequences. The skeptics had been proven wrong.

Capecchi and Smithies were eager to take the next step, to see whether foreign DNA could be targeted to insert at a specific gene site in embryonic stem cells. Evans provided both of them with mouse embryonic stem cells from his laboratory, and they independently set to work, in both cases focusing on the same gene, HPRT, located in the X-chromosome. Defects of HPRT cause Lesch-Nyhan syndrome, also known as juvenile gout, a rare condition that affects male children, causing excessive uric acid to build up in all body fluids. The disease leads to intellectual disability, reduced muscle tone, and other serious kidney and neurological issues, including self-mutilation.

The embryonic stem cells that Smithies had obtained from Evans had a defective version of the HPRT gene, where the first part of the gene had been deleted. Using a DNA fragment containing the missing pieces of the HPRT gene, the team led by Smithies introduced the DNA into the cells by electroporation. In this technique, the cells are subjected to high-voltage pulses that temporarily cause their cell membranes to become porous, allowing the DNA to enter the cells. The experiment was a success; they repaired the gene via homologous recombination [11].

In contrast, Capecchi used a process called microinjection, in which the DNA was injected directly using tiny glass tubes to pierce the cell membrane and inject the DNA. Of the two techniques, microinjection was more efficient, but much more labor-intensive. Also in contrast to Smithies, Capecchi and his partner, Kirk R. Thomas, at the University of Utah, sought to inactivate the HPRT gene in embryonic stem cells possessing a normal version of the gene. Their injected DNA fragment contained the neo gene sequence inserted inside part of the sequence of the HPRT gene. Their experiment was also a success [12]; their inserted DNA had inactivated the HPRT gene and also made the cells resistant to neomycin (actually, they used a drug related to neomycin, geneticin, also known as G418).

Thomas and Capecchi speculated that their transformed stem cells containing the inactivated gene could be injected into mouse embryos that could then develop into mice containing the targeted gene inactivation in only one copy of their two chromosomes, thus avoiding the immediate effects of lethal recessive mutations. They wrote, "If successful, this technology will be used in the future to dissect the developmental pathway of the mouse as well as to generate mouse models for human disease." Their words proved prophetic.

There was one final hurdle before their transformed cells could be used to create chimeric mice. After treatment with G418, the cultured stem cells that had not inserted the foreign DNA construct had no resistance and died. However, of the cells that survived, only one in 1000 actually had the DNA inserted in the target gene. Instead, in 99.9% of cases, the DNA was inserted at random places in the genome. Making chimeric mouse embryos, impregnating foster mice mothers, and waiting for them to come to term was time-consuming, not to mention labor-intensive. No one would do all that work and waiting if there were only 0.1% chance of success. What Capecchi's team needed was a way to select cells that had inserted the foreign DNA *and* had it inserted in the right place.

By 1988, they had the answer, and it was ingenious. Capecchi and his team added the tk gene at the end of their DNA construct, which they got into the cell using electroporation (they had switched to the less labor-intensive method by this time) [13]. With homologous recombination, the cell could insert a foreign DNA sequence into the target gene if it was sandwiched between DNA that matched part of the target gene sequence, and they chose the neo gene as the foreign DNA sequence. If the DNA was inserted by homologous recombination, only the neo gene and the flanking DNA homologous to the target gene would be inserted at the target site. On the hand, if the DNA construct was inserted at a random site in the genome, the full construct, including the tk gene, would be inserted.

So what is the big deal about the tk gene? Recall that it is an enzyme from a virus. Mice have their own version of the enzyme, but it is sufficiently different from the viral enzyme that it does not bind to the drug ganciclovir. Does the drug sound familiar? It is used to treat opportunistic eye infections caused by cytomegalovirus, and it is closely related to drugs used to treat herpes. In fact, the tk gene that Capecchi's team used was from the herpes simplex virus. The viral tk enzyme converts ganciclovir into a cell toxin. By putting the tk gene at the end of their foreign DNA construct, they could treat their cell cultures with ganciclovir, and any cell that had incorporated the construct by random insertion would have the tk enzyme, which would convert the ganciclovir into its toxic form, and the cell would die. Capecchi called his new strategy "positive-negative" selection. By treating their cell cultures with both G418 and ganciclovir, only the cells with the targeted DNA insertion would survive.

By 1989 the floodgates had opened, with several laboratories creating chimeric mice using stem cells created with the new gene-targeted DNA technique. They then bred the chimeras, and a fraction of their offspring would be heterozygous for the inserted DNA: that is, with one copy of the inserted DNA in every somatic cell in their bodies. Then, by breeding the heterozygous mice, they could obtain homozygous mice in which both copies of the targeted gene had the inserted DNA. The inserted DNA sequences could be used to repair damaged genes or introduce foreign genes, but the most common use of the new technique was to create mice in which the target gene had been inactivated, so-called knockout mice. Soon, scientists in every biological field were clamoring for mice with their particular gene of interest knocked out.

At first, it was thought that the cost of creating and maintaining the knockout mice strains would be prohibitive, and the clamoring scientists would just have to be patient. However, where interest is focused, money is funneled, and laboratories and facilities were created with the sole purpose of creating and maintaining the ever-increasing numbers of knockout strains. By the 2000s, thousands of strains with different genes knocked out had been created. To save money, they would create new knockouts, characterize any effect on the mice, and then freeze their eggs and sperm, so that they did not need to maintain the live mice. When a scientist needed the particular knockout strain, they could thaw and fertilize the eggs and impregnate foster mice mothers to recreate the desired knockout mouse line. In 2007, the International Knockout Mouse Consortium was created, connecting the various laboratories with the goal of creating a knockout for every gene in the mouse genome. At the time of writing, they have nearly achieved this, creating over 20,000 different knockout strains.

The same year the consortium was created, Capecchi, Evans, and Smithies were awarded the Nobel Prize in medicine. In the 20 years between their breakthrough work and the award, surprisingly little progress had been made in making knockouts of other species. Rats were an especially unhappy surprise; the techniques that had worked so well for mice either performed poorly or not at all for rats, despite how closely the species were related. Primates were seemingly a pipe dream; they could not even be cloned. Nevertheless, eventually, and ironically, knockout rats were created in the same year as the knockout mouse Nobel Prize [14]. New advances in gene-targeting technology, DNA sequence-targeting zinc-finger nucleases called TALENs, for example, have helped with the effort.

For primates, the tide might finally be turning. In 2015, the first knockout rhesus monkeys were created in China using yet another new DNA sequence-targeting technique, the CRISPR/Cas9 system, this time adapted from the immune systems of bacteria (yes, bacteria have immune systems). The CRISPR/Cas9 system helps protect bacteria from them one of their major nemeses, bizarre-looking viruses called bacteriophages (do a Web search on them!). With the new technology and direct DNA sequencing, the old multistep selection process for knockout stem cells is no longer necessary, and the new technology is termed "gene editing."

Using this new gene-editing technology on whole embryos, Chinese scientists introduced mutations into the dystrophin gene, which is mutated in the disease Duchenne muscular dystrophy [15]. Unlike knockouts in mice, the introduced mutations were not uniform throughout the tissue of the affected monkeys, and whether the affected monkeys can pass their gene-targeted knockout mutations to their offspring has not yet been reported. Still, even partial knockouts in primates will be invaluable to medicine. In 2018, China achieved two more milestones. They successfully cloned primates for the first time by nuclear transfer [16], the same technique used to create Dolly the Sheep. Moreover, opening a Pandora's box of ethical issues, He Jiankui of the Southern University of Science and Technology, China, shocked and angered the science world by announcing the creation of two CRISPR gene-edited human babies. That bombshell will no doubt be featured in many books to come.

In 2000, the first results on alpha-synuclein knockout mice were published by a team led by Asa Abeliovich at UCSF [17]. Finally, the moment of truth: I would get to learn why alpha-synuclein was so important that it had been conserved all the way back to our last common ancestor with the coelacanth. And the answer was—nothing. Well, practically nothing. The mice were perfectly healthy with no "gross morphological or behavioral abnormalities." I had expected major defects, profoundly aberrant movement, or at the very least some pronounced neurological deficits—but, nothing. I was dumbfounded.

The brains of the alpha-synuclein knockout mice looked completely normal; however, the researchers saw some subtle differences, particularly in the level and release of the neurotransmitter dopamine in specific parts of the brain. That was something; dopamine is the neurotransmitter that is deficient in Parkinson's patients. The drug amphetamine works by increasing dopamine levels in the brain, so the researchers tested its effect on the mice and saw that their locomotion was less stimulated than control mice. Aha! locomotion—at least I had a straw to grasp.

Another paper came out in 2001, by Christian G. Specht and Ralf Schoepfer, on mice that had spontaneously lost their alpha-synuclein gene [18]. This paper was even more underwhelming than the first. In 2002, yet another alpha-synuclein knockout paper came out, by Deborah E. Cabin et al. at the National Human Genome Research Institute at NIH [19]. This team had inactivated *SNCA* by inserting DNA in a different part of the gene. They confirmed the results of the first paper and saw that the dopamine-releasing neurons had less dopamine in reserve compared with normal mice. They also observed a subtle behavioral difference: in an open field (actually a small, walled area in the laboratory), the knockout mice reared less on their hind legs and spent more time near the walls. Perhaps this was important evolutionarily somehow? Rearing up can help one scope out potential mates and rivals better. On the other hand, the evolutionary consequences are quite different if one is asking, "Is that an owl?"

In the interest of full disclosure, at the time I read the alpha-synuclein knockout papers, I had been studying the protein for less than a year, and I came late to the scene, starting only in 2010. At that time, I was not actually dumbfounded by the knockout results, just mildly puzzled. It took several more years of studying the synucleins, their protein interactions, and their sequence conservation across species for me to go from puzzlement to dumbfundity (okay, not a real word, but I am playing my artistic license card). It was because of this mystery that I became obsessed with alpha-synuclein, trying to reconcile its apparent unimportance with its central role in Parkinson's disease, and from that obsession, this book was born.

Also, I have to apologize for not mentioning earlier that there are entire orders of mammals that appear to have lost the alpha-synuclein gene. No alpha-synuclein gene has yet been found in marsupials and monotremes, although they have beta- and gamma-synuclein genes. One placental mammal, the common shrew *Sorex araneus* also appears to lack alpha-synuclein. The evidence that alpha-synuclein is expendable was already there in nature. What sort of evolutionary selection could have kept the alpha-synuclein sequence so conserved in tetrapods, and yet could have allowed their loss in some mammals? In a way, this wrinkle makes the synuclein sequence conservation question even more intriguing, and we will return to this evolutionary paradox; but first, there is more of the synuclein knockout story to tell.

As research continued on alpha-synuclein knockout mice, the situation became more muddled, as other laboratories, some that had generated their own versions of the knockout mice, could not reproduce the subtle effects seen by others. Perhaps the precise location where the DNA was inserted into the *SNCA* gene mattered, or maybe the lineage of the strain of mice used had an influence on the results. Some argued that to see clear-cut results, the best way to proceed was to also knock out the other synuclein genes. They reasoned that because there are three synuclein genes, there could be redundancy in their function, so that in alpha-synuclein knockout mice, beta- and gamma-synuclein were able to fill in and carry out the most important roles of alpha-synuclein in the neurons.

By 2004, a team of scientists led by Thomas C. Südhof at the Howard Hughes Medical Institute at the University of Texas, Dallas, had created a double knockout mouse with both alpha- and beta-synuclein inactivated [20]. And the result was—nothing. The mice seemed completely normal with normal brains, body, and behavior, although with a small reduction in dopamine levels reminiscent of what other laboratories had seen with the single alpha-synuclein knockout. Südhof's team also tested the levels of 36 neurologically relevant proteins and found that levels of several of them were significantly altered. In particular, gamma-synuclein levels were 50% higher in the double knockout mice than in controls. This lent credence not only to the idea of overlapping functionality of the synucleins, but also that neurons could increase the level of the remaining synuclein to compensate for the synuclein that was knocked out.

Obviously, the next step was to make the triple knockout, with all three synuclein genes inactivated, but for years this did not happen. Why? It might have something to do with another type of transgenic mouse being created at almost the same time as the alpha-synuclein knockout, the human alpha-synuclein mouse. Two papers came out in 2000, one by Herman van der Putten et al. working with a team of scientists in Basel, Switzerland [21] and the other by Philipp J. Kahle et al. working with a joint German—Swiss team [22]. In this mouse, or I should say mice, because multiple versions were created, DNA containing the human *SNCA* sequence, along with neural specific promoter DNA, was microinjected into mouse eggs, where it was hoped that the DNA would be incorporated into the genomic DNA by random insertion. Unlike gene-targeted insertion by homologous recombination, the chance for random insertion into egg embryonic cell genomic DNA was high enough that they could use this more direct approach.

The large majority of injected embryos developed into mice with no obvious differences from control mice; either the microinjected DNA had failed to be incorporated or it had been incorporated, but in such a way that the DNA had no noticeable effect on the mice. However, in a small percentage, the effect was dramatic. These mice had significant motor and other neurological deficits, and on a cellular level, clumps of alpha-synuclein were often present, reminiscent of the pathology seen in Parkinson's disease. The clumps were smaller, and the pattern of central nervous system neurons affected differed somewhat from that in Parkinson's patients, more in the spinal cord and less in the brain, but it did not matter. Finally, the scientists had a mouse in which they could test hypotheses involving alpha-synuclein pathology, a mouse model of Parkinson's disease. Whereas the other scientists had their embarrassingly normal synuclein knockout mice, these scientists had mice that were actually sick. Guess who had an easier time getting their research funded?

Since the initial two papers in 2000, there has been an unending stream of papers on human alpha-synuclein transgenic mice. Most of the scientists focused on normal human versus mutant human versions of the alpha-synuclein mice, although a few also looked at the effects of the other two synucleins, beta and gamma [23,24]. The multitude of papers confirmed the primary finding of moderate to severe motor deficits in the mice, and typically found that the mutant version of alpha-synuclein was different or worse in some way, although details of the neural deficits and pathology in the transgenic mice often varied depending on the study.

Later studies looked more carefully at the expression levels of human alpha-synuclein in varying regions of the mouse brain, and how they correlated with symptom severity [25]. Typically, more significant deficits were correlated with higher levels of transgenic alpha-synuclein expression. Intriguingly, beta-synuclein transgenic mice typically did not have motor problems. Beta-synuclein even had a protective effect; for transgenic mice that had both beta- and alpha-synuclein inserted in their genomic DNA, much less motor impairment was seen than for alpha alone [23]. On the other hand, overexpression of gamma-synuclein produced motor deficits similar to, but not identical to, alpha-synuclein overexpression [24].

The correlation between higher transgenic alpha-synuclein levels and more severe symptoms jibed with what was already known from Parkinson's patients: namely, that having alpha-synuclein gene duplication or triplication was just as bad as having mutant alpha-synuclein. Mice overexpressing normal human alpha-synuclein displayed the same symptoms as those overexpressing mutant versions [26]. Indeed, later scientists generated mice from embryos injected with mouse alpha-synuclein DNA. These mice, now overexpressing their own version of alpha-synuclein, also developed neural pathology, although following the general trend in the transgenic synuclein studies, the symptoms differed somewhat from the various symptoms seen in mice overexpressing human synuclein [27].

In retrospect, it is hard to say whether the results showing more severe or altered symptoms for mutant alpha-synuclein were really due to the mutations. Perhaps some of the mutant transgenic mice had the DNA inserted in such a way as to produce more of the protein, and the researchers happened to focus on these mice because their symptoms were more severe. This, in fact, is one of the central debates addressed in this book. Are mutant alpha-synuclein molecules somehow toxic to neurons in and of themselves, or do the mutations somehow lead to higher levels of alpha-synuclein in neurons? Is it the higher level of alpha-synuclein, normal or mutated, which ultimately leads to Parkinson's disease?

After a hiatus of 6 years, in 2010, a paper finally appeared on a triple-knockout synuclein mouse, written by a team of scientists headed by S.S. Chandra at Yale [28]. Alpha/beta-synuclein double knockouts had been bred to gamma-synuclein knockout mice (which were also perfectly normal, of course). And the result was—pronounced neural deficits! Okay,

I admit it is creepy to cheer for the creation of mice suffering from debilitating neurological issues. But the synucleins had to be important! Why else had they been so conserved for hundreds of millions of years? Otherwise, the theory of evolution made no sense, and if evolution made no sense, then none of the rest of biology did, either. Evolution is the paradigm, the foundation on which all of the life sciences are built! Whew, deep breath, science existential crisis averted.

Back to the mice. The brain development of the triple synuclein knockouts was normal, with no gross anatomical defects, but the neuron synapses were different. They were smaller, yet despite this, the synapses of the younger knockout mice were actually more active than controls. As the mice aged, however, synaptic activity steadily declined, eventually becoming much slower than in control mice. The life span of the triple knockouts was shortened, too. And the biggest surprise? As the mice aged, they became blind. Wow! Didn't see that one coming. I was curious how they could tell a mouse was blind, so I combed through the methods section of the triple-knockout paper. The test consisted of picking up the mouse by the tail and then bringing a rod toward it. If the mouse reached for the rod, then it was not blind.

Although the synuclein triple knockout proved that synucleins were important, this still did not explain why the alpha- and beta-synuclein DNA and protein sequences had been so conserved throughout tetrapod evolution. If they could be deleted with no obvious consequences, then why should it matter so much when a single base of the gene or single amino acid of the protein was mutated? One possibility was that, somehow, changing even a single position in the gene resulted in toxicity to the neuron, so that any mutations were strongly selected against, even though inactivation of the gene produced no ill effects in the single knockout mice.

The transgenic synuclein mice results seemed to support the idea that single mutations result in toxicity, with several reports finding that transgenic mice with mutant human synuclein had more severe deficits than those with normal, nonmutant human synuclein. However, I wonder whether the reports of sicker mutant human synuclein transgenic mice might reflect researcher bias. Here is how I think it might have played out. A collection of transgenic mice was produced in each study by random insertion of human synuclein DNA with the mutation of interest. For control mice, they could also produce transgenic mice with normal human synuclein or obtain them from other laboratories that had already made them. They would, of course, screen the transgenic mutant synuclein mice for those that were the most affected. As for the control transgenic mice, all that mattered for publication was that the controls were transgenic for the normal human synuclein DNA with the same neural promoter sequence used for the transgenic mutant mice. In other words, there was no incentive to choose sicker control mice.

Obviously, I am not convinced by the studies showing transgenic mice with mutant human synuclein were sicker than controls, although I admit that this is just my opinion, and I could be wrong. What would it take to convince me? Knock-in mice. Using the same tricks that generate knockouts, mice can be created with mutant human synuclein DNA inserted at target sequences in the mouse genome via homologous recombination. With knock-in mice an extra step is often added, called floxing. In this technique, the extra DNA used for in vitro selection, such as the *neo* gene, can be cut out after the selection step. For floxed knock-in mice, everything thing is nearly exactly the same, with the foreign DNA inserted in exactly the same place, the alpha-synuclein gene in this case, and with practically the same sequence except for the mutation of interest.

Has a mouse knock-in for alpha-synuclein ever been created? Out of the more than 1000 papers on mouse knock-ins, I have found only one that involved synuclein, in a Polish journal in 2008 by a team of Estonian scientists [29]. They looked at the A30P mutation (alanine in position 30 to proline) in alpha-synuclein; this mutation was the second of the familial mutations in humans to be linked to Parkinson's disease. In these floxed knock-in mice, everything was the same as in the controls except for the change to the amino acid proline in position 30. The mice were generated by breeding parents that had just one copy of the A30P mutation. Roughly half of their offspring had just one copy of the A30P mutation, like their parents, and roughly one-quarter of the offspring had two copies of A30P. The remaining offspring had two copies of the normal version of the gene, and they were used as the controls in the study.

As the mice aged, their motor performance was tested. On the whole, those that had inherited the A30P mutation moved around just the same as their littermates who had inherited two normal versions of the alpha-synuclein gene, but there were a few differences. On two tests in particular, the A30P mice showed pronounced differences. In one test, the ink test, they would paint the rear paws of mice in one color of paint and their forepaws in another, and then they would let the mice run across a piece of paper. By analyzing the prints, they could determine the step size and other features of their stepping patterns. At first, all the mice performed the same on all tests, but by 13 weeks of age, the mice with the A30P mutation developed a more uneven gait, taking smaller steps on average. At the same age, the A30P mice also started to slip more than their normal littermates on a test in which they had to traverse a thin rod.

Curiously, the A30P alpha-synuclein knock-in mice were reminiscent of synuclein knockout mice; they also responded less to amphetamine and had a moderately lower dopamine levels in their brains. Perhaps the A30P mutation made the protein so nonfunctional that it was similar to having no alpha-synuclein at all. The Estonian study reported on the A30P knock-in mice only up to 16 weeks old. What happened after that? Did their symptoms continue to worsen? Unfortunately, I could find no subsequent reports. I did find, however, that both mouse and rat knock-ins with the A53T mutation are now commercially available. A53T was the first familial alpha-synuclein mutation identified for Parkinson's disease.

In fact, a report describing a rat knock-in of another Parkinson's disease—related alpha-synuclein mutation, G51D, came out in the form of a doctoral thesis by Stephen West at the University of Edinburgh as I was putting the finishing touches on this book [30]. West observed no difference in neural morphology in the G51D rats, but the mutant alpha-synuclein was distributed differently in the neurons, with less in presynaptic boutons compared with controls and, intriguingly, more localized to mitochondria. The rats displayed no differences in movement or behavior, although they did not appear to have undergone the more rigorous testing described in the Estonian study.

As journals were being flooded with reports on alpha-synuclein transgenic mice, a trickle of papers on alpha-synuclein knockouts continued. In one such paper, a team led by Robert L. Nussbaum at the National Human Genome Research Institute at NIH observed changes in the chemical makeup of fat molecules in the brains of the knockout mice: in particular, a fat molecule called cardiolipin, which is important for the cellular machinery that uses oxygen to create energy needed for cells to function [31]. This machinery is located in compartments, or organelles, within the cell called mitochondria. In the knockout mice, the energy production of their neurons was reduced, although the effect was subtle. It turns out that many of the damaging aspects of Parkinson's disease pathology are also associated with the energy-producing machinery of neurons, as we will explore in coming chapters.

One of the most interesting studies on alpha-synuclein knockouts came from a group from the Institute of Physiologically Active Compounds,

Russian Academy of Sciences [32]. This study focused on learning tasks involving movement. Earlier studies had seen no difference between knockouts and controls, but those studies had been conducted using younger mice. The Russian study tested older mice in the prime of their adulthood, at 10 and 14 weeks, which corresponds roughly to 21 and 25 human years. Recall, this is about the same age at which deficits in A30P knock-in mice became apparent. First strength and locomotion of the knockouts and controls were compared. No difference in strength was observed, but there were slight differences in locomotion using the open field test, reminiscent of what Cabin had seen in 2002, although not in all respects. The knockouts reared up on the hind legs slightly less, as seen before, but they were also more what one might term fidgety: that is, they moved around more, but covered less distance overall.

The additional learning tasks that the knockout mice underwent were not pleasant. In the passive avoidance test, the mouse was placed in a platform enclosed under a glass dome, and the platform sat on a grid that could deliver an electric shock. The first day had a single learning session; the glass dome would be removed, and after about 4 s on average, the mouse would step off the platform and an electric shock would be delivered. The next day, the procedure was the same. On the second day, the control mice waited 27 s on average before chancing escape from the platform, but the average time the knockout mice waited was only 8 s.

There was also an active avoidance test. The mouse was placed in a narrow enclosure divided in the middle by a barrier with a door. Initially, both sides of the enclosure would be dark. Then, the light would be turned on the side the mouse was placed, and the door to the other side opened. The mouse had 6 s to escape through the door, else it would receive shocks through the floor until it did. After it ran through the door, it was given 10 s of rest in the dark, and then the light on its side was turned on and the opposite side turned off, and the trial repeated. The first 2 days were for learning, and the trial was repeated five times each day. For the next 6 days after this, the trial was repeated 15 times each day, and the number of successful escapes was recorded each day for the control mice and knockouts. At first, the knockouts had fewer successful escapes, about 4% of the time, but not significantly less than the controls, at about 8%. By day 8, however, the control mice had learned to escape before the shock with a 45% success rate; for the knockouts, the success rate had risen only to 10%.

Finally, there was the Morris water maze test. Originally designed for rats by Richard G. Morris, now at the University of Edinburgh, a smaller version is used for mice. The maze actually consists of a round pool filled with water with a small nonvisible platform in one quadrant, submerged just below the surface. For rats, the test is not so bad; they are natural swimmers in the wild. For mice, the test is more stressful; they normally avoid swimming whenever possible. In the test, the mouse is released in the water near the wall of a quadrant not containing the platform. It is allowed to swim freely for a minute, and if it has not found the platform by then, the researcher uses his or her hand to guide the mouse to it gently. Outside the pool are external cues serving as landmarks to the swimming mouse, although the Russian study did not specify what the external cues were.

For 5 days, the knockout mice and control mice were tested one at a time. Each day, two release points were chosen near the wall in quadrants not containing the platform. Every mouse was first released twice from the first release point and twice from the second. The average time it took the mice to find the platform was averaged over the four tries. At first, all of the mice performed similarly, with most swimming the full 60 s before being guided to the platform, but by day 4, the performance of the control mice had improved dramatically, taking only 30 s on average. In contrast, the alpha-synuclein knockouts improved only slightly, still taking over 50 s on average. In their discussion of the results, the Russian group noted that a previous version of the test had failed to show an effect because the platform was too wide. Apparently, they had spent some effort optimizing at least some of these learning tasks to get clearer results. Being a laboratory mouse sucks, but even more so in Russia, it would seem.

In the Russian study, the adult alpha-synuclein knockout mice clearly showed deficits in working and spatial memory that were not apparent in younger mice. Although they did not mention whether they had done their specific versions of the tests on younger mice, I would not be surprised if they had, and seeing lack of deficits, they had chosen not to bother trying to publish results already reported by others. Were the age-dependent deficits in working and spatial memory direct consequences of knocking out alpha-synuclein? Perhaps not; other, more nuanced hypotheses are conceivable. The tests were stressful, and perhaps alpha-synuclein somehow helps mice deal with stress better, thus allowing better learning under stressful conditions. Another hypothesis is that alpha-synuclein does not necessarily affect learning in general, but instead is involved more specifically in conscious suppression of random motions that interfere with achievement of the task.

The problem with these alternate hypotheses is that they do not explain why the effect of knocking out synuclein is age-dependent. Age-dependent effects were also seen for the synuclein triple-knockout mice, as well as the A30P knock-in mice. When learning tasks, relevant neural connections in the brain can be strengthened, others weakened, new connections formed, and others removed. The simplest explanation of the effect of alpha-synuclein on working and spatial memory, in my opinion at least, is that the protein maintains neural cells in a state at which they can be effectively remodeled when exposed to the stimuli of learning tasks. The developing neurons in young knockout mice can obviously form and remodel their connections without alpha-synuclein, as the earliest papers showed, but it is as if something runs out in adulthood for the knockout mice.

One possibility, a speculative one that we will explore in later chapters, is that this something is related to gangliosides, special fat molecules especially abundant at neural connections. In early childhood, the fat molecules needed for brain development come from mother's milk. After weaning, fat must come from other dietary sources, and the variety of fat molecules can be different, possibly affecting the synthesis of gangliosides in neurons.

Alpha-synuclein can potentially affect the balance of gangliosides and other fat molecules in the cell, such as cardiolipin in mitochondria. Indeed, alpha-synuclein's impact on mitochondria, which is bad when too much is present, might actually be favorable at normal levels. Whatever the benefit, it does not have to do with the number of mitochondria; aging knockout mice appear to have the same number of mitochondria in neurons as age-matched controls [31]. Perhaps alpha-synuclein helps maintain normal function of mitochondria through adulthood, so that they can continue to provide the energy required for neurons to fire and remodel themselves in response to external stimuli. In this putative scenario, neurons in adult synuclein knockout and A30P knock-in mice become moribund to varying degrees, with their mitochondria unable to produce the chemical energy needed for neural rewiring.

The studies of transgenic mice have confirmed that too much alpha-synuclein causes clear neuronal deficits, whereas knocking out alpha-synuclein or knocking in mutant alpha-synuclein produces much more subtle effects that do not manifest until adulthood. We are still left with the puzzle of why the synucleins, especially alpha and beta, have been so conserved in the hundreds of millions of years since the common ancestor of lobe-finned fishes and tetrapods branched off the vertebrate evolutionary tree, even while alpha-synuclein has been lost in marsupials and monotremes. How can this have come about?

The Russian team used difficult tests optimized to highlight the subtle deficits in the locomotion and learning of alpha-synuclein knockout mice versus controls. However, nature is the harshest test of all; if you fail, your genes die with you. An interesting experiment would be to release a population of alpha-synuclein knockouts and controls, which are genetically identical except for the knocked-out gene, into a simulated wild, a large area where they would have to compete for food and mates, avoid being prey, and survive environmental pathogens, as much as possible like being in their natural habitat. I suspect the slight edge the adult control mice have over the knockouts would lead to the knockout population being selected out, eventually dwindling away. In other words, in highly competitive environments, genes that provide an advantage, even just a slight one, can be strongly selected for.

What of marsupials and monotremes? I wonder if they might have lost alpha-synuclein owing to a founder effect in an initially low-competition environment. The founder effect occurs when a new population is established by a small number of individuals from a larger population. Mutations carried by these individuals, perhaps rare in the population from which they came, become common in their descendants. Perhaps some of the founding individuals of the protomarsupial or protomonotreme populations had a mutation where alpha-synuclein was deleted. Then, when they arrived at a locale only recently having become habitable after some natural disaster, their population rapidly expanded and speciated to fill the available ecological niches.

As we have seen for mice, alpha-synuclein knockouts are viable, and with little competition, there should be no natural selection to remove them from the expanding founder population. Many of the new species could, by chance, be founded by individuals with no alpha-synuclein. In fact, only three marsupials, an opossum, the Tasmanian devil, and the koala, have had their genomes published so far, and only one monotreme, the platypus. It could be that there are other marsupial and monotreme species that still have the ancestral alpha-synuclein gene. Moreover, when it comes to species number, geographic extent, and invasive species, placental mammals are clearly winning the evolutionary race. There are myriad reasons for the front-runner status of placentals, but it is not inconceivable that alpha-synuclein, or its lack, is part of the explanation.

In a way, the synucleins are Goldilocks proteins, especially alpha-synuclein. Not having enough can cause deficits, whereas having too much is even worse. To help understand why this is and the connection to Parkinson's disease, we need to explore their molecular roles in neurons, and the good, the bad, and the ugly details of these curious proteins.

References

[1] Steensma DP, Kyle RA, Shampo MA. Abbie Lathrop, the "mouse woman of Granby": rodent fancier and accidental genetics pioneer. Mayo Clinic Proceedings November 2010;85(11):e83.

[2] Crow JFCC. Little, cancer and inbred mice. Genetics August 1, 2002;161(4):1357—61.

[3] Rutstein DD. An open letter to Dr. Clarence Cook little. The Atlantic Monthly October 1957;10.

[4] Jaenisch R, Mintz B. Simian virus 40 DNA sequences in DNA of healthy adult mice derived from preimplantation blastocysts injected with viral DNA. Proceedings of the National Academy of Sciences United States of America April 1974;71(4):1250—4.

[5] Papaioannou VE, et al. Participation of cultured teratocarcinoma cells in mouse embryogenesis. Journal of Embryology and Experimental Morphology 1978;44:93—104.

[6] Evans MJ, Kaufman MH. Establishment in culture of pluripotential cells from mouse embryos. Nature 1981;292:154—6.

[7] Martin GR. Isolation of a pluripotent cell line from early mouse embryos cultured in medium conditioned by teratocarcinoma stem cells. Proceedings of the National Academy of Sciences United States of America 1981;78:7634—8.

[8] Bradley A, et al. Formation of germline chimaeras from embryo-derived teratocarcinoma cell lines. Nature 1984;309:255—6.

[9] Folger KR, et al. Patterns of integration of DNA microinjected into cultured mammalian cells: evidence for homologous recombination between injected plasmid DNA molecules. Molecular and Cellular Biology November 1982;2(11):1372—87.

[10] Smithies O, Gregg RG, et al. Insertion of DNA sequences into the human chromosomal beta-globin locus by homologous recombination. Nature September 1985;317(6034):230—4.

[11] Doetschman T, et al. Tragetted correction of a mutant HPRT gene in mouse embryonic stem cells. Nature December 10-16, 1987;330(6148):576—8.

[12] Thomas KR, Capecchi MR. Site-directed mutagenesis by gene targeting in mouse embryo-derived stem cells. Cell November 6, 1987;51(3):503—12.

[13] Mansour SL, Thomas KR, Capecchi MR. Disruption of the proto-oncogene int-2 in mouse embryo-derived stem cells: a general strategy for targeting mutations to non-selectable genes. Nature November 24, 1988;336(6197):348—52.

[14] Kitada K, et al. Transposon-tagged mutagenesis in the rat. Nature Methods February 2007;4(2):131—3.

[15] Chen Y, et al. Functional disruption of the dystrophin gene in rhesus monkey using CRISPR/Cas9. Human Molecular Genetics July 1, 2015;24(13):3764—74.

[16] Liu Z, et al. Cloning of Macaque monkeys by somatic cell nuclear transfer. Cell February 8, 2018;172(4):881—7.

[17] Abeliovich A, et al. Mice lacking alpha-synuclein display functional deficits in the nigrostriatal dopamine system. Neuron January 2000;21(1):239—52.

[18] Specht CG, Schoepfe R. Deletion of the alpha-synuclein locus in a subpopulation of C57BL/6J inbred mice. BioMedCentral (BMC) Neuroscience 2001;2:11.

[19] Cabin DE, et al. Synaptic vesicle depletion correlates with attenuated synaptic responses to prolonged repetitive stimulation in mice lacking alpha-synuclein. Journal of Neuroscience October 15, 2002;22(20):8797−807.
[20] Chandra S, et al. Double-knockout mice for alpha- and beta-synucleins: effect on synaptic functions. Proceedings of the National Academy of Sciences United States of America October 12, 2004;101(41):14966−77.
[21] van der Putten H, et al. Neuropathology in mice expressing human alpha-synuclein. Journal of Neuroscience August 15, 2000;20(16):6021−9.
[22] Kahle PJ, et al. Subcellular localization of wild-type and Parkinson's disease-associated mutant alpha -synuclein in human and transgenic mouse brain. Journal of Neuroscience September 1, 2000;20(17):6365−73.
[23] Hashimoto M, et al. beta-Synuclein inhibits alpha-synuclein aggregation: a possible role as an anti-parkinsonian factor. Neuron October 25, 2001;32(2):213−23.
[24] Ninkina N, et al. Gamma-synucleinopathy: neurodegeneration associated with overexpression of the mouse protein. Human Molecular Genetics May 15, 2009;18(10):1779−94.
[25] Chesselet MF. In vivo alpha-synuclein overexpression in rodents: a useful model of Parkinson's disease? Experimental Neurology January 2008;209(1):22−7.
[26] Kirik D, et al. Parkinson-like neurodegeneration induced by targeted overexpression of alpha-synuclein in the nigrostriatal system. Journal of Neuroscience April 1, 2002;22(7):2780−91.
[27] Rieker C, et al. Neuropathology in mice expressing mouse alpha-synuclein. PLoS One September 26, 2011;6(9):e24834.
[28] Greten-Harrison B, et al. αβγ-Synuclein triple knockout mice reveal age-dependent neuronal dysfunction. Proceedings of the National Academy of Sciences United States of America November 9, 2010;107(45):19573−8.
[29] Plaas M, et al. Alpha-synuclein A30P point-mutation generates age-dependent nigrostriatal deficiency in mice. Journal of Physiological Pharmacology June 2008;59(2):205−16.
[30] West SA. Phenotypic analysis of the CRISPR-engineered G51D alpha-synuclein rat. Edinburgh: s.n.; 2019.
[31] Ellis CE, et al. Mitochondrial lipid abnormality and electron transport chain impairment in mice lacking alpha-synuclein. Molecular and Cellular Biology November 2005;25(22):10190−201.
[32] Kokhan VS, Afanasyeva MA, Van'kin GI. α-Synuclein knockout mice have cognitive impairments. Behavioural Brain Research May 16, 2012;231(1):226−30.

CHAPTER 5

Shapeshifter – The molecular structure and normal biological role of synuclein in neurons

"Elusive," "mysterious," a protein "Bermuda Triangle:" such descriptions pervade the scientific literature on synuclein proteins, alpha-synuclein in particular. The irony is that alpha-synuclein is one of the most studied proteins in science. There are literally thousands of other proteins about which *less* is known. Its physical properties have been described from every angle in painstaking detail. It is, however, a protein of fascinating peculiarity.

The structure of alpha-synuclein is protean; in water it appears completely disordered, like a wiggly piece of spaghetti. Add lipid membranes to the water, and the first part of the protein changes completely, ordering itself into an extended alpha-helix. Make the water more acidic, and the structure changes again, the protein molecules slowly condensing together into long filaments called fibrils. Most terrible, it can also adopt structures intermediate to these, called oligomers, which are highly toxic to neurons. But I am getting ahead of the story. What's a lipid membrane? What's an alpha-helix? I'm so glad you asked!

Synuclein and the Coelacanth
ISBN 978-0-323-85707-9
https://doi.org/10.1016/B978-0-323-85707-9.00011-3

The fats, or lipids, in our diet, as well as that stored in our fat cells, are primarily in triglyceride form; that is, they consist of a glycerol molecule with three fatty acid hydrocarbon tails attached. The fatty acid hydrocarbon tails are made of chains of carbon atoms with one or two hydrogen atoms attached to each carbon, except for the end carbon, which has three. Chains that have a neighboring pair of carbons with just one hydrogen attached to each are monounsaturated, and two or more such pairs are polyunsaturated. If all of the carbons have two hydrogens (except the last, which always has three), the chain is saturated; that is, it has the maximum allowed number of attached hydrogens.

The lipids that form lipid membranes are different; they are diglycerides, with only two long hydrocarbon chains attached to the glycerol, although the chains can be either saturated or unsaturated. The more saturated the chains, the stiffer the membrane. Instead of the third hydrocarbon chain, nearly all diglycerides have a phosphorus-containing molecule attached that carries a negative electric charge, a phosphoryl group, making what is called a phospholipid. The glycerol—phosphoryl bit is called the head group.

The negative electric charge of the phosphoryl group makes this part of the lipid hydrophilic, preferring to interact with water. The hydrocarbon chains are hydrophobic, preferring to avoid water. As with proteins, the trick is to arrange the phospholipids so that the hydrophilic bits are exposed to water and the hydrophobic bits are hidden away. One way to do this is to divide the lipid molecules into two layers. The top layer has all of the lipids aligned with their phosphorus bits pointing one way and hydrocarbon chains the other, whereas the bottom layer is arranged pointing the opposite way, so that when the layers are brought together, packing the tail ends of the hydrocarbon chains together, all of the hydrophobic bits are hidden in the middle and the phosphorus bits are exposed to water. This bilayer arrangement of phospholipids is a lipid membrane, and it is central to life. The contents of our cells are held in by lipid membranes; without them, our cells would just dribble away.

All cells of all creatures are surrounded by cell membranes. All plants and fungi and most bacteria also have a cell wall made of linked carbohydrates outside the membrane, but we animals make do without. Inside cells are more lipid membranes enclosing various compartments called organelles, with diverse functions such as the nucleus where DNA is located, and mitochondria, which have two membranes, an inner and an outer one. Small lipid membrane-enclosed compartments called vesicles store neurotransmitters that are released at synapses when neurons fire. These vesicles are where synucleins do their most important work, helping to regenerate, maintain, and regulate them.

Let me paint the synucleins' role more vividly. The neuron, which receives a barrage of chemical signals from the mesh of neurons connected to it, passes a point of no return, sending an avalanche of charged ions flooding into the cell. This flood flows in a wave of electric voltage coursing down the neuron's main conduit, the axon, branching treelike into thousands of synapse-tipped tendrils. The entirety of the axon branching structure is called an axonal arbor. Upon reaching these synaptic terminals, or in neurology jargon, presynaptic boutons, the ion avalanche triggers the release of their neurotransmitters, sending them across the synaptic cleft to chemically signal the attached neurons.

Synucleins moderate the final step in the firing neuron, making sure the axonal regions releasing neurotransmitters do not get carried away and release too many vesicles [1]. These regions are generally known as axonal varicosities, which include boutons, as well as en passant synapses, which form at swellings along axon branch fibers. Then, the synucleins help tidy up the aftermath after firing, getting the next set of vesicles properly outfitted and in place, ready for the next avalanche [2]. At least, that seems to be the consensus view based on experimental results to date. It is always possible that some brilliant new set of experiments will turn this picture on its head.

Synucleins, especially alpha and beta in the central nervous system, are found in varicosities of multitudinous neurons, excitatory in some brain regions, inhibitory in others, varying in proportion and protein levels depending on the specific region and neural environment [3,4]. Excitatory neurons send chemical signals to other neurons that stimulate them to fire, whereas inhibitory ones send different signals that suppress firing. Importantly for alpha-synuclein, one type of neuron in which it is found releases the neurotransmitter dopamine. The death of these dopaminergic neurons causes many of the symptoms of Parkinson's disease, especially movement-related ones. Because of the importance of these neurons, researchers have sought animals to serve as model systems in which they can study these neurons in more detail.

One such model animal for the study of the synucleins is the mouse, which was the subject of the last chapter. Another is the zebra finch, which is native to Australia and parts of Indonesia. Although the overall architecture of the avian brain is different from that of mammals, nearly all of the neuronal circuits are analogous. Early in the developing zebra finch brain, just like our brains and those of all vertebrates, axons grow and branch and form synapses. The postsynaptic parts of these synapses typically form on dendrites, branches emerging from the main body of the neuron distinct from the axon and its branches. The dendrites are where the neuron receives the bulk of the chemical signals from the neurons connected to it.

Later in the development of the zebra finch brain, alpha-synuclein arrives at the presynaptic boutons of the already formed synapses [5]. The zebra finch studies occurred in the mid-1990s, before the name "synuclein" had become standard, and in zebra finches, alpha-synuclein was first known as synelfin: *syn* for synapse and *elf* from the German word for 11, for the characteristic 11-amino acid repeats in the protein sequence.

As the young male zebra finch grows, alpha-synuclein continues to be present at high levels in the dopamine-releasing boutons until a critical time is reached. Just before the young male is about to undergo the most important stage in learning bird song, which is perhaps his most challenging motor skill, alpha-synuclein levels begin to fall in the dopamine-releasing neurons. The drop first occurs in the main part of the neuron, the cell body, where synuclein is synthesized. A drop in synuclein levels in the boutons follows, but more slowly over several days, perhaps reflecting the normal half-life of synuclein in cells. After the drop in alpha-synuclein, half of the presynaptic boutons of the dopamine releasing neurons are lost, whereas the remaining boutons grow to twice their previous size. Interestingly, in synuclein knockout mice, presynaptic boutons were underdeveloped [6]. Clearly, this remodeling process, or synaptic pruning, appears to be connected somehow with alpha-synuclein levels, but I am getting ahead of the story again. We still need to understand how the synucleins interact with membranes.

So, what is going on with alpha-synuclein at the molecular level? When synucleins interact with vesicles, they adopt a structure called an alpha-helix. To explain this structure, we need to learn a bit more about amino acids. Recall that each of the 20 types of amino acid has a particular side chain that determines its properties, whether it likes water, whether it is charged, and so on. Each amino acid also has a peptide backbone, which, with one exception, is the same for all of the amino acids.

The peptide backbone forms the chain part of the protein, with the peptide carbon atom chemically bonded to the peptide nitrogen of the next amino acid in the sequence. The peptide nitrogen of the first amino has no preceding peptide to which to bond; thus, the first residue is called the N-terminus, where N is the chemical symbol for nitrogen. For the final peptide, the carbon has no subsequent peptide to which to bond, and the last amino acid is called the C-terminus, where C is the chemical symbol for carbon. The N-terminus is considered the first amino acid in the chain because this is the first amino acid read from the messenger RNA by the ribosome when it synthesizes the protein.

In addition to bonding to the nitrogen of the next amino acid, the peptide carbon of each amino acid is bonded to an oxygen atom. Similarly, the peptide nitrogen, with one exception, is also bonded to a hydrogen atom in addition to the carbon of the preceding residue. The peptide oxygen has a slight negative charge, whereas the peptide hydrogen has a slight positive charge. Because opposite charges attract, the peptide backbone prefers structural arrangements in which peptide oxygens and hydrogens are close; such oxygen—hydrogen pairs are called hydrogen bonds.

Hydrogen bonds are much weaker than normal chemical bonds, only about 1/20th as strong. The alpha-helix is a tight spiraling structure that was first proposed in the 1930s by William Astbury based on x-ray studies of hair [7], but it was not until 1951 that chemist Linus Pauling and colleagues described the hydrogen bonding arrangement responsible for its structure [8]. The one amino acid that is an exception is proline, which has a side chain carbon attached to nitrogen instead of a hydrogen, and thus its peptide nitrogen cannot form a hydrogen bond.

So, if peptide oxygens and hydrogens like alpha-helices, why do the synucleins not adopt this ordered structure in water? The answer is entropy. Nature favors disorder: "Alpha-helix beware, Nature abhors thee!" For synucleins in a watery environment, the hydrogen bonds are too weak on their own to overcome the tendency toward disorder: hence their wiggly spaghetti structure. Technically speaking, alpha-synuclein is what is called an intrinsically disordered protein. However, by interacting with a membrane surface, this tendency toward disorder can be overcome.

The amino acid sequences of the synucleins include hydrophobic side chains that can burrow into the lipid bilayer and other side chains that prefer water. Still other side chains, in particular the side chains of the amino acid lysine, prefer to lie parallel to the lipid surface, interacting with both the hydrocarbon tails and their positive charged ends interacting with the negatively charged phosphoryl part of the diglycerides. But to really understand what is special about the helical structure formed upon interaction with the lipid membrane, one must understand the periodic nature of the alpha-helix structure, and how its structure repeats itself.

In the alpha-helix, the peptide oxygen forms a hydrogen bond with the peptide hydrogen that comes four amino acids later. Imagine the alpha-helix as a cast iron spiral staircase going up counterclockwise, with each amino acid as a step and the hydrogen bonds as vertical supports. Going around once takes only 3.6 steps (a steep, narrow staircase). The reason why it is only 3.6 steps is that the hydrogen bond goes up from oxygen at the end of one amino acid's peptide to the hydrogen at the beginning of the peptide four amino acids later, not quite four full steps.

The alpha-helix staircase repeats itself every 18 steps. If the bottom step points at 12 o'clock, you have to go up 18 steps to get to the next one that points exactly at 12 o'clock; this is because 18 is the first whole number multiple of 3.6. Now, visualize the amino acid side chains sticking out to the side of each step of the staircase. If you have an amino acid sequence that repeats itself every 18 amino acids, each of those 18 will have its side chain pointing the same way as the next 18 amino acids.

The N-terminal part of the amino acid sequences of the synucleins has a periodicity, too, but repeating every 11 amino acids, not 18. The repeats do not have the same sequence; they are imperfect repeats but are still similar; alpha- and gamma-synuclein have seven repeats and beta-synuclein has six. To accommodate the 11—amino acid periodicity, the alpha-helix changes its twist slightly, going from 3.6 to 3.67 steps per turn. This way, three turns take just 11 amino acids.

This small tweak to 3.67 steps per turn also causes the helix to curve in an arc called a superhelical twist. Work at the laboratory of Ralf Langen at the University of Southern California has shown how the repeating amino acids in the curved helices are arranged so that the hydrophobic ones stick out from the inside part of the curve, whereas the hydrophilic ones stick out on the outside of the curve, and the ones that like to interact with both the chains and phosphoryl groups of lipids stick out at right angles, between the hydrophobic and hydrophilic ones [9]. Because of the curve of the helix, the synucleins prefer to bind lipid membrane surfaces of similar curvature, in particular, to the convex side of the curved membranes, so that the hydrophobic side chains are pointed inside the lipid membrane.

Synaptic vesicles are some of the smallest vesicles in the cells. The high curvature of their membranes makes them unstable. Left to their own devices, the lipid membranes would be much happier in larger, less curved structures. As the neuron is flooded with calcium ions during a neuron firing event, a calcium binding protein called synaptotagmin causes a collection of proteins at the vesicle surface and synapse cell membrane known as a SNARE[a] complex to change their structure, causing their alpha-helices to intertwine. This structure change forces the vesicle and cell membranes together. After this, the vesicle rapidly fuses with the cell

[a] SNARE stands for "**S**oluble ***N***-ethylmaleimide-sensitive factor **A**ttachment protein **RE**ceptor"

membrane and its highly curved structure relaxes into the relatively flat cell membrane, releasing its cargo of neurotransmitters into the synaptic cleft. This vesicle release process is known as exocytosis, and for synaptic vesicles, their membrane typically fully fuses with the cell membrane. Exocytosis also occurs for other types of vesicles and does not always need to be complete, so-called "kiss and run" events.

One of the presumed roles of the synucleins is to brace synaptic vesicles, binding to their highly curved membrane surface with their curved alpha-helices, stabilizing them until the moment of release. Proving this experimentally has been problematic. Add more synuclein and fewer vesicles are released, take synuclein away and more are released, at least in some of the mouse knockout experiments [1]. This would seem to support their putative stabilizing role; yet after purifying the vesicles from the neurons, numerous studies failed to find any synucleins bound [10]. Perhaps they fell off during the purification. Indeed, other studies found that small amounts of synucleins do copurify with synaptic vesicles, but most of the synucleins end up in the unbound fraction [11]. If their function really is to stabilize the high curvature of the vesicle, why do they seem to fall off so easily? I suspect part of the explanation might be the protein calmodulin, which is especially abundant in presynaptic boutons.

Unless the neuron is firing, calcium is kept at a low concentration inside the cell membrane, 20,000 times lower than outside. When the neuron fires, special protein gates at the cell membrane open, flooding the neuron with calcium ions, which bind to a number of proteins including calmodulin, the ubiquitous, highly conserved protein introduced in the first chapter. After firing, protein pumps at the neuron cell membrane transport the calcium back out until the low concentration is reached again. Calmodulin directly interacts with the N-terminal amino acids of alpha-synuclein, the same amino acids that are important for membrane binding, but only when calmodulin is in its calcium bound state. This interaction could be important in firing neurons, with calmodulin helping to remove synucleins from vesicles fusing with the cell membrane, so that the synucleins can then help get the next set of vesicles get ready for the next firing event. When purifying synaptic vesicles, it is necessary to break the cell membrane, likely exposing the vesicles to elevated calcium levels and enabling calmodulin to remove the bound synuclein.

I determined the structure of calmodulin bound to the first 19 amino acids of alpha-synuclein [12] while in the laboratory of Nico Tjandra, a leading expert in advanced nuclear magnetic resonance (NMR) techniques (see "The nuclei sing"). It was one of the most challenging structures on which I have ever worked. Part of the difficulty was my fault. To get a high-quality structure, I needed to have the 19—amino acid alpha-synuclein peptide synthesized using the isotopes of carbon and nitrogen necessary for NMR spectroscopy. This was during the years of federal budget sequestration, and so to save money, I chose a cheaper synthesis in which only a subset of the amino acids had the necessary isotopes. Eventually I muddled through, although the months of extra struggle were definitely not worth the roughly $1000 saved.

The nuclei sing — NMR spectroscopy

All atoms consist of a dense nucleus surrounded by a diffuse cloud of one or more electrons. For some elements, the nuclei have a special property called magnetic spin. These nuclei are like little bar magnets. Placed in a magnetic field, these nuclei prefer to align their spin with the direction of the field. Something marvelous happens to these nuclei when irradiated with a brief pulse of radio waves. Their alignment to the magnetic field gets perturbed, and their spins, with their own little magnetic fields, end up rotating like tops, or gyroscopes, around the axis of the external magnetic field, rotating with particular frequencies called resonances.

Using miniature antennas and electronics tuned to the frequency of the nuclei, radio waves generated by the rotating nuclear magnetic fields can be measured. The measurement of these radio waves is called NMR spectroscopy. The electromagnets used to generate the magnetic fields for NMR are over 100,000 times as strong as the earth's magnetic field. The electromagnets generate their fields using massive currents flowing through ribbons of superconducting metal chilled to a few degrees above absolute zero using liquid helium. The stronger the field, the stronger the NMR signal.

The frequency of a nuclear spin depends on the element or, more precisely, on the isotope of the atom. Hydrogen nuclei have a totally different frequency than phosphorus nuclei, for instance. In addition, the chemical environment an atom in a molecule will alter its frequency very slightly, by less than one-thousandth of 1%. The hydrogen atom in the methyl group at the end of a lipid chain will have a different frequency from that of the hydrogen attached to the backbone nitrogen of a peptide chain. In a complex molecule such as a protein, local differences in the magnetic fields surrounding a nucleus can further alter the nuclear frequency. Because of this, the frequency of a hydrogen in one amino acid in the chain will typically be different from that of the corresponding hydrogens in the other amino acids.

The NMR frequencies for a small molecule are like a simple musical chord, whereas those of a protein are like a symphony. At least, that is what I imagined until I actually listened to NMR signals. The range of NMR frequencies spans several kilohertz, which can be converted directly to audio frequencies. The smallest molecules sound like bells, moderately sized ones can sound like UFOs or alien cicadas, but proteins just sound like a dull *thunk*. So much for my glorified imaginings. You can listen to the sounds yourself; just search online for "NMR signal sound" and listen to the videos.

The best nuclei for NMR spectroscopy are ones that have spin of ½. I will say more on what spin ½ means in just a moment. Luckily, for scientists like me in the biological sciences, hydrogen, the most common atom in biological molecules, has spin ½. Unfortunately, the most abundant isotopes of carbon (^{12}C) and oxygen (^{16}O) have spin 0, whereas the common form of nitrogen (^{14}N) has spin 1. Nuclei with spin 0 have no magnetic resonance signals, whereas those with spins of 1 or greater interact strongly with local electric fields, which makes their signals too short-lived to be useful for most applications.

The spin of the nuclei arises from how the protons and neutrons that comprise them interact. The "12" of the isotope ^{12}C means that the total number of protons plus neutrons in the nucleus is 12, ^{16}O has 16, and so on. Protons and neutrons also have spin ½, and depending on how they interact, their spins can add or cancel. Nuclei with an even total of protons and neutrons can have only zero or even spin, whereas those with an odd total can have only spins of ½, 3/2, 5/2, and so on. The highest nuclear spin for the known elemental isotopes is 8, an honor shared by ^{90}Nb, ^{180}Ta, and ^{258}Md.

In NMR studies of proteins, we boost the number of frequencies we measure by replacing the ^{12}C and ^{14}N nuclei with the less common but nonradioactive isotopes ^{13}C and ^{15}N, both of which have spin ½. We do this by using *E. coli* bacteria genetically engineered to make the desired protein, and then feed them only ^{13}C glucose and ^{15}N ammonium chloride as their sole carbon and nitrogen sources. The bacteria grow poorly on this strict diet, but we just grow larger volumes of them to compensate. Because nuclei with spin ½ are largely unaffected by electric fields, their signals last much longer, which is critical for discriminating between resonance frequencies that can differ by as little as a few millionths of 1%. For a protein, the NMR signals typically last a fraction of a second, providing a frequency discrimination, or resolution, of about 5–10 Hz.

After the radio-frequency pulse, the perturbed nuclei seek to return their magnetic fields back into alignment with the external magnetic field. One of the ways they return, or relax, back into alignment is through magnetic interactions with their neighboring nuclear spins. Advanced NMR techniques take advantage of these spin–spin interactions, using series of radio-frequency pulses to manipulate the nuclear spins. One type of experiment provides information on which nuclear spins are close to each other through chemical bonds, whereas another type shows which hydrogens are close together in space. The resulting data are like a jigsaw puzzle, figuring out which resonance goes with which nucleus. Finally, the spatial data can be used to figure out the three-dimensional structure of the molecule, even for proteins with hundreds of amino acids.

The 19 amino acids of the alpha-synuclein peptide whose structure I studied included the first eight amino acids of the protein plus the first of the seven 11–amino acid repeats. It turns out that for many proteins, including the synucleins, the N-terminus nitrogen is attached to something preceding it, not an amino acid, but something called an acetyl group, chemically related to acetic acid (i.e., vinegar). The acetyl group at the beginning of alpha-synuclein makes the N-terminal amino acids more helical and makes alpha-synuclein bind to calmodulin more tightly. In fact, bound to calmodulin, the first eight amino acids were even more helical than the following 11. N-Terminal acetylation also increases the strength of alpha-synuclein interaction with membranes, probably also owing to the increased preference of the N-terminal amino acids for the helical state. The first eight amino acids appear to be the first part of alpha-synuclein to interact with the membrane when binding to vesicles.

Alpha-synuclein is important for recruiting the SNARE protein synaptobrevin-2 to synaptic vesicles [2]. Recall that the SNARE proteins are responsible for vesicle fusion with the cell membrane. Recruitment of the synaptobrevin, also called VAMP2 (Vesicle Associated Membrane Protein 2), protein requires the tail of alpha-synuclein, the negatively charged amino acid region at the end of its chain. The first parts of the synucleins share similar sequences, but their tails differ. The synucleins display individuality with their tails, which leads to differing interactions with their respective partners. At least, that is the picture that study of alpha-synuclein has suggested; the less notorious siblings, beta- and gamma-synuclein, have not yet received the same level of detailed scrutiny.

Synaptic vesicles are regenerated from the cell membrane of presynaptic boutons and other axonal varicosities in a small region surrounding the active zone where the vesicles are released. The process is called endocytosis. Numerous proteins participate in endocytosis. One of these, clathrin, is clearly important. Clathrin forms a cage of interlocking hexagons and pentagons, like a soccer ball, scaffolding the newly forming vesicle, budding inward from the cell membrane. Because endocytosis occurs so near the site of exocytosis, many of the synaptic vesicle proteins that became part of the cell membrane upon fusion are recovered. New proteins also arrive via the axon from the cell body, to replace those lost owing to cellular wear and tear. Once the vesicle has formed enough to pinch off from the cell membrane, the clathrin cage must be removed. One of the proteins involved in uncoating clathrin is auxilin, another protein whose structure I have studied (more alpha-helices!) [13].

Auxilin is also known by its gene name, *DNAJC6*, mutations of which cause a rare juvenile form of Parkinson's disease. Another protein associated with newly forming vesicles and their transport is leucine-rich repeat kinase 2 (LRRK2), and mutations of LRRK2, mentioned in Chapter 2, are one of the most common genetic causes of Parkinson's disease. At some point during vesicle formation, alpha-synuclein presumably has its first interaction with the vesicle membrane, stabilizing its structure and helping to recruit proteins such as synaptobrevin to the vesicle surface. Given the participation of auxilin, LRRK2, and alpha-synuclein, synaptic vesicle endocytosis would seem to be a molecular hot spot for Parkinson's disease pathology.

The interaction of alpha-synuclein and synaptobrevin was detected by an experimental technique called a pull-down assay, which usually employs tiny beads covered in the protein of interest (i.e., alpha-synuclein or synaptobrevin in this case). The beads are mixed with preparations derived from cells that have been broken open, or lysed. The beads are isolated, and the proteins that are bound by the protein of interest are identified. The attached proteins are said to be pulled down by their interaction with the target protein. These proteins might directly interact with target protein, or their interaction might be mediated by other proteins as part of larger complexes.

I used NMR to check whether synaptobrevin and alpha-synuclein interact directly, but I saw no interaction, much to my disappointment. My experiment did not include vesicles, but experiments by others have included them, and still no direct interaction was observed [14]. Nevertheless, remove the negatively charged tail of synuclein, and it no longer pulls down synaptobrevin-2 [2]. One possibility is that the tail of alpha-synuclein interacts with another vesicle protein, which then interacts with synaptobrevin.

LRRK2 can also pull down alpha-synuclein [15]. LRRK2 is a large protein consisting of several domains with different functions. The *K* in LRRK2 stands for kinase, which is a type of enzyme that can attach phosphoryl groups to other proteins, a process called phosphorylation, as well as attach phosphoryl groups to itself, called, as you might guess, autophosphorylation. LRRK2 is capable of phosphorylating alpha-synuclein, so one might say that these proteins directly interact, at least long enough for the reaction to occur. Whether this interaction, with the accompanying phosphorylation of alpha-synuclein, is biologically relevant, involved in Parkinson's pathology, or just an experimental artifact is hotly debated. However, both alpha-synuclein and LRRK2 are elevated in the brains of Parkinson's disease patients, and both proteins are toxic to neurons at high levels, so this is an important debate, and we will revisit LRRK2 in later chapters.

Another protein that has been shown to interact directly with alpha-synuclein in neurons is the dopamine transporter protein (DAT), which transports dopamine back into dopamine-releasing neurons after a firing event [16]. Pull-down experiments using various regions of alpha-synuclein and DAT have localized the interaction to the middle region of alpha-synuclein and the C-terminal region of DAT. Another protein that directly interacts with the DAT C-terminal region is calmodulin-dependent kinase II (CAMKII). In a protein ménage-a-trois, CAMKII also interacts with the C-terminal tail of alpha-synuclein.

Perhaps CAMKII is the intermediary in a functional interaction between DAT and alpha-synuclein, or perhaps they compete for interaction with the C-terminal region of DAT. Interestingly, as its name implies, the kinase activity of CAMKII depends on its binding to the calcium-bound form of calmodulin, which, as mentioned, can also bind the N-terminal amino acids of alpha-synuclein. Thus, it could be that the dopamine transport function of DAT involves an interplay of all four of these proteins along with calcium ions.

Pull-down experiments have identified dozens of potential partners of alpha-synuclein, but most of these have not been confirmed using techniques that detect direct interactions. One such partner, calmodulin, has already been described, and direct interactions with DAT have also been confirmed. Alpha-synuclein can pull down hemoglobin in brain neural tissue samples. This was surprising to me because I thought hemoglobin was found only in red blood cells. Hemoglobin helps manage iron levels and appears to have a biological role inside mitochondria of neurons, transporting oxygen and perhaps interacting with free radicals [17].

A particularly interesting partner with proven direct interaction is the protein tau [18]. Neurodegenerative diseases known as tauopathies involve abnormal deposits of hyperphosphorylated tau protein, called neurofibrillary tangles, which are commonly seen inside neurons of Alzheimer's and less commonly in Parkinson's disease patients. In the normal cellular setting, tau binds to microtubules, protein tubes that run inside axons and serve as tracks for cellular cargoes coming and going from the cell body to the axon termini. The negatively charged tail of alpha-synuclein interacts with tau, interfering with its binding to microtubules [19], and alpha-synuclein can bind to microtubules, as well [20].

Direct interactions can be measured by many techniques. One clever approach is the yeast two-hybrid system, in which, when two proteins interact, the two components of the GAL4 protein are brought together, generating a chemical reaction that can be detected. Using this technique, direct interaction of alpha-synuclein has been shown with the protein synphilin-1; in particular, the middle domain of synphilin was shown to interact with the first 65 amino acids of alpha-synuclein [21]. Like so many proteins, the function of synphilin, which is found in the brain, endometrium, and ovaries, is largely unknown, though it might be involved in targetting proteins for degradation [28].

Another alpha-synuclein interacting protein discovered using the two-hybrid technique is cytochrome C oxidase, the mitochondrial protein in which oxygen is used to generate chemical energy [22]. Cytochrome C oxidase, which is located spanning the inner membrane of the mitochondrion, is a protein critical for oxygen-based life. In a cascade of chemical reactions, electrons are transferred from food molecules through the protein complexes of the mitochondrion, and ultimately to the oxygen bound to cytochrome C oxidase, producing water and carbon dioxide. The carbon dioxide binds to hemoglobin and is transported to the lungs for release.

Pull-down experiments have identified many more mitochondrial proteins that might interact with alpha-synuclein. However, with the exception of cytochrome C oxidase, no direct interaction been observed between alpha-synuclein and these mitochondrial proteins. The synucleins interact directly with the lipid cardiolipin, however, and numerous cardiolipin molecules form integral parts of mitochondrial protein complexes. We will return to the mitochondrion in later chapters, because this is where perhaps the most damaging events occur in Parkinson's disease pathology that lead to neuronal death.

Using NMR, direct interactions were observed between beta-synuclein and alpha-synuclein in the laboratory of Jean Baum at Rutgers University [23]. They also observed that alpha-synuclein directly interacts with copies of itself, whereas copies of beta-synuclein are less inclined to interact with each other. One way to detect the direct interaction between alpha-synuclein molecules is to use fluorescence techniques. In fluorescence measurements, the sample is illuminated with a particular wavelength of light, and molecules, or parts of molecules known as fluorophores, absorb the incoming light and reemit, or fluoresce, longer-wavelength, lower-energy light. In one of the simplest experiments, a fluorescent dye molecule, the fluorophore, is attached to alpha-synuclein in one test tube, and a molecule called a quencher, which prevents fluorescence if it comes in contact with the dye molecule, is attached to alpha-synuclein in another test tube. When the two are mixed, any direct interactions that bring the dye and quencher in contact cause a reduction in the observed fluorescence.

Fancier fluorescent experiments can reveal additional details about the protein interactions. The technique called FRET (Förster Resonance Energy Transfer) uses two fluorescent dye molecules: one that emits at the same wavelength the other absorbs. The two dyes are attached to the two proteins of interest, and by quantifying the amount of transferred fluorescence, the distance between the dye molecules, and thus the proteins, can be estimated. Using this technique, the lysosomal enzyme glucocerebrosidase, the main subject of Chapter 9, was observed to interact directly with the C-terminal amino acids of alpha-synuclein in the absence of vesicles [24]; and in the presence of vesicles, it interacts with N-terminal amino acids as well [25].

My first experience with alpha-synuclein was in collaboration with Jennifer C. Lee at the National Institutes of Health (NIH), who heads a laboratory specializing in fluorescence techniques. Knowing about previous fluorescence studies showing direct self-interaction in alpha-synuclein, the researchers were curious whether NMR could confirm this result. The signals in NMR are exquisitely sensitive to any changes in the environments of the hydrogen atoms in the molecule being studied, especially changes in structure and intermolecular interactions. Even tiny changes in the sample acidity or salt concentration can easily be detected. I performed a standard experiment for intermolecular interaction, a dilution series. As the sample is diluted to lower concentrations, intermolecular interactions are less likely to occur; that is, alpha-synuclein is less likely to bump into and interact with other alpha-synuclein molecules. The hydrogens at the interaction interface between the interacting alpha-synuclein molecules will experience less of the interaction as the concentration goes down, and the change this causes to their environment should be reflected in a change in their frequencies.

Starting with alpha-synuclein at the highest concentration, I measured the spectrum, and then diluted the sample and measured again. I made sure the acidity and salt concentration would not change, to avoid spurious changes in the spectra not associated with alpha-synuclein interaction. I continued diluting and measuring, going all the way down to the lowest concentration that could be measured on the NMR spectrometer. I saw no evidence of interaction, with no change in frequencies even at 1/100th my starting concentration. My conclusion? The fluorescence results had to be wrong! NMR was too sensitive not to be able to see the interaction the fluorescence people claimed. There must be some flaw in their experiments.

I decided to learn more about fluorescence spectroscopy, seeking to help my colleagues by showing how their technique had produced an erroneous result. The more I learned, reading the relevant literature, the more my confidence was eroded. I realized that there was no way their results could be wrong; it had to be the NMR results that were the problem. But how could that be? If there were any change to the environment of the hydrogens at the interface or the interacting alpha-synuclein molecules, NMR would see it. Even with my eroded confidence, I could not accept that NMR would not be sensitive to a change in the environment of the hydrogens at the interaction interface. Then it dawned on me. There was no change in their environment. There were interactions, but the loss of intermolecular interactions upon dilution had been perfectly balanced with intramolecular interactions: that is, by interactions within a single alpha-synuclein molecule.

To explain these balancing interactions, one needs to know about the charge distribution in alpha-synuclein. The N-terminal regions of the synucleins have a slight excess of positively charged amino acids, whereas the C-terminal regions have more negatively charged amino acids. When two alpha-synuclein molecules interact, they tend to do so in a head-to-tail fashion; that is, the N-terminal region of one interacts with the C-terminal region of the other, and vice versa. Because the structure of alpha-synuclein in water is so dynamic and flexible, its N- and C-terminal regions can just as easily interact within the same molecule in a manner that is indistinguishable from the intermolecular interaction. Alpha-synuclein is a tricky protein! At least, it sure fooled me.

In cells, there is evidence that alpha-synuclein might interact not just with one copy of itself, but with three other copies, a complex of four proteins called a tetramer [26]. A single protein, or any molecule for that matter, on its own is called a monomer, two molecules interacting is called a dimer, three a trimer, and so on. When brain tissue is treated with a chemical that bonds particular amino acid side chains together, a process called cross-linking, the tetramer complex can be detected experimentally using antibodies to alpha-synuclein [27]. More tetramer is observed than dimer or trimer, which suggests that tetramer is the predominant form in neurons. Using antibodies for beta-synuclein, cross-linked tetramers have also been observed, but whether these tetramers consist only of beta-synuclein has not been determined. In a test tube, beta-synuclein prefers to interact with alpha-synuclein rather than itself. Thus, one might expect the observed beta-synuclein—antibody-binding tetramer to be two alpha—beta synuclein dimers interacting to form a tetramer. Unfortunately, these putative tetramers are unstable outside the cells, and although there are hints that they adopt at least some alpha-helix structure, attempts to characterize their structure more clearly have not produced consistent results.

There is something else curious about the NMR spectra of alpha-synuclein, this time in the presence of vesicles. In watery solutions with no vesicles, the signals from alpha-synuclein are sharp. That is, each signal is spread over a narrow range of frequencies, much narrower than for other proteins of comparable size. The sharp signals are a consequence of the rapid, dynamic tumbling of its wiggly spaghetti structure. When a protein interacts with another protein, or any molecule for that matter, in which the interaction is dynamic (that is, exchanging between free and bound forms), the range of environments for the nuclei whose signals are being measured becomes more varied. As a consequence, the range of frequencies for each signal widens and the signals becomes less sharp. However, when interacting with vesicles, the signals for the synucleins stay sharp; instead, they just weaken in intensity. Ad Bax of the NIH, one of the world's premier NMR scientists, first brought this odd behavior to my attention.

When the vesicles are first added at low concentration, the alpha-synuclein signals of the first amino acids at the N-terminal weaken first. This is because the N-termini have the highest affinity for the vesicles, and because there are so few vesicles, there is insufficient room for all of the synuclein molecules to bind their full membrane-binding regions. Then, as more vesicles are added and more vesicle surface is available, additional amino acids farther from the N-terminus can bind, weakening their intensities, until finally at the highest vesicle concentration, all 90 or so N-terminal amino acids that make up the membrane-binding domain reach their weakest intensity.

The only explanation I can come up with for this weakening in the absence of signal broadening is that alpha-synuclein binding to the vesicle membrane is a slow stepwise process. First, binding starts at the N-terminus and additional amino acids bind as more vesicle surface becomes available. This process must take place slowly, more slowly than the time it takes to measure an NMR spectrum, about 1 s; otherwise, significant signal broadening would be observed.[b] Because the vesicles are much larger than alpha-synuclein, the vesicle-bound amino acids tumble slowly and their

[b] Technically speaking, it is not possible for no signal broadening to be present, but it is possible that the signal is the sum of strong, sharp and weak, broad components. If the population of vesicle bound synuclein giving rise to the weak, broad component for a particular amino acid signal is small enough, the strong, sharp component can effectively obscure the weak component to the point where it can no longer be detected experimentally.

signals become too broad to be seen; that is, they effectively disappear, causing no detectable broadening of the NMR signal. The sharp signals that are seen correspond to the fraction of alpha-synuclein amino acids not interacting with membrane.

No one expected this result, at least among the NMR colleagues with whom I have talked, including David Eliezer at Weill Cornell Medical College, New York. Instead, we imagined that once the first N-terminal residues had bound and adopted their helical conformation, the rest of the membrane-binding domain, with its seven curved alpha-helical 11—amino acid repeats, would rapidly follow, like a row of collapsing dominoes. In other words, we expected the NMR signals of the entire membrane-binding region to weaken together, along with some broadening from exchange between bound and unbound forms. This rapid process, called cooperative folding, is the general rule for alpha-helices in protein folding.

Instead, what we saw with binding vesicles is that alpha-synuclein is like the reluctant swimmer, who, fearing *curglaff* (cool word, look it up!), slowly puts her toes in first, waits for them to get used to the water, then puts in her feet, waits, next her legs, and so forth. Is what we see in the test tube what actually happens in the cell? Perhaps not. There might be the molecular equivalent of the bully who pushes the swimmer in, where "in" in this case means onto the synaptic vesicle; or more likely, a thoughtful friend already bound to the vesicle helps to shorten the agony by pulling alpha-synuclein on.

Our tricky protein is something of a paradox. Experiments suggest that synucleins have roles in recruiting proteins to synaptic vesicles and moderating the release of their neurotransmitter cargo when the neuron fires. Both of these results are consistent with synucleins binding and stabilizing the highly curved vesicle surface with their naturally curving alpha-helices while bringing various proteins to the vesicles via their individualistic C-terminal tails. However, when synaptic vesicles are isolated from neurons, practically no synucleins seem to be present, and in test tubes, they bind vesicles with an unexpected stepwise slowness.

In the developing zebra finch brain, a drop in alpha-synuclein level precedes synaptic pruning of dopamine-producing neurons. Some might propose that the high level of synuclein primes the synapses for remodeling, and the drop is necessary to lock in the circuitry of the subsequently learned birdsong. Others might propose the opposite, that high synuclein prevents pruning by strengthening the synapse, and that the drop is necessary to allow the pruning accompanying birdsong learning. Because alpha-synuclein protein levels tend to rise in the brains of human adults as they age, I suspect the latter proposal is closer to the truth: the older the brain, the more it is set in its ways. In any case, with regard to normal function in the brain, alpha-synuclein presents a multifaceted and not altogether clear picture, but more tricks are in store.

References

[1] Nemani VM, et al. Increased expression of alpha-synuclein reduces neurotransmitter release by inhibiting synaptic vesicle reclustering after endocytosis. Neuron January 14, 2009;65(1):66–79.
[2] Burré J, et al. Alpha-synuclein promotes SNARE-complex assembly in vivo and in vitro. Science September 24, 2010;329(5999):1663–7.
[3] Li J, Henning Jensen P, Dahlström A. Differential localization of alpha-, beta- and gamma-synucleins in the rat CNS. Neuroscience 2002;113(2):463–78.
[4] Taguchi K, et al. Expression of α-synuclein is regulated in a neuronal cell type-dependent manner. Anatomical Science International October 25, 2019;94(1):11–22.
[5] George JM, et al. Characterization of a novel protein regulated during the critical period for song learning in the zebra finch. Neuron August 15, 1995;15(2):361–72.
[6] Greten-Harrison B, et al. αβγ-Synuclein triple knockout mice reveal age-dependent neuronal dysfunction. Proceedings of the National Academy of Sciences, USA November 9, 2010;107(45):19573–8.
[7] Astbury WT, Street A. X-ray studies of the structures of hair, wool, and related fibres. I. General. Philosophical Transactions, Series A 1932;230:75–101.
[8] Pauling L, Corey RB, Branson HR. The structure of proteins: two hydrogen-bonded helical configurations of the polypeptide chain. Proceedings of the National Academy of Sciences, USA April 1, 1951;37(4):205–11.
[9] Jao CC, et al. Structure of membrane-bound α-synuclein from site-directed spin labeling and computational refinement. Proceedings of the National Academy of Sciences, USA December 2008;105(50):19666–71.
[10] Burré J, Volknandt W. The synaptic vesicle proteome. Journal of Neurochemistry 2007;101:1448–62.
[11] Kahle PJ, et al. Subcellular localization of wild-type and Parkinson's disease-associated mutant alpha -synuclein in human and transgenic mouse brain. Journal of Neuroscience September 1, 2000;20(17):6365–73.
[12] Gruschus JM, et al. NMR structure of calmodulin complexed to an N-terminally acetylated α-synuclein peptide. Biochemistry May 21, 2013;52(20):3436–45.
[13] Gruschus JM, et al. Structure of the functional fragment of auxilin required for catalytic uncoating of clathrin-coated vesicles. Biochemistry March 23, 2004;43(11):3111–49.
[14] DeWitt DC, Rhoades E. α-Synuclein can inhibit SNARE-mediated vesicle fusion through direct interactions with lipid bilayers. Biochemistry April 9, 2013;52(14):2385–7.
[15] Guerreiro PS, et al. LRRK2 interactions with α-synuclein in Parkinson's disease brains and in cell models. Journal of Molecular Medicine April 2013;91(4):513–22.
[16] Wersinger C, et al. Modulation of dopamine transporter function by alpha-synuclein is altered by impairment of cell adhesion and by induction of oxidative stress. Federation of American Societies for Experimental Biology Journal November 17, 2003;17(14):2151–3.
[17] Freed J, Chakrabarti L. Defining a role for hemoglobin in Parkinson's disease. NPJ Parkinson's Disease. 2016;2(1). https://doi.org/10.1038/npjparkd.2016.21.
[18] Qureshi HY, Paudel HK. Parkinsonian neurotoxin 1-methyl-4-phenyl-1,2,3,6-tetrahydropyridine (MPTP) and alpha-synuclein mutations promote Tau protein

phosphorylation at Ser^{262} and destabilize microtubule cytoskeleton in vitro. Journal of Biological Chemistry February 18, 2011;286(7):5055–68.
[19] Dasari AKR, et al. Tau interacts with the C-terminal region of α-synuclein, promoting formation of toxic aggregates with distinct molecular conformations. Biochemistry June 25, 2019;58(25):2814–21.
[20] Alim MA, et al. Tubulin seeds alpha-synuclein fibril formation. Journal of Biological Chemistry January 18, 2002;277(3):2112–7.
[21] Neystat M, et al. Analysis of synphilin-1 and synuclein interactions by yeast two-hybrid beta-galactosidase liquid assay. Neuroscience Letters June 7, 2002;325(2):119–23.
[22] Elkon H, et al. Mutant and wild-type alpha-synuclein interact with mitochondrial cytochrome C oxidase. Journal of Molecular Neuroscience June 2002;18(3):229–38.
[23] Janowska MK, Wu KP, Baum J. Unveiling transient protein-protein interactions that modulate inhibition of alpha-synuclein aggregation by beta-synuclein, a pre-synaptic protein that co-localizes with alpha-synuclein. Scientific Reports October 19, 2015;5:15164.
[24] Yap TL, et al. Alpha-synuclein interacts with Glucocerebrosidase providing a molecular link between Parkinson and Gaucher diseases. Journal of Biological Chemistry August 12, 2011;286(32):28080–8.
[25] Yap TL, et al. Membrane-bound α-synuclein interacts with glucocerebrosidase and inhibits enzyme activity. Molecular Genetics and Metabolism January 2013;108(1):56–64.
[26] Bartels T, Choi JG, Selkoe DJ. α-Synuclein occurs physiologically as a helically folded tetramer that resists aggregation. Nature August 14, 2011;477(7362):107–10.
[27] Dettmer U, et al. n vivo cross-linking reveals principally oligomeric forms of α-synuclein and β-synuclein in neurons and non-neural cells. Journal of Biological Chemistry March 1, 2013;288(9):6371–85.
[28] Krüger R. The role of synphilin-1 in synaptic function and protein degradation. Cell Tissue Res 2004;318(1):195–9.

CHAPTER 6

A New Kind of Pathogen – Prion diseases and amyloid fibrils

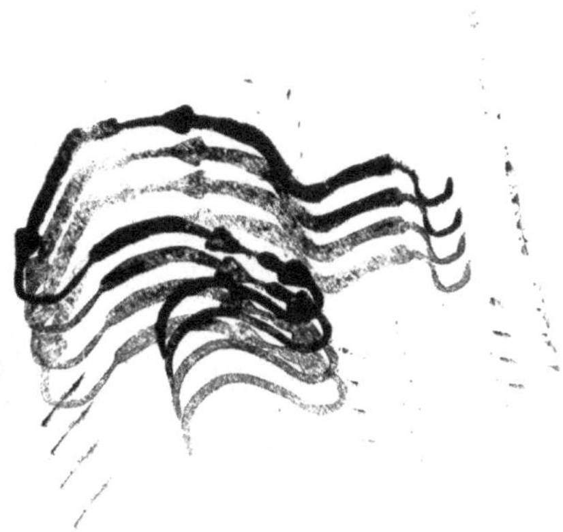

In 1994, in Wiltshire, England, something started to change for Stephen Churchill, a healthy young student doing his A-levels and preparing to join the RAF. Losing weight, feeling miserable, schoolwork flagging, he started to develop an unsteady gait. Over the next months, he developed hallucinations and lost the ability to walk, even to communicate. In May 1995, 19-year-old Stephen Churchill died.

Stephen Churchill was the first victim of a variant form of Creutzfeldt-Jakob disease (CJD), a rare, fatal neurodegenerative disease. In previous decades, the disease typically did not strike people until their sixth or

Synuclein and the Coelacanth
ISBN 978-0-323-85707-9
https://doi.org/10.1016/B978-0-323-85707-9.00002-2

seventh decade, but this changed in the 1990s. Churchill contracted the disease by eating beef tainted with bovine spongiform encephalitis, better known as mad cow disease. The infectious agent was not a virus or a bacterium, but a type of protein called a prion.

A prion is a protein that has misfolded in a way that has insidious consequences. When the prion form of a particular protein encounters another molecule of the same protein in its normal folded form, the prion can cause the normal protein molecule to change into the misfolded prion form. The converted proteins go on to transform more of the normal molecules of the protein, and the prions can accumulate and condense into long fibers, called amyloid fibrils. In variant CLD (vCJD), the amyloid fibrils collect in plaque deposits in the brain while nearby neurons die, leaving holes in the brain tissue, giving it a spongy appearance. Once the symptoms appear, the disease is invariably fatal [1].

Mad cow disease and vCJD are caused by the prion form of a particular protein called prion protein (PrP), an unfortunate name choice that can lead to some confusion. To reiterate, in the field of molecular biophysics, a prion is considered to be any protein that can convert its corresponding correctly folded protein into the misfolded, toxic prion form. PrP, a protein of unknown function found on the neuron cell surface, is just one example of a protein that can become a prion.

The prion form of PrP is called PrP scrapie (PrPsc). Scrapie is the sheep and goat version of mad cow disease. In fact, British cows might first have caught the disease in the 1970s from being fed processed remains of sheep, which have had outbreaks of scrapie for centuries. However, subsequent to the initial infection(s) in cows, the disease was mostly spread by cows being fed processed remains of infected cows, and it is also possible the initial bovine case(s) were not contracted from sheep, but occurred spontaneously in the cows, as in human CJD. Similar diseases, collectively called transmissible spongiform encephalopathies (TSEs), occur in minks, cats, antelopes, and deer [2]. Interestingly, dogs do not appear to be affected, perhaps owing to protective amino acid mutations in their version of PrP [3].

Another human TSE involving PrP is the disease kuru. Kuru, from the Fore word *kuria,* "to shake," occurred among members of the Fore tribe of Papua New Guinea who practiced funerary cannibalism. After the death of a family member, the body was cooked and eaten, and those who ate the brain, usually women, could be exposed to PrPsc [4]. The disease could take 10–50 years to strike. By 1960, medical authorities had realized that funerary cannibalism was spreading the disease, and the already declining practice was abandoned. The disease steadily declined in the following decades, and no cases of kuru have occurred since 2009 [5]. Curiously, comparing kuru and vCJD, there are differences in the two diseases; autopsies have also shown differences in pathology, such as the presence of PrPsc in lymphatic tissue in vCJD, but not kuru [6]. This raises the possibility that there are different strains of PrPsc: that is, different misfolded prion forms of the same protein, leading to different disease courses.

So what do prions have to do with the synucleins? It turns out that alpha-synuclein undergoes a transformation similar to prions, also condensing into amyloid fibrils. Unlike mad cow disease, there is no evidence (yet) that Parkinson's disease is transmissible by eating material tainted with alpha-synuclein in any form, whether it is in the form of amyloid fibrils, in its normal, unstructured monomer form, or in any form in between. People with Parkinson's disease can donate blood. Among hemophiliacs, who must regularly receive human blood or blood products, there is no evidence of an increased occurrence of Parkinson's disease [7].

There are other differences between Parkinson's disease and mad cow disease, along with its human form, vCJD [1]. Instead of forming extracellular plaques outside neurons, alpha-synuclein amyloid forms deposits inside neurons, large spherical deposits called Lewy bodies, smaller deposits in branches projecting from the cell body called Lewy neurites, and countless tiny deposits throughout the cell. Lewy bodies were discovered by Frederic Henry Lewey, a Jewish American neurologist who had worked in the laboratory of Alois Alzheimer before fleeing Nazi Germany [8]. He changed the spelling of his last name after coming to America; the protein deposits are known by the original German spelling.

Although there is no evidence of Parkinson's disease being transmitted among humans, perhaps it is wise not to let down our guard just yet. As we saw in an earlier chapter, mice have been engineered to make high levels of human alpha-synuclein. When alpha-synuclein prepared in a test tube in its amyloid fibril form is injected into these transgenic mice, the injected fibrils can transform the alpha-synuclein in the mouse brain into the amyloid form, and the mice develop progressively more severe movement deficits reminiscent of Parkinson's disease [9]. The first study showing this effect injected the preformed amyloid fibrils directly into the brain, but later studies showed similar results from injections into the bloodstream and even into the hind limb of the mice [10]. Somehow, the pathology was able to enter from the hind limb muscle into the sciatic nerve and then on to the brain, because cutting the nerve slowed or prevented the pathology. A similar experiment was also performed, but with the stomach of the mice injected with fibrils, and cutting the vagus nerve prevented the pathology reaching the brain stem [11].

Since the 1980s, over 300 Parkinson's patients have undergone a controversial procedure. Fetal brain tissue has been transplanted into the brain to replace the dopamine-producing neurons destroyed by the disease [12]. For some, improvement after the procedure was dramatic; for others, not so much. The improvements tended to fade over the course of a few months, and many developed pronounced side effects: in particular, involuntary writhing, bobbing, and fidgeting movements known as dyskinesia. Dyskinesia also occurs as a well-known side effect of drugs used to treat Parkinson's disease, especially the L-dopa—based drugs. Ultimately, those who underwent the procedure fared no better on average in the long run than those who did not [13]. Yet, those lucky few who did have longer-term benefits inspired variations in the experimental procedure, tweaking the parameters and modulating neuron dosage, with the hope of maximizing benefits while reducing the side effects [14]. Even today, the effort continues, either by injecting dopamine-producing neurons derived from stem cells from the patient or by transforming astrocytes already in the substantia nigra into dopamine-producing neurons [15,16].

These medical procedures have had a macabre dividend, something that has added to our understanding of the underlying pathology of Parkinson's disease. In the years after the first transplants, some of the patients died, not necessarily from Parkinson's disease—related causes. Brain autopsies were performed, and, at first, most tissue implants looked healthy, with no obvious signs of Parkinson's disease pathology in the transplanted cells. Because Parkinson's generally does not manifest until late in life, it was hoped that it would also take just as long (i.e., a human lifetime) before similar problems appeared in the transplanted cells. Unfortunately, as the years passed and patients died who had had implant procedures performed over a decade earlier, Lewy body pathology began to be seen in the transplanted tissue [17]. As in the genetically engineered mice, it appeared that alpha-synuclein in its amyloid form could be transferred from cell to cell in humans, triggering new amyloid formation on its way.

More evidence that alpha-synuclein might act like a prion surfaced as more autopsies were performed, this time not on tissue transplant patients, but on Parkinson's patients who had died at various stages in the disease progression [18]. In the earliest stages, Lewy bodies and neurites were found only in two regions of the brain: in the olfactory bulb, which is connected to the nerves in the nose, and in the medulla of the brain stem, in a region connected to the vagus nerve, the nerve that runs to most organs, including the lungs, stomach, and intestines. This earliest phase of Parkinson's pathology is called Braak stage 1, the first of six stages. The stages are named for Heiko Braak, who, with his wife and other scientific colleagues, performed the pathology studies and proposed the staging scheme, inspired by their earlier work describing Alzheimer's disease progression.

The brain locations showing pathology in the first stage struck an ominous chord; Parkinson's patients often experience from a loss of smell (anosmia), and issues regarding the digestive tract, such as constipation, a decade or more before movement-related symptoms appear [19]. Furthermore, epidemiologists had noticed that farmers were more prone to Parkinson's disease, and they suspected pesticide and herbicide exposure was to blame [20]. The nose, lungs, and the digestive system are possible routes for pesticides to enter the body. Finally, in patients who had undergone a now outdated procedure in which the vagus nerve was cut to treat peptic ulcers, it was discovered that their odds of developing Parkinson's disease was reduced [21].

In the second Braak stage, the pathology moves further up the brain stem into specific regions of the pons. It has been hypothesized that Lewy body—associated pathology in at least one of these regions might be responsible for rapid eye movement (REM) sleep behavior disorder, which is another early symptom in patients with Parkinson's disease [22]. In REM sleep behavior disorder, or RBD, patients move during REM sleep, acting out the movements in their dreams. It can be a distressing disorder, even more so for the patient's bedmate, who gets struck by a flailing arm or kicking leg in the middle of the night.

In the third Braak stage, the pathology spreads up into the midbrain. It is here where the most serious damage is done in Parkinson's disease [18]. The pathology attacks several regions, most importantly a region called the substantia nigra pars compacta (SNc). By the time the signature symptoms of Parkinson's disease appear, tremor, rigidity, and bradykinesia, about 30% of neurons in the substantia nigra pars compacta have already died [23].

The SNc delivers the neurotransmitter dopamine to neurons higher up in the brain, in the striatum. Two types of neurons predominate in the striatum: ones that dopamine stimulates and others that it suppresses [24]. The ones that dopamine stimulates are part of the direct motor pathway. When you decide to make a movement, neurons in the motor cortex of the cerebrum signal these neurons in the striatum. When enough dopamine is present, the signal from the motor cortex causes these neurons to inhibit other neurons that normally inhibit motion, which causes a strengthening the motor cortex signal, and the voluntary motion occurs. The neurons that dopamine suppresses are part of the indirect motor pathway. The indirect pathway, when not inhibited by dopamine, suppresses movement. Thus, lack of dopamine leads to the inability to perform voluntary motions and contributes to rigidity.

In the fourth Braak stage, additional regions are affected in the middle region of the brain, including the retrorubral and ventral tegmental (VTA) areas. Retrorubral neurons deliver dopamine to a brain region called the globus pallidus, and some have hypothesized that damage to the retrorubral area might develop into the characteristic Parkinson's tremor [25]. Neurons in the ventral tegmentum area supply dopamine to the reward circuitry neurons of the brain. In the fourth Braak stage, pathology also spreads to the amygdala, which is involved in emotional response [26]. Pathology in these regions, especially the VTA, is thought to contribute to the depression and apathy commonly experienced by Parkinson's patients [27]. As if the other symptoms were not depressing enough?

Also in the fourth Braak stage, the pathology first starts to spread further upward toward the neocortex, the part of the brain where we ideate movement, interpret senses, and plan for the future. This spread enters the neocortex in the fifth Braak stage. A common symptom that can appear at this stage is visual hallucinations and delusions, such as thinking someone is there when he or she is not. As the pathology worsens, dementia often sets in. Symptoms at the onset of dementia are often not the same as those seen with Alzheimer's disease. Instead, the patient might experience episodes of an inability to pay attention, zoning out, or loss of executive function, such as being unable to exit the expressway while driving, but initially without the short-term memory loss typical of Alzheimer's.

In the sixth Braak stage, the pathology is widespread throughout the affected areas of the cortex. However, even at this stage, some patients still show no obvious signs of dementia [28]. Curiously, primarily in those who have dementia, amyloid plaques are often observed in the affected striatal cortical regions, like the plaques seen in Alzheimer's disease, although not as widespread [29]. This observation suggests that Lewy pathology, or whatever its underlying cause is, might also trigger Alzheimer's pathology, further contributing to neuronal dysfunction, and thus to the dementia.

The Braak staging scheme has proven to be accurate for describing a large subset of Parkinson's disease patients, namely those who have idiopathic Parkinson's disease. Strictly speaking, the medical term "idiopathic" means that the cause is unknown, but in the case of Parkinson's disease, what is actually meant is that the cause is not obviously genetic; that is, those who have the disease have no known mutation linked to Parkinson's and no family history of the disease. The movement-related symptoms seen for Parkinson's disease can also be seen in other diseases, such as supranuclear palsy, in which the neuron-level pathology manifests as neurofibrillary tangles consisting of the protein tau. The symptoms can also be caused by vascular problems, such as a stroke that happens to affect the SNc, as well as by neurotoxins targeting the same region. Cases such as parkinsonism without Lewy pathology are obviously excluded from the Braak staging scheme.

In cases of Parkinson's disease with an obvious genetic cause, the Braak staging scheme seems to break down [28]. For patients with mutations in the protein glucocerebrosidase, the cortical progression of the Lewy pathology seems to progress more rapidly, blurring the boundaries between Braak stages [30]. For those with alpha-synuclein mutations, including cases in which the whole gene has been duplicated or even triplicated (that is, in which two or three copies of the gene are present on the chromosome instead of just one), the Lewy pathology is so pronounced and widespread that upon autopsy, nearly all cases look like stage 6, with no way to confirm whether earlier Braak stages occurred in the usual order. At the other end of the spectrum, for those with Parkinson's disease with mutations in the LRRK2 or parkin proteins, sometimes there are no Lewy bodies, even with near complete loss of neurons in the SNc [28].

Another confusing aspect of the Braak staging scheme is that the various brain regions displaying Lewy pathology often do not have direct synaptic connections to each other. For example, the olfactory bulb, lying above the nasal cavity, lacks direct connections to the dorsal motor nucleus of the vagus nerve in the brain stem, yet both regions are included in Braak stage 1. On the other hand, maybe it is this separation of the vulnerable regions that gives rise to the distinct stages of the Braak scheme. That is, once a region vulnerable to Lewy pathology takes up alpha-synuclein in its prion form, the pathology can quickly spread throughout the region, but to proceed to the next stage, alpha-synuclein in its prion form must somehow be taken up by the next vulnerable region in the Braak staging scheme, a chance process that might take years to occur.

There is one more important set of cases that are excluded from the Braak staging scheme: patients who have dementia with Lewy bodies (DLB). In DLB, patients have most of the same symptoms of Parkinson's disease, but the dementia appears before or no more than a year after movement-related symptoms. In other words, some patients seem to show a staging of symptoms in the reverse chronological order given by the Braak staging scheme [31]. Some researchers have suggested that DLB might be more common for those whose pathology originates in the olfactory bulb, which is closer to the cortical and limbic parts of the brain, especially the amygdala, which is involved in DLB [32]. In fact, a new staging scheme seeking to integrate the different patterns of Lewy pathology seen for DLB with those of idiopathic Parkinson's disease has been proposed [33]. Most DLB cases are also idiopathic, although intriguingly, a few DLB patients have mutations in beta-synuclein [34].

Patients with DLB also have a poorer prognosis; they often die of the disease within 5 years of diagnosis. Could it be, like PrP in kuru and vCJD, that there are different strains of alpha-synuclein prion in Parkinson's disease? Are they the same prion, emerging in different regions of the brain, or do the prion molecular structures differ in the olfactory bulb and the vagus nerve? Is the prion hypothesis for Parkinson's disease even correct? To explore these questions further, we need to learn more about the molecular structure of alpha-synuclein in its amyloid form.

Protein molecular structure is hierarchical, with four levels. The first level, called primary structure, is simply the sequence of amino acids; each of the 20 types of amino acids has the same chemical structure for its peptide backbone atoms, except for proline, and each of the 20 has a different chemical structure for its particular side chain. The next level, called secondary structure, consists of the amino acids arranged in structures like the alpha-helix described in the last chapter. The other main type of secondary structure is called beta-sheet, which is the element necessary for us to understand amyloid structure. Most protein structure is basically the secondary structural elements, alpha-helices and beta-sheets, folding together in a particular way, with each of the elements connected by sequences of intervening amino acids called loops.

The particular way the secondary structural elements come together is called the tertiary structure, or fold, of the protein molecule; that is, the tertiary structure is synonymous with the actual three-dimensional (3D) structure into which the sequence of amino acids folds. Finally, there is the quaternary structure, the structure that results when the folded protein molecules come together to form complexes. Many proteins have no quaternary structure; they function fine with just their tertiary structures. Quaternary structure can involve complexes of a protein interacting with one or more different proteins, or it can involve interactions with copies of itself. Amyloid fibrils are nearly always composed of just one kind of protein, so the quaternary interactions involve interprotein interactions between copies of the amyloid protein.

One of nature's aims in folding a protein is to find a structure for which the hydrophilic bits of the peptide backbone, namely, the peptide hydrogen and oxygen atoms, are either exposed at the protein surface or involved in hydrogen bonds. The alpha-helix is one of the ways in which the peptide backbone can provide hydrogen bonding to itself. The other major secondary structural element, the beta-sheet, also involves peptide hydrogen bonds. In a beta-sheet, the peptide chain is extended so that the peptide hydrogen and oxygen atoms both point to the right for one amino acid, both point to the left for the next amino acid, right for the next, and so on. In this secondary structural motif, another extended peptide chain lies alongside the first, forming hydrogen bonds with it, and more chains can be added to either side, forming more hydrogen bonds, each chain forming a strand in the beta-sheet structure.

The adjacent strands can be arranged in two ways. If the strand runs in the same N-terminal to C-terminal direction as the first chain, they are said to form a parallel beta-sheet. If the other chain runs in the direction opposite the first, they form an antiparallel beta-sheet. The beta-sheet structure can be built up further, with additional extended chain segments lying parallel or antiparallel alongside the first two chains, continuing the sheet structure to the left and right. Thus, beta-sheets can be all parallel, all antiparallel, or a mixture of both. The beta-sheet is like a mesh with all of the chains aligned in one direction and the hydrogen bonds connecting them at right angles to the chains, but in most cases, the sheet is not flat. Instead, the sheets typically have a slight to moderate twist when viewed along the direction of the hydrogen bonds.

In protein tertiary structures, about 80% of the beta-sheets are antiparallel. It could be that antiparallel beta-sheets are more energetically favorable, as some computational studies have suggested [35]. Another possibility is that antiparallel sheets are easier to form. That is, it is easy after the first chain of the sheet to turn around 180 degrees with a short loop and run the next chain down the opposite direction. This is the simplest beta-sheet structure, called a beta hairpin, and is one of the most common beta-sheet structures. In contrast, to add another chain segment running parallel to the first, the loop would need to arc all the way back over the first chain segment so that the continuing chain can run in the same direction. This does happen; a common structural motif is for the intervening amino acids between segments of parallel sheet to form an alpha-helix; such protein folds are called alpha/beta.

Here is why understanding the details of the beta-sheet structure is so important to understanding amyloids. One of the peculiarities of amyloid fibrils is that, with few exceptions, their beta-sheets are parallel [36]. Even more peculiar is that these sheets are formed not from the chain of a single protein molecule, as is typical for most protein structures, but from the chains of multiple molecules of the same protein. In other words, the sheets are elements of the quaternary structure, as opposed to being elements of the tertiary structure. What can explain this peculiar structure? The answer lies in one more peculiar aspect of the amyloid structure: namely, the chains forming the parallel beta-sheets are almost always in register. That is, the strands in each sheet are composed of the same stretch of amino acids in the protein sequence so that each amino acid in a particular beta strand lies next to the same amino acid in the neighboring strands.

The in-register arrangement has advantages. Hydrophobic side chains pack together, helping each other to avoid contact with water, for instance. Hydrophilic side chains without a net electric charge can form hydrogen bond ladders, each side chain hydrogen bonding with its like neighbors. Charged side chains are problematic, however. Like electric charges repel, and being in register places charge next to like charge. Sometimes the amyloid structure handles a charged side chain by having an opposite charged side chain close by, forming a salt bridge, so-called because salt is made of crystals of oppositely charged atoms or molecules lying next to each other in such a way that the attractive forces outweigh the repulsive ones. For some types of amyloid, they can form only when changing solution conditions result in the neutralization of charges of certain side chains. For example, this can happen when the acidity of the solution is increased, a situation that happens inside particular organelles within the cell.

Another way in which alpha-synuclein can form amyloid is through interactions with negatively charged lipid membranes. Recall from the last chapter that alpha-synuclein first binds to lipid membrane vesicles via its first few N-terminal amino acids. At low vesicle concentrations, the alpha-synuclein molecules crowd onto the vesicles with just their N-termini bound. This arrangement greatly increases the odds of in-register, parallel chain interactions between the crowded chains, accelerating the initiation of amyloid structure. Ironically, at high vesicle concentrations, the situation is reversed [37]. With all of the extra vesicle surface area, crowding is eliminated, and the alpha-synuclein can fully bind its first 100 or so amino acids to the vesicles, including the amyloid core non—amyloid-beta component (NAC) region amino acids, keeping them out of mischief.

Amyloid fibrils have not yielded the fine details of their structures without experimenters having to overcome substantial challenges. When viewed in an electron microscope, fibrils purified from cells almost never have a uniform appearance. Amyloid deposits such as senile plaques in Alzheimer's disease and Lewy bodies in Parkinson's disease contain all sorts of proteins, lipids, and other compounds in addition to the amyloid fibrils [38]. Even fibrils grown in the laboratory can adopt different shapes, with different widths, most often appearing as multiple fibrils twisted together. X-ray diffraction studies also hinted at unusual structures at the atomic level. Unfortunately, x-ray crystallography, the workhorse of molecular structure determination techniques, was not initially an option for amyloid fibrils, because the fibrils could not form into a crystal lattice structure necessary for the technique (see the sidebar on Crop Rows). Instead, the diffraction pattern was smeared out into a powder pattern, so named because it is the same type of pattern produced by a powder of randomly oriented crystalline grains.

Crop rows — X-ray crystallography

When I was a kid, we went on a lot of road trips. I remember passing through the agricultural regions of California, seeing fields of various crops go by, strawberries, artichokes, most I could only guess at. Driving by at 55 miles per hour (the speed limit had been lowered to improve fuel efficiency because of the oil embargo of 1973), lines of crops would spring into sight for a split second, to be replaced by new ones a split second later. The lines radiated not just along the furrows, but at diagonals, too, with the regularly spaced plants lining up one behind another. X-ray crystallography relies on such alignments, in which the crystals are a 3D analog of crop fields, each molecule like a plant in the field. Because they are three-dimensional instead of flat fields, the alignments become planes through the crystal, each reflecting the incoming x-ray beam in a unique way.

X-rays are a high-energy form of light, and they interact with atoms, in particular, with the electron clouds surrounding the atomic nuclei. The heavier the atom, the more electrons in its surrounding cloud. The electrons respond to the x-ray beam, feeling the force of its oscillating electric field, scattering a tiny bit of the x-ray beam in all directions. For molecules lying in a crystal plane, their scattering of the x-ray beam combines in a special way in a particular direction, bouncing away from the plane at an angle equal to that at which the beam hit it, like a cue ball banking off the side of a pool table. The intensity of the reflection depends on this angle and the plane spacing. Also important is the wavelength of the beam, which is inversely related to the frequency of its oscillating field via the speed of light; the shorter the wavelength, the higher the frequency. Most important for structure determination, each electron's contribution to the reflection depends on its spatial location within the molecule.

When measured, originally using x-ray—sensitive films, but these days, using electronics, the result of this scattering is a patterned array of dots of varying intensity around the strong signal in the center corresponding to the main x-ray beam. The pattern and spacing of the array yield important information, such as the spacing between molecules and exactly how they are packed together into a crystal. Growing the crystal is an art unto itself, especially for biomolecules such as proteins. For proteins, which are chiral molecules, there are exactly 65 different ways to pack them into a crystal: that is, 65 different ways of spacing and orienting each molecule relative to its neighbors to generate a regularly repeating, symmetric, 3D array. To find out more about chirality, you will have to wait until the section "Twisting Light," but for now, just know that it has to do with right- versus left-handedness.

The structure of the molecules is encoded in the varying intensities of the array of dots. To maximize the amount of structural information, the crystal is rotated to different orientations and each new pattern is measured. To determine the structure of the small molecule, or just to get the crystal symmetry group's and its molecular spacing, a relatively small tabletop experimental setup using an x-ray source similar to that used for medical x-ray images is adequate. For large molecules such as proteins, however, the more powerful the x-ray beam, the better. The stronger the beam, the more the weaker dots at larger angles to the beam become visible, providing higher-resolution data on the structure. The strongest beams are produced by circular particle accelerators called synchrotrons, which range in size from a few tens of meters to over 26 km in circumference. The beams are so intense that they can tear apart the molecular structure; to slow this process, the crystals are cooled in a jet of liquid nitrogen.

X-rays are waves, with crests and troughs. When the crests of the scattered x-rays from the electrons in various regions of the molecules align with each other along a particular reflection angle, the intensity of the dot is increased, but when a crest scattered from one part of the molecules aligns with the trough from another part, they cancel, reducing the intensity. Complicating matters, knowing just the intensity of the dots is not sufficient to determine the structure of the molecule. The phase of the dots is required, too, in which "phase" means the timing with which the x-ray wave crests hits the measuring device relative to the waves forming the other dots.

There are two ways to deal with the phase problem. The first way is more direct, using heavier atoms incorporated into the crystal structure that strongly scatter the x-beam. Some heavier atoms scatter even more strongly when tuned to a particular frequency. Next, one measures the intensities of the dots measured with and without the heavy atoms, or alternatively, with the x-ray beam at the special frequency and then with the beam well away from the special frequency. Comparison of these data sets provides enough missing phase information to begin the structure determination process.

The other way to obtain phase information, called molecular replacement, requires no incorporation of heavier atoms, but it does require prior knowledge regarding the molecular structure. Using an approximate molecular structure, or at least part of it, the pattern of dot intensities it would produce can be compared and adjusted until it best matches the measured data. Both approaches provide initial, approximate values of the phases. Using both the intensities and approximate phases, a mathematical analysis called a Fourier transform can convert the data into a density representing the electron cloud.

Finally, in an iterative process called refinement, the structures of the individual components of the molecule, amino acids in the case of proteins, are fitted into the calculated electron density, yielding an idealized, model electron density. The model is used to generate better approximations of the phases, producing a more accurate calculated density, and the model density is readjusted to fit it. This back and forth process continues until no further improvement can be made in the match between the calculated and model densities, yielding the final molecular structure.

In 1968, the first x-ray diffraction patterns produced by aligned amyloid fibrils were reported by E. D. Eanes and G. G. Glenner at the National Institutes of Health (NIH) [39]. The patterns were reminiscent of those seen for DNA 15 years earlier. They revealed that along the axis of the fibril, the atoms were spaced about 0.49 nm (billionths of a meter) apart, a distance that corresponds to the separation of chain segments in a beta-sheet. This type of structure, with the beta-sheet chain segments perpendicular to the fibril axis, and the beta-sheet they form running parallel to the fibril axis, is called a cross-beta structure. Subsequent studies have shown additional signals in fibril powder patterns. Many show a signal indicating an additional separation of about 1.0 nm, corresponding to the distance between two beta-sheets stacked on top of each other, whereas powder patterns for other fibrils have indicated different separations or lacked these signals altogether [40]. Even at this crude level, differences in amyloid fibril atomic structures were already apparent.

By employing amino acid segments much shorter than original amyloid forming proteins, x-ray crystallographic techniques were finally able to reveal details of fibril structure at the atomic level, confirming the cross-beta structure and identifying features such as the hydrogen bonding ladders mentioned earlier [41]. However, for the longer protein chains that form amyloid fibrils in the cellular milieu, in particular those involved with Alzheimer's and Parkinson's diseases, a different technique was required. That technique was solid-state nuclear magnetic resonance imaging (NMR), a version of NMR spectroscopy that was introduced in the last chapter.

In solid-state NMR, to obtain sharp signals, a tiny, hollow ceramic tube packed with the amyloid fibrils must be spun at speeds of thousands of rotations per second. The angle at which the sample is spun relative to the axis of the magnetic field in the NMR spectrometer is critical. The angle is 54.7 degrees, the magic angle corresponding to the body diagonal of a cube. By rotating the sample at this angle, with the magnetic field parallel to the cube edge, the interactions that broaden the NMR signals of the atomic nuclei average to zero, producing sharp signals at the corresponding frequencies of each nucleus.

Using careful protocols to produce samples containing just one type of fibril shape, or in amyloid jargon, morphology, magic angle spinning solid-state NMR studies were performed on amyloid beta, the protein found in senile plaques of Alzheimer's patients. These studies, which were performed in the laboratory of Robert Tycko at NIH, were the first to provide a detailed atomic look at different strains of amyloid fibrils arising from the same protein [42,43]. The less twisted, striated ribbon fibrils had a stacked beta-sheet structure, consistent with that predicted by the early x-ray diffraction studies. The twisted fibrils had a more complicated structure, with three stacked beta-sheets arranged in a triangle, with the line joining the middle of the triangles running more or less parallel to the fibril axis. These structures proved that amyloid fibrils from the same protein could adopt different structures even at the atomic level.

The first alpha-synuclein amyloid atomic level structure was determined by a team including Julia M. George, who conducted the zebra finch studies mentioned in the last chapter, and solid-state NMR spectroscopist Chad M. Rienstra [44]. It has the same cross-beta structure as all other amyloids, but with its chain in a more complex shape known as a Greek key motif. This motif owes its name to the interlocking geometric patterns that were commonly used to decorate ancient Greek pottery. Another name for this type of motif is the jelly roll fold, because its pattern is reminiscent of the pastry made from a layer of jelly rolled up between two sheets of dough. The region comprising the amyloid structure consists of amino acids 46—96 of the 140-amino acid—long protein, including the NAC region, mentioned in Chapter 1, which makes up the most hydrophobic region of the protein. The structure contains side chain hydrogen bond ladders and packed hydrophobic side chains typical of amyloid structures, and one well-defined salt bridge. Not all amino acids in the amyloid adopt a well-defined structure. Amino acids 55—62 form a disordered loop, possibly a consequence of the four charged side chains in this stretch, making an ordered, in-register structure energetically unfavorable.

In addition to the disordered loop, the amino acids before and after the amyloid structure region also do not adopt a well-defined structure. I have performed many successful solution NMR studies looking at alpha-synuclein and other disordered proteins interacting with much larger molecules and structures [45]. One such study was the interaction with lipid vesicles described in the last chapter [46]. Because of this, I looked with a fair amount of optimism at the solution NMR spectrum of pure, isotopically labeled amyloid fibrils prepared in a test tube by a standard protocol.

Much to my disappointment, I saw none of the sharp signals for alpha-synuclein to which I was accustomed; in fact, I saw practically no signal at all from the amyloid sample except for a few amoeba-esque blobs peeking just above the level of experimental noise. Perhaps my disappointment was unwarranted; negative results are just as important to science as positive ones. The problem with negative results is that to publish, one usually needs to show that the negative result is always obtained even after varying all relevant parameters, a higher level of rigor than usually required for positive results. To my discredit, because of the extra rigor required, I opted to drop the project and pursue experiments more likely to produce positive results that were easier to publish.

Why were there no signals in the solution NMR spectrum of the alpha-synuclein fibrils? Perhaps the reason is the interactions between the N- and C-terminal regions similar to those seen for alpha-synuclein in its nonamyloid state, described the last chapter. A variety of studies, using antibodies, for instance, suggest that the C-terminal region tends to be more exposed in fibrils [47]. It could be that the N-terminal region lies wrapped around the amyloid core, with the C-terminal region then wrapping around the N-terminal amino acids, restricting the motion of the terminal regions and broadening their signals.

At the microscopic level, different morphologies of alpha-synuclein amyloid fibrils abound, with various descriptors such as right- and left-handed twisted helices, ribbons, and, my favorites, spaghetti and linguine (though not the wiggly spaghetti of the unstructured monomer, which is too small to see by microscopic techniques) [48]. Two more alpha-synuclein structures were determined in 2018 [49,50]. Using a technique called cryoelectron microscopy (cryo-EM) (see Ghost in the Static), the researchers observed the same Greek key motif structure, but the fibril morphology was different, and many atomic level details differed, as well.

For both structures, instead of the single filament morphology of the fibrils used for the solid-state NMR structure, they saw two filaments twisted around each other. Instead of a disordered loop for amino acids 55—62, these amino acids were ordered. Indeed, amino acids 50—57 were at the interface between the filaments. This stretch of amino acids includes 50, 51, and 53, sites where mutations cause familial Parkinson's disease. Mutations at these sites could potentially destabilize the two-filament structure. A host of other experiments also support the idea that different atomic structures underlie the various morphologies. For instance, one such experiment partially digests fibrils using protein-cutting enzymes called proteases, and different fibril shapes yield different digestion patterns, which suggests that different parts of the alpha-synuclein amino acid chain are more exposed to the proteases, depending on the fibril shape [51].

Ghost in the static — Cryo-EM

Did you ever take a page of text and copy it on an old copying machine, and then copy the copy, and so on, until the text became unreadable? Now, imagine that you start over with the original page and repeat the process, copying the copies, and you start over again, and so on, until you have 100 different unreadable pages. Here's the really cool part. One can recover the original text (that is, reduce the noise obscuring it) by averaging the images of the unreadable pages.

The reason this is possible is because the noise obscuring the text is random (well, for the sake of the argument, let us assume the noise is random, and not due to dirty plates or mechanical imperfections). Say we have digitized the 100 unreadable pages, and let us consider a particular pixel location in the images. The noise at that location might make the pixel darker in one image and brighter in another; on average, the noise would average to zero if you had enough images. The reduction in noise obeys an inverse square root relationship; that is, if you average 100 images, the noise becomes reduced by a factor of 10 (the square root of 100). If you had 400 images, the noise would be reduced by a factor of 20.

Structure determination by cryo-EM relies on this noise reduction technique. Taking advantage of the quantum weirdness of wave-particle duality, electrons can be treated like waves and used to create a microscope capable of resolving images much smaller than possible with visible light, even down to the levels of individual protein molecules. To image protein molecules in a water solution, they have to be frozen, to stifle the quantum jitteriness of Brownian motion; the colder, the better, and hence the "cryo" part. Protein molecules are at the limit of what can be observed, and their images are noisy.

One more challenge to tackle is that the protein molecules being imaged are in random orientations. The images have to be sorted, clumping images of similar appearance into sets corresponding, it is hoped, to similar orientations. Sometimes hundreds of thousands of these images have to be sorted through. Each set of images gets averaged together, yielding a 2D projection of the molecule from the viewpoint of each orientation. Then, starting with a crude model for the 3D structure, a computer program compares the 2D projections the model yields with the actual projections, iteratively improving the 3D model to yield better agreement with the data. There are many ways of doing this, but all require massive amounts of computer time.

One thing that helps reduce the complexity of the 3D reconstruction is when the molecule has some sort of symmetry. For instance, say a protein forms a symmetric hexamer, six equivalent protein molecules arranged in a complex that, when rotated 60 degrees around the axis of symmetry, yields the same structure, like a pinwheel with six vanes or a six-petaled flower. This reduces the complexity of 3D model reconstruction by a factor of six. Furthermore, the projection image with sixfold symmetry must correspond to the projection along the axis of symmetry, giving a leg up on creating and refining the initial model.

The regularly repeating structures of amyloid fibrils make them ideal for structure determination by cryo-EM. Their monomers must be stacked about 0.49 nm (49% of a billionth of a meter) apart, perpendicular to the fibril axis, a requirement imposed by their cross-beta structure. The regularly repeated twist pattern typically seen for most fibrils makes it straightforward for experimenters to align and average images from different fibrils with each other, even with images from different parts of the same fibril. Some of the highest-resolution cryo-EM structures to date are of amyloid fibrils, rivaling the atomic resolution of x-ray and NMR structures.

There is evidence that mice inoculated with different alpha-synuclein fibrils, either purified from the brains of patients or grown in a test tube, show different symptoms and pathology. However, additional work is needed to determine whether this is the result of differing fibril morphology or other factors present in the purified fibrils [52].

In one experiment, fibrils were injected purified from patients who had had the rare synucleinopathy known as multiple system atrophy (MSA), which affects glial cells in the brain called oligodendrocytes. The mice developed alpha-synuclein pathology in their neurons similar to that seen by other experimenters using fibrils from Parkinson's patients, but they also developed glial cell pathology reminiscent of MSA [53]. Glial cells support neurons; they have roles in remodeling neuronal connections, scavenging intercellular debris, and for oligodendrocytes in particular, coating axons with myelin, a lipid membrane sheathing that helps speed electrical signals traveling down axons. Normally, oligodendrocytes do not produce alpha-synuclein, at least not in significant amounts, so the alpha-synuclein they accumulate in MSA pathology is assumed to come from enveloping extracellular synuclein somehow secreted from surrounding neurons.

The laboratory experiments overwhelmingly support the idea that alpha-synuclein fibrils can act like prions. Despite this experimental evidence, autopsy results from actual patients present a contradictory picture. Lewy bodies and neurites have also been seen in brains of patients who displayed no obvious symptoms of Parkinson's disease or DLB, yet with pathology just as extensive as for patients with symptoms [28]. Perhaps these patients had some sort of benign strain of alpha-synuclein fibrils, but so far, there appear to be no obvious structural differences between the fibrils from these patients and those from patients with Parkinson's disease. It is possible that there are differences at the molecular level too subtle to be detected by standard microscopic techniques, small differences in their amyloid cross-beta structures, or chemical modifications such as phosphorylation.

Cells have various machinery for regulating protein turnover, by recycling old proteins by breaking their peptide chemical bonds, an enzymatic process called hydrolysis, generating amino acids to be reused to make new proteins. There are two major locations for protein degradation in the cell. One is the lysosome, an acidic organelle that breaks down not just proteins, but also carbohydrates and lipids. The other location is the proteasome. Proteasomes break down only proteins, and the cell has a way of targeting a protein for proteasomal degradation, by attaching a chain of small proteins called ubiquitin to the protein. No cell types have been found that do not express ubiquitin; hence its name from the word "ubiquitous." Lewy bodies have high amounts of ubiquitin, much of which is attached to alpha-synuclein, and they have other proteins as well that are parts of the proteasomal and lysosomal machinery [54]. It is as if the cell were trying to degrade alpha-synuclein via the proteasomal and lysosomal pathways, but the processes got gummed up somehow and the resulting mess ended up trapped in the Lewy body.

The alpha-synuclein in Lewy bodies is also hyperphosphorylated; that is, it has more phosphoryl groups attached than does alpha-synuclein in healthy neurons [55]. Some researchers have speculated that the phosphorylation increases the propensity of alpha-synuclein to aggregate into amyloid, but it is unclear whether this actually happens in the cell. Recall that charged groups can be problematic for amyloid formation because of their electrostatic repulsion of each other, especially for in-register, parallel beta-sheet structure. Personally, I lean toward an alternate theory, that alpha-synuclein becomes hyperphosphorylated after amyloid formation. The cell is full of phosphorylating enzymes, or kinases. The longer a protein lingers in the cell, the greater the chance that some of the less discriminating kinases might accidently attach a phosphoryl group, even if alpha-synuclein is not their ideal substrate. Lewy bodies are likely to be months or even years old by the time of death, plenty of time for accidental phosphorylation to occur.

The presence of Lewy bodies and neurites in the autopsies of patients who showed no symptoms of Parkinson's or dementia have led some authors to propose an alternate hypothesis with regard to alpha-synuclein amyloid in disease: to wit, that amyloid formation could actually have some sort of protective role [56]. The idea is that Lewy pathology is not actually pathological, but rather a way for the cell to corral misfolded alpha-synuclein, to keep it tucked away where it cannot do damage. Some have proposed that a Lewy body might start as something called an aggresome, also known as a juxtanuclear quality control compartment, because they are found next to cell nuclei [57]. When the cell's proteasomal degradation pathway becomes clogged with too many ubiquitinated proteins, the cell tucks them away in the aggresome, perhaps to await another degradative pathway called macroautophagy. What is not known is why this supposed macroautophagy pathway seemingly fails to degrade the aggregated alpha-synuclein and its associated lipids and proteins, resulting instead in a Lewy body.

In parallel with this hypothesis that Lewy bodies might be protective, scientists have discovered amyloid proteins that appear to serve normal biological, beneficial roles in organisms ranging from bacteria to humans [58,59]. In humans, amyloids have numerous roles, from protein hormone storage to oocyte fertilization. I studied one such functional amyloid with Ryan McGlinchey in the laboratory of Jennifer C. Lee. The protein PMEL forms amyloid under acidic conditions and serves as a scaffold for melanin pigment synthesis; then, it dissolves under neutral pH conditions [60]. My NIH lab mate Daniel L. Morris, in the laboratory of Nico Tjandra, is studying another peptide, humanin, made by mitochondria, that appears to form amyloid fibrils that help to prevent cell death [61].

Even if Lewy bodies and neurites can have a protective role, sequestering misfolded alpha-synuclein, I doubt such a role could have arisen owing to natural selection. The large, dense Lewy bodies seen in the SNc are unique to humans; they are not seen in any of our great ape relatives. With few exceptions, Lewy bodies develop decades after our peak reproductive years, so it is hard to imagine a scenario in which the propensity to form Lewy bodies could have been selected for. A likelier possibility is that Lewy pathology is a relatively benign side effect of whatever the true neurotoxic agent is in Parkinson's disease, and that the neurotoxicity can vary from person to person based on genetics and particular environments.

Nevertheless, experiments in laboratory animals have shown that alpha-synuclein fibrils are the most efficient at inducing alpha-synuclein pathology capable of spreading from the site of inoculation. The Braak staging scheme, which accurately describes most nongenetically related Parkinson's disease, is consistent with the prion hypothesis: namely, that misfolded synuclein in the gut, lungs, or nose can travel to the brain stem and olfactory bulb, where the pathology can spread further to vulnerable regions in the basal ganglia, limbic regions, and cortex.

On the other hand, DLB and Parkinson's disease resulting from to genetic mutations do not generally follow the Braak staging scheme, and asymptomatic patients with Lewy pathology call into question the main premise of the scheme, that Lewy pathology mirrors Parkinson's pathology. Finally, I reiterate that there is no evidence that Parkinson's disease has ever been transmitted among humans. I suspect that Parkinson's disease is a kind of hybrid with prion disease aspects, especially with regard to environmental factors, and nonprion aspects with regard to genetic influences. The idea that Lewy pathology might just be a side effect begs the question of what the true neurotoxic agent in Parkinson's disease really is. But just because Lewy bodies might be relatively benign, do not think this means that alpha-synuclein is off the hook.

References

[1] Trevitt CR, Singh PN. Variant Creutzfeldt-Jakob disease: pathology, epidemiology, and public health implications. The American Journal of Clinical Nutrition September 1, 2003;78(3):651S−6S.

[2] Williams ES, Miller MW. Ransmissible spongiform encephalopathies in non-domestic animals: origin, transmission and risk factors. Revue Scientifique et Technique (Paris) 2003;22(1):145−56.

[3] Fernández-Borges N, et al. Unraveling the key to the resistance of canids to prion diseases. PLoS Pathogens 2017;13(11):e1006716.

[4] Whitfield JT, et al. Mortuary rites of the South Fore and kuru. Philosophical transactions of the Royal Society of London. Philosophical Transactions of the Royal Society Series B. Biological sciences 2008;363(1510):3721−4.

[5] Whitfield JT, et al. Cultural factors that affected the spatial and temporal epidemiology of kuru. Royal Society Open Science 2017;4(1):160789.

[6] Hilton DA, et al. Specificity of lymphoreticular accumulation of prion protein for variant Creutzfeldt-Jakob disease. Journal of Clinical Pathology March 2004;57(3):300−2.

[7] Khleif AA, et al. Multiple comorbid conditions among middle-aged and elderly Hemophilia patients: prevalence estimates and implications for future care. Journal of Aging Research September 7, 2011:985703.

[8] Holdorff B. Friedrich Heinrich Lewy (1885−1950) and his work. Journal of the History of the Neurosciences 2002;11(1):19−28.

[9] Luk KC, et al. Intracerebral inoculation of pathological α-synuclein initiates a rapidly progressive neurodegenerative α-synucleinopathy in mice. Journal of Experimental Medicine May 7, 2012;209(5):975−86.

[10] Sacino AN, et al. Intramuscular injection of α-synuclein induces CNS α-synuclein pathology and a rapid-onset motor phenotype in transgenic mice. Proceedings of the National Academy of Sciences United States of America July 22, 2014;111(29):10732−7.

[11] Kim S, et al. Transneuronal propagation of pathologic α-synuclein from the gut to the brain models Parkinson's disease. Neuron August 21, 2019;103(4):627−41.

[12] Clarkson ED. Fetal tissue transplantation for patients with Parkinson's disease: a database of published clinical results. Drugs and Aging 2001;18(10):773−85.

[13] Freed CR, et al. Transplantation of embryonic dopamine neurons for severe Parkinson's disease. New England Journal of Medicine March 8, 2001;344(10):710−9.

[14] Hallett PJ, et al. Long-term health of dopaminergic neuron transplants in Parkinson's disease patients. Cell Reports June 2014;7(6):1755–61.
[15] Xiao B, et al. Induced pluripotent stem cells in Parkinson's disease: scientific and clinical challenges. Journal of Neurology, Neurosurgery, and Psychiatry July 2016;87(7):697–702.
[16] Qian H, Kang X, Hu J, et al. Reversing a model of Parkinson's disease with in situ converted nigral neurons. Nature 2020;582:550–6.
[17] Kordower JH, et al. Lewy body-like pathology in long-term embryonic nigral transplants in Parkinson's disease. Nature Medicine May 2008;14(5):504–6.
[18] Braak H, et al. Staging of brain pathology related to sporadic Parkinson's disease. Neurobiology of Aging March–April 2003;24(2):197–211.
[19] National Institute of Aging. Parkinson's Disease. https://www.nia.nih.gov/health/parkinsons-disease.
[20] Gorell JM, et al. The risk of Parkinson's disease with exposure to pesticides, farming, well water, and rural living. Neurology May 1, 1998;50(5):1346–50.
[21] Svensson E, et al. Vagotomy and subsequent risk of Parkinson's disease. Annals of Neurology October 2015;78(4):522–9.
[22] Boeve BF. Idiopathic REM sleep behaviour disorder in the development of Parkinson's disease. The Lancet Neurology 2013;12(5):469–82.
[23] Cheng HC, Ulane CM, Burke RE. Clinical progression in Parkinson disease and the neurobiology of axons. Annals of Neurology June 2010;67(6):715–25.
[24] Freeze BS, et al. Control of basal ganglia output by direct and indirect pathway projection neurons. Jounal of Neuroscience 2013;33(47):18531–9.
[25] Helmich RC, et al. Cerebral causes and consequences of parkinsonian resting tremor: a tale of two circuits? Brain 2012;135(11):3206–26.
[26] Papapetropoulos S, et al. Cortical and amygdalar Lewy body burden in Parkinson's disease patients with visual hallucinations. Parkinsonism and Related Disorders May 2006;12(4):253–6.
[27] Alberico SL, Cassell MD, Narayanan NS. The vulnerable ventral tegmental area in Parkinson's disease. Basal Ganglia August 2015;5(2–3):51–5.
[28] Burke RE, Dauer WT, Vonsattel JP. A critical evaluation of the Braak staging scheme for Parkinson's disease. Annals of Neurology November 2008;64(5):485–91.
[29] Dugger BN, et al. Presence of striatal amyloid plaques in Parkinson's disease dementia predicts concomitant Alzheimer's disease: usefulness for amyloid imaging. Journal of Parkinson's Disease 2012;2(1):57–65.
[30] Neumann J, et al. Glucocerebrosidase mutations in clinical and pathologically proven Parkinson's disease. Brain July 2009;132(7):1783–94.
[31] Donaghy PC, McKeith IG. The clinical characteristics of dementia with Lewy bodies and a consideration of prodromal diagnosis. Alzheimer's Research and Therapy July 21, 2014;6(4):46.
[32] Rey, et al. The olfactory bulb as the entry site for prion-like propagation in neurodegenerative diseases. Neurobiology of Disease January 2018;109(Pt B):226–48.
[33] Adler CH, et al. Unified staging system for Lewy body disorders: clinicopathologic correlations and comparison to Braak staging. Journal of Neuropathology and Experimental Neurology September 5, 2019:nlz080.
[34] Ohtake H, et al. Beta-synuclein gene alterations in dementia with Lewy bodies. Neurology September 14, 2004;63(5):805–11.
[35] Perczel A, Gáspári Z, Csizmadia IG. Structure and stability of beta-pleated sheets. Journal of Computational Chemistry August 2005;26(11):1155–68.

[36] Groveman BR, et al. Parallel in-register intermolecular β-sheet architectures for prion-seeded prion protein (PrP) amyloids. Journal of Biological Chemistry August 29, 2014;289(35):24129–42.
[37] Galvagnion C. The role of lipids interacting with α-synuclein in the pathogenesis of Parkinson's disease. Journal of Parkinson's Disease 2017;7(3):433–50.
[38] Shahmoradian SH, et al. Lewy pathology in Parkinson's disease consists of crowded organelles and lipid membranes. Nature Neuroscience July 2019;22(7):1099–109.
[39] Eanes ED, Glenner GG. X-ray diffraction studies on amyloid filaments. Journal of Histochemistry and Cytochemistry November 1968;16(11):673–7.
[40] Serpell LC, et al. Fiber diffraction of synthetic alpha-synuclein filaments shows amyloid-like cross-beta conformation. Proceedings of the National Academy of Sciences United States of America April 25, 2000;97(9):4897–902.
[41] Nelson R, et al. Structure of the cross-beta spine of amyloid-like fibrils. Nature June 9, 2005;435(7043):773–8.
[42] Petkova AT, et al. A structural model for Alzheimer's beta -amyloid fibrils based on experimental constraints from solid state NMR. Proceedings of the National Academy of Sciences United States of America December 24, 2002;99(26):16742–7.
[43] Tycko R. Solid-state NMR studies of amyloid fibril structure. Annual Review of Physical Chemistry 2011;62:279–99.
[44] Tuttle MD, et al. Solid-state NMR structure of a pathogenic fibril of full-length human α-synuclein. Nature Structural and Molecular Biology May 2016;23(5):409–15.
[45] Yap TL, et al. Alpha-synuclein interacts with Glucocerebrosidase providing a molecular link between Parkinson and Gaucher diseases. Journal of Biological Chemistry August 12, 2011;286(32):28080–8.
[46] Jiang Z, et al. Segmental deuteration of α-synuclein for neutron reflectometry on tethered bilayers. Journal of Physical Chemistry Letters January 5, 2017;8(1):29–34.
[47] Almandoz-Gil L, et al. Mapping of surface-exposed epitopes of in vitro and in vivo aggregated species of alpha-synuclein. Cellular and Molecular Neurobiology October 2017;37(7):1217–26.
[48] Melki R. Role of different alpha-synuclein strains in synucleinopathies, similarities with other neurodegenerative diseases. Journal of Parkinson's Disease 2015;5(2):217–27.
[49] Guerrero-Ferreira R, et al. Cryo-EM structure of alpha-synuclein fibrils. eLife July 3, 2018:36402.
[50] Li Y, et al. Amyloid fibril structure of α-synuclein determined by cryo-electron microscopy. Cell Research September 2018;28(9):897–903.
[51] Tanaka G, et al. Sequence- and seed-structure-dependent polymorphic fibrils of alpha-synuclein. Biochimica et Biophysica Acta - Molecular Basis of Disease June 1, 2019;1865(6):1410–20.
[52] Peelaerts W, Baekelandt V. α-Synuclein strains and the variable pathologies of synucleinopathies. Jounral of Neurochemistry 2016;139(Suppl. 1):256–74.
[53] Prusiner SB, et al. Evidence for α-synuclein prions causing multiple system atrophy in humans with parkinsonism. Proceedings of the National Academy of Sciences United States of America September 22, 2015;112(38):E5308–17.
[54] Wakabayashi K, et al. The Lewy body in Parkinson's disease and related neurodegenerative disorders. Molecular Neurobiology April 2013;47(2):495–508.
[55] Fujiwara H, et al. alpha-Synuclein is phosphorylated in synucleinopathy lesions. Nature Cell Biology February 2002;4(2):160–4.
[56] Kramer ML, Schulz-Schaeffer WJ. Presynaptic alpha-synuclein aggregates, not Lewy bodies, cause neurodegeneration in dementia with Lewy bodies. Journal of Neuroscience February 7, 2007;27(6):1405–10.

[57] Raiss CC, et al. Functionally different α-synuclein inclusions yield insight into Parkinson's disease pathology. Scientific Reports March 17, 2016;6:23116.
[58] Bergman P, et al. Amyloid formation: functional friend or fearful foe? Journal of Internal Medicine August 2016;280(2):139–52.
[59] Jackson MP, Hewitt EW. Why are functional amyloids non-toxic in humans? Biomolecules September 22, 2017;7(4):E71.
[60] McGlinchey RP, et al. Probing fibril dissolution of the repeat domain of a functional amyloid, Pmel17, on the microscopic and residue level. Biochemistry December 13, 2011;50(49):10567–9.
[61] Morris DL, et al. Humanin induces conformational changes in the apoptosis regulator BAX and sequesters it into fibers, preventing mitochondrial outer-membrane permeabilization. Journal of Biological Chemistry November 2019. jbc.RA119.011297.

CHAPTER 7

Addicts — Neurotoxins that cause parkinsonism and the search for analogues in the environment

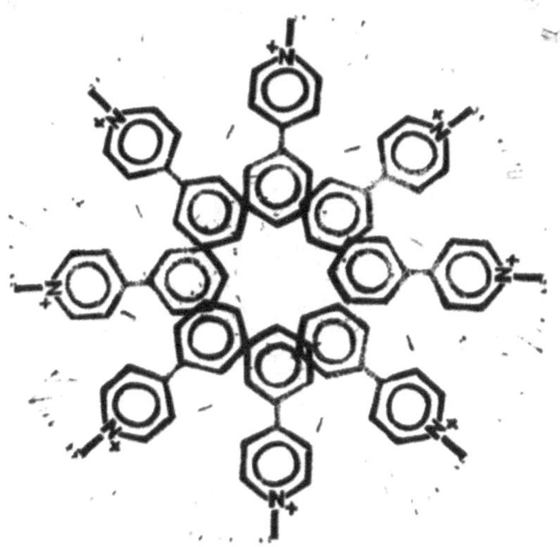

As I write this, America is experiencing yet another drug crisis, the opioid epidemic, the latest chapter in the unending saga of substance abuse. It certainly is one of the deadliest; more people die daily in the United States from opioid overdoses than from AIDS at its peak. One reason for the increased mortality is the rise in synthetic opioids, fentanyl compounds in particular, which are much more potent than heroin [1]. The synthetic compounds are diluted with inert fillers or added to enhance low-grade heroin, resulting in potencies that can vary greatly from dealer to dealer and batch to batch. This week's dose might yield a disappointing high, but next week's might kill you. On top of that, heroin overdose intervention drugs need

Synuclein and the Coelacanth
ISBN 978-0-323-85707-9
https://doi.org/10.1016/B978-0-323-85707-9.00013-7

to be administered at higher doses when fentanyl compounds are involved, a complication that further increases the overdose death rate.

Type "synthetic opioid" into a website such as Wikipedia and over a 100 different compounds will be listed, and these are just the more common ones. Although many were first synthesized in pharmaceutical laboratories in the so far fruitless search for opioids without addictive properties, many new ones are designer drugs, synthesized by recreational drug enthusiasts and dealers to get around Drug Enforcement Agency—controlled substance schedules. One such enthusiast was Barry Kidston, 23-year-old Maryland college graduate student in chemistry. After reading a 1947 paper describing the synthesis of a new opioid, desmethylprodine (MPPP), a compound closely related to Demerol, he synthesized it himself using a home chemistry kit.

For a few months the precocious addict synthesized and injected his homemade MPPP; then something went wrong. Perhaps hoping to speed up a particular step, he heated the mixture instead of keeping it at room temperature. He knew something was wrong when he injected the resulting substance; it produced a burning sensation unlike previous batches. Within 3 days he had become paralyzed and unable to speak. After unsuccessful treatments with antipsychotic medication and electroconvulsive therapy, a neurologist, Ramon B. Jenkins, examined him and recognized his symptoms as parkinsonism.

Dr. Jenkins treated Barry with L-dopa, and like the patients in Oliver Sacks' book *Awakenings*, he unfroze. Realizing how important this "synthesis gone wrong" case could be to Parkinson's disease research, Jenkins had Barry Kidston transferred to the National Institutes of Health (NIH). Researchers there had little success with him, unfortunately, and Kidston continued to abuse drugs, even the L-dopa they gave him, for the initial rush. Two years later, he walked onto the NIH campus, sat under a tree, and overdosed on cocaine. His parents allowed his brain to be autopsied. The neurons of his substantia nigra had been destroyed, presumably by a toxin in the bad batch of MPPP.

Barry Kidston's experience is not a recent episode in the opioid epidemic; it happened in 1976. His story can be read in J. William Langston's and Jon Palfreman's book, *The Case of the Frozen Addicts*. Designer drugs have been around for decades, as have horror stories involving botched syntheses. In the case of Barry Kidston's bad batch, Sanford Markey, a chemist at the National Institute of Standards and Technology, found two compounds in addition to MPPP [2]. One of these, 4-phenyl-1,2,3,6-tetrahydropyridine (MPTP), was ultimately found to be the compound responsible for causing parkinsonism, although not until many years later, after more cases of frozen addicts appeared.

Why did it take so many years to find the neurotoxin that destroyed Barry Kidston's substantia nigra? Rats. The go-to laboratory animal for toxicology studies in the 1970s was the brown rat *Rattus norvegicus*, though their popularity has waned a bit since knockout mice came onto the scene. When Sanford Markey injected rats with MPPP and the other compounds he had isolated, including MPTP, the rats would become temporarily paralyzed, what Markey and colleagues jokingly called "ratatonia," but they would fully recover with no neural damage, even after repeated injections [3].

When more victims of drug-induced parkinsonism surfaced in California in 1982, an investigation found that the drug the victims had thought was heroin was actually MPPP tainted with MPTP. When NIH scientist Stan Burns, who had flown out to California to examine the victims, learned about this, he knew what to do. He already knew about Markey's attempts using rats, which had failed to cause lasting neural damage. Rats are not humans and their neurochemistry is not identical. He arranged to have MPTP tested in primates.

Burns' first test of MPTP in a monkey was a success: in fact, far too much so. The monkey rapidly developed a tremor and then died. For the next monkey, Burns used much lower doses injected over a number of days. Eventually, the monkey developed mild parkinsonian symptoms. He then tried a moderately increased dosage over several days on a third monkey. The third monkey developed full-blown parkinsonian symptoms, tremor with severe bradykinesia and muscle rigidity. When he administered L-dopa to the affected monkey, the symptoms disappeared for a time, the same as it did in addicts and in people with classic Parkinson's. After euthanizing the monkey, an autopsy showed that the monkey's substantia nigra had been destroyed. Burns had proven that MPTP was the neurotoxin responsible for the parkinsonian symptoms [4].

Once the news had reached the scientific community that MPTP was the neurotoxic agent, researchers jumped to discover its mechanism of action. Early research revealed that in addition to humans and monkeys, MPTP caused varying neurotoxicity in other mammals, affecting dogs more than cats, for example [5]. A team of scientists led by J. William Langston, the doctor who first diagnosed the frozen addicts in California, discovered that frogs were sensitive to MPTP [6]. Another team including Sanford Markey, discoverer of the MPTP in Barry Kidston's botched synthesis, found that goldfish were also affected [7]. MPTP neurotoxicity has even been studied in the fruit fly, an invertebrate [8]. In rodents, some species such as rats were discovered to be much more resistant to MPTP than others [9]. Even within a species, variations were seen; certain strains of laboratory mice were found to be more sensitive to MPTP, including the black 6 strain used to generate knockout mice lines. This was a lucky thing for researchers, because mice cost much less than monkeys to maintain, and research burgeoned, with thousands of studies done in the decades after the discovery of MPTP.

One of the first things the early MPTP studies revealed was that MPTP was not the active form of the neurotoxin; instead, its metabolite, MPP+, was the primary culprit [10]. Once in the brain, an enzyme called monoamine oxidase B converts MPTP into MPP+. This occurs primarily in brain cells called astrocytes, star-shaped cells that help maintain neurons. The plus sign in MPP+ means that the compound carries a positive charge. MPTP can also be converted into MPP+ in the liver, but the positive charge prevents MPP+ from crossing the blood—brain barrier, unlike MPTP, which is neutral and can cross it. One reasons why some species might have resistance to the neurotoxic effects of MPTP is that their livers more efficiently metabolize MPTP, resulting in less of it reaching the brain.

One might think that because MPP+ is created in astrocytes, they should experience its toxic effects, but this is not the case. Apparently, once made, MPP+ does not linger inside the astrocytes. The neurotoxin diffuses through the extracellular medium between brain cells until it encounters a particular protein found in abundance on the cell membranes of dopaminergic neurons, the dopamine transporter (DAT) [11]. Normally, after a synapse fires, the DAT protein helps end the dopamine neurotransmitter signal by transferring excess extracellular dopamine back into the cell, allowing it to be recycled for subsequent firing events. Unfortunately, it also transports MPP+ into the dopaminergic neurons, and the MPP+ blocks dopamine from reentering, so that more dopamine builds up outside the cell.

Amphetamine drugs act on neurons in much the same way as MPP+, becoming transported inside by DAT and blocking the reuptake of dopamine [12]. The excess dopamine this causes outside the cell increases and prolongs its interaction with other neurons, which helps cause the high from the drug. One of the frozen addicts in California, a drug dealer, described the high from the MPTP-laced heroin as "terrific stuff," unlike anything he had used before, despite the burning it caused when injected. Perhaps the amphetamine-like effect of MPP+, added to the euphoria from the synthetic MPPP opioid, created the unique high. Amphetamine drug users have a three times greater risk for developing Parkinson's disease [13]. However, even in habitual amphetamine users, there are normally few long-term neural effects, just some synaptic alterations, with no obvious loss of dopaminergic neurons.

There are a variety of amphetamine drugs, such as Adderall and methamphetamine, or crystal meth. If you compare the structure of an amphetamine drug molecule with a dopamine molecule, the resemblance is obvious. They both have a ring of carbon atoms at one end attached to a short chain of carbon atoms that ends in a nitrogen atom that can be positively charged. It is unsurprising that amphetamines can interact with the same protein complexes as dopamine. The resemblance between MPP+ and dopamine is a bit more of a stretch, however. MPP+ has a ring at one end, like dopamine, but then instead of a chain, the other end is made of another ring with a positively charged nitrogen as part of the ring structure, an arrangement different from that of dopamine. I would never have had predicted that MPP+ would be able to interact with the same proteins as dopamine, yet another lesson from nature always to question my assumptions.

Once transported inside dopaminergic cells, MPP+ mucks up more cellular machinery, competing with dopamine to enter presynaptic vesicles [14]. Because of this, dopamine levels in the cellular fluid rise. The fluid inside cells that surrounds all of the organelles and vesicles is called the cytosol. Unlike most dopaminergic and related neurons in the rest of the brain, those in the substantia nigra lack the monoamine oxidases needed to break down dopamine [15]. Once the level becomes high enough, dopamine starts to leak out of the neuron, ironically, via DAT, the same protein that normally transports it inside [16]. Somehow, the presence of MPP+ and the elevated dopamine level allows the DAT protein to reverse dopamine transport across the cell membrane. The effect is even more pronounced with amphetamines. Dopamine is not a particularly stable chemical, and elevated levels in places it is not supposed to be, such as the cytosol, is definitely not a good thing, but MPP+ does not stop there. The most damaging thing MPP+ does is to mess with the energy-producing machinery in the cell, inhibiting a huge complex of 44 proteins found in mitochondria called complex I [17].

What is complex I and how does MPP+ mess with it? When functioning normally, complex I shuttles electrons and protons (hydrogen nuclei) from the inner compartment of the mitochondrion, called the matrix, across the inner mitochondrial membrane. The electrons and protons come from metabolism of carbohydrates and lipids, primarily from glucose in brain cells. The pathway the electrons follow takes them through several more complexes on the inner membrane, and they ultimately end up being transferred to oxygen molecules, which becomes converted into water. The mitochondrion uses the excess protons outside the inner membrane as an energy source, like water behind a dam, allowing them to flow back into the matrix through another protein complex, driving the synthesis of the chemical energy molecule adenosine triphosphate (ATP). Cells use ATP to carry out thousands of different enzymatic processes throughout the cell. The series of complexes that shuttle the electrons is called the respiratory chain, and the entire process, from shuttling the electrons to creating the ATP, is called oxidative phosphorylation.

MPP+ blocks the final step of the transport of the electrons by complex I [17]. The electrons are supposed to be transferred to a molecule called ubiquinone, which transports them to the next complex in the respiratory chain, which is called complex III (there is a complex II, which involves electron transfer, but at an earlier step in metabolism). By preventing the electrons from reaching ubiquinone, MPP+ forces them to find other places to end up, creating damaging reactive oxygen species, such as hydrogen peroxide [18]. Normally, small amounts of the damaging molecules are created by mitochondria as side products of normal oxidative phosphorylation, and the cell has enzymes to detoxify them. However, with MPP+ present, the levels rise above what the enzymes can handle. To make matters worse, dopamine has a high propensity to react with the damaging molecules, and with the elevated level of dopamine in the cytosol caused by MPP+, the detoxifying enzymes cannot get to the damaging molecules in time [19]. The reacted dopamine compounds are themselves toxic, capable of chemically bonding to proteins and other molecules in the cells, damaging their function.

In addition to the flood of damaging reactive oxygen species, oxidative phosphorylation cannot be maintained without the normal flow of electrons and protons across the mitochondrial inner membrane, ATP levels drop, and enzyme function starts to shut down in the cell [20]. Meanwhile, glial cells surrounding the neuron sense the elevated levels of reactive oxygen species, both directly and indirectly through other chemical messengers, starting the inflammation process [21,22]. If glial cells are stimulated above a certain point, they send signals to the neuron for it to commit a special form of cell suicide called apoptosis.

In apoptosis, the cell releases enzymes that break down the cellular infrastructure and chromosomes, and the cell breaks into small, lipid membrane enclosed pieces called blebs that are easily taken up and recycled by the glial cells. Unfortunately, ATP levels in the MPP+-affected neuron are often too low to carry out apoptosis successfully, and a different kind of cell death occurs: necrosis [23]. In necrosis, the cell's machinery can no longer maintain the integrity of the cell membrane, and rather than forming blebs, the membrane ruptures and the interior of the cell leaks into the extracellular fluid. This much messier situation triggers more inflammation, which can put neighboring neurons at risk for cell death, too.

MPP+ selectively kills dopaminergic neurons because the DAT protein at their cell membranes concentrates MPP+ inside the neuron, where it inhibits complex I in mitochondria, generating damaging reactive oxygen species and reducing ATP levels. The reactive oxygen species trigger an inflammatory response from neighboring glial cells, and the neuron dies either by apoptosis owing to the inflammatory response or by necrosis owing to low ATP levels.

The discovery in the early 1980s that the neurotoxin MPTP could selectively destroy the substantia nigra caused a great deal of excitement in the Parkinson's disease research field. Around this time, the first twin studies had been done, which seemed to show, erroneously as we now know, that the great majority of Parkinson's cases resulted from nongenetic factors such as exposure to toxic chemicals. With a bona fide parkinsonism-causing toxin in hand, the race was on to find similar toxins in the environment.

Over the next two decades, the race slowed to a crawl. Only a few candidate neurotoxins were discovered, and at the same time, mounting evidence of the role of genetics has in Parkinson's disease was stealing the spotlight. Pesticides and herbicides have already been touched on in Chapter 2 [24], and they seem to account for only about 2% of Parkinson's cases in the United States. One of these, the herbicide paraquat, has a structure similar to that of the toxic MPTP metabolite MPP+. Paraquat also interferes with complex I, but by a different mechanism. Instead of blocking transfer of the electrons to ubiquinone, paraquat can steal the electrons directly from complex I, becoming a free radical that can react with other molecules in the cell, leading to the creation of damaging reactive oxygen species. In addition, paraquat carries two positively charged nitrogen atoms, and thus, it should not be able to cross the blood–brain barrier. However, several studies in rats have shown that at least some paraquat is able to enter their brains [25,26]. On the other hand, later studies on monkeys found that when injected into the blood, there was no evidence of significant amounts of paraquat crossing the blood–brain barrier [27].

Another pesticide known to increase the risk for developing Parkinson's disease is rotenone, which is also a potent inhibitor of complex I, much more so than MPP+. When injected into rats, rotenone can cause dopaminergic neuron loss [28]. However, the molecular structure of rotenone is different from MPP+, larger and more complicated. Rotenone is an uncharged molecule capable of passing easily and indiscriminately into neural and nonneural cells. Exposure to rotenone in humans is primarily in the respiratory tract or gut, because the chemical is inhaled or swallowed; it does not pass through the skin. With no mechanism for targeting and concentrating rotenone specifically in dopaminergic neurons in the brain, it does not seem plausible that rotenone directly causes the specific pattern of neuron loss seen in Parkinson's disease. Instead, the damage should be primarily in the cells rotenone encounters first, in the gut and respiratory tract.

In the West Indies, mostly on the island of Guadeloupe, a cluster of atypical cases of Parkinson's disease as well as a tauopathy called progressive supranuclear palsy were tied to eating or drinking an herbal tea made from a fruit in the pawpaw family *Annonaceae* [29]. The fruit contains an isoquinoline compound chemically similar to MPTP, which has been shown to be neurotoxic in laboratory animals. Other potential toxins chemically similar to MPTP and MPP+ came under scrutiny, such as the beta-carbolines, compounds found in various flora and fauna throughout nature. Beta-carbolines can be fluorescent; one such compound causes scorpions to glow in UV light. The psychedelic drug ayahuasca derives its effect from beta-carbolines in the liana vine *Banisteriopsis caapi*. Some beta-carbolines are even found naturally in the human brain, and their levels can be elevated in alcoholics and drug users, as well as in those with neurological disorders, such as essential tremor and Parkinson's disease, although for Parkinson's disease the increase was found to be relatively small [30]. Overall, in the end, the quest to discover environmental or endogenous neurotoxins was a failure. After decades of searching, it is safe to say that the vast majority of Parkinson's cases have nothing to do with neurotoxins resembling MPTP.

Although the focus in Parkinson's research had swung from environmental causes to genetics, the MPTP story was not quite finished. In 2002, a team at Columbia University led by William Dauer, now at the University of Michigan, reported that alpha-synuclein knockout mice were resistant to MPP+ toxicity [31]. In other words, alpha-synuclein seemed to enhance or complement the damage caused by MPP+. Researchers focused on three points in the MPP+ neurotoxicity pathway where alpha-synuclein could potentially lead to enhanced damage.

The first point to receive special scrutiny was the entry of MPP+ into the dopaminergic neuron via the DAT protein. By 2002, it was known that alpha-synuclein directly interacts with DAT, and this interaction appeared to make more DAT available at the cell surface [32]. With more DAT at the cell surface, MPP+ could enter the neuron in greater quantity and build to toxic levels more quickly. The Columbia research team tested whether this might be true in knockout mice by using an amphetamine drug, reasoning that if the knockout mice had less functional DAT, they would be less affected by the drug than controls, because the drug also uses DAT to enter the neuron. They saw no significant difference between the knockouts and controls.

On the other hand, other studies on alpha-synuclein knockout mice have seen significant, albeit not large, differences in amphetamine response [33]. Perhaps differences in dosing protocols or particular mouse strains used might explain why the Columbia team saw no effect. Thus, it remains a possibility that alpha-synuclein enhances MPP+ toxicity by causing more DAT to be available at the cell surface. Another possibility the Columbia team tested is that alpha-synuclein reduces the amount of dopamine that later flows back out via DAT. Here, they saw a large effect, with 60% less dopamine flowing back out in the knockout mice.

The second point in the pathway was the interference of MPP+ with delivery of dopamine into synaptic vesicles. In addition to blocking dopamine delivery, MPP+ itself gets transported inside the vesicles. On the one hand, blocking delivery of dopamine inside vesicles is bad because it leads to higher dopamine levels in the cytosol. On the other hand, sequestering MPP+ inside vesicles keeps it away from complex I in mitochondria. The Columbia team saw no difference in dopamine delivery to vesicles between the knockouts and controls, but they acknowledged that alpha-synuclein could influence the delivery process and vesicle dynamics in ways their tests could miss. Another thing to consider is that in younger alpha-synuclein knockout mice, their dopaminergic neurons were seen to be more active than those of controls, releasing more vesicles during firing events [34]. Thus, it could be that in knockout mice MPP+ is transferred back out of the neurons more quickly than in controls owing to enhanced vesicle release.

The third point in the pathway was the inhibition of complex I by MPP+. Complex I activity is considerably reduced in dopaminergic neurons in the substantia nigra of Parkinson's disease patients. In addition, pull-down assays have seen interactions between alpha-synuclein and components of complex I in brain samples from Parkinson's patients [35]. This suggests that alpha-synuclein might also inhibit complex I, and perhaps it acts in concert with MPP+ to enhance its toxicity. In cells overexpressing alpha-synuclein, the protein can even be seen at the mitochondrial inner membrane where complex I is located [36]. On the other hand, in cells expressing normal levels of alpha-synuclein, it does not appear to be localized at the inner membrane; instead, it appears to be localized to where mitochondria contact another organelle, the endoplasmic reticulum [37]. Furthermore, some studies of alpha-synuclein knockout mice, although not all, found that the mitochondria of dopaminergic neurons were *less* active than controls, which suggests that alpha-synuclein might actually stimulate complex I activity [38].

One possible explanation to explain these discrepancies is that the elevated dopamine levels and reactive oxygen species caused by MPP+ cause individual alpha-synuclein molecules to aggregate into oligomers, and the oligomeric form of alpha-synuclein enhances neurotoxicity. These oligomers are the topic of the next chapter, but before we move on, I would like to continue exploring the potential interaction between mitochondria and alpha-synuclein monomers in healthy cells. Knockout mice had less mitochondrial activity, which suggests that under normal conditions, alpha-synuclein somehow stimulates activity. If so, where might one expect to find alpha-synuclein in the mitochondria?

The amino acids at the N-terminal end of alpha-synuclein resemble the N-terminal sequences of proteins that become targeted by the cell to the inner mitochondrial membrane, and it is reasonable to assume that alpha-synuclein might use the same machinery to cross the outer membrane to reach the inner one [35]. The mitochondrial inner membrane is highly wrinkled, consisting of deep, regularly spaced folds called cristae (singular form, crista). The synucleins prefer to bind highly curved, convex membrane surfaces. They also prefer to bind to negatively charged phospholipids. The mitochondrial inner membrane contains cardiolipin, which, unlike most phospholipids, has four hydrophobic lipid tails instead of two, and two negative charges instead of one. Normally, most of the cardiolipin is on the inner side of the inner membrane, but some is on the outer side, too, where alpha-synuclein can interact with it. The hydrophilic part of the lipid, the part with the phosphorus atom(s), is called the head group. Because the head group of cardiolipin is relatively small, and because of how its four lipid tails are attached to it, cardiolipin prefers to be located at concave membrane surfaces [39].

There is only one type of location on the mitochondrial inner membrane where alpha-synuclein and cardiolipin can interact and still satisfy their membrane curvature preferences: saddle points. These are places on the membrane surface where along one direction the membrane slopes up on both sides, and along the direction at a right angle, the membrane slopes down on both sides. This membrane shape resembles a saddle (hence the name), but I prefer to think of saddle points as mountain passes. The path itself slopes down on both sides of the pass, whereas left and right of the path the ground slopes up to the mountain peaks. The membrane-binding helical region of alpha-synuclein prefers to lie along the path, whereas the cardiolipin prefers to array itself in the rising slopes to the sides. Most saddle points on the inner membrane are located at the bottom of the cristae. Perhaps the relative sparseness of such points is why experiments with alpha-synuclein at normal levels failed to observe it colocalizing with the mitochondrial inner membrane, although this is just speculation on my part.

To speculate further, it could be that alpha-synuclein does not need to interact directly with complex I, but instead interacts with cardiolipin at saddle points, and thereby stabilizes and/or modifies the structure of the cristae in a way that enhances mitochondrial respiratory chain activity. The respiratory chain complexes, including complex I, all incorporate cardiolipin in their structures, and perhaps this cardiolipin is what binds alpha-synuclein in the pull-down assays mentioned earlier. The N-terminal sequences of beta and gamma synuclein are similar to that of alpha-synuclein, so one would expect they might also reach and interact with the mitochondrial inner membrane in a similar way, although I have not found any studies that address this. With their differing C-terminal tails, the three synucleins could potentially have different interactions with the proteins making up the respiratory chain complexes. Synucleins are primarily expressed in neurons, and neurons consume much more energy than other cells [40]. Maybe one function that evolved for the synucleins is to modify or rev up the output of mitochondria in energy-hungry neurons, to get more bang for the mitochondrial buck.

In reading up on mitochondrial energetics in alpha-synuclein knockout mice, I came across a graph that gave me pause. It was in a paper that claimed that the energetics in the mitochondria of the knockouts did not significantly differ from that of controls. I will not name the authors, because I wish to use what I read as a cautionary tale (and it is also possible I somehow misinterpreted what the authors meant). The graph displayed the mitochondrial activity in knockout mice dopaminergic neurons compared with controls over time as stimulatory and inhibitory chemicals were applied. At every time point, the average activity of the knockout neurons was less than that of the controls. The experiment was done simultaneously on several knockout samples and several control samples, and in addition to the average for each time point, the graph included error bars reflecting how much the measurements on individual samples typically differed from the average, a number called the standard deviation.

From the size of the error bars, which reflect how much the individual measurements varied, it was evident that at each time point, a minority of knockout samples had activities higher than those of the control samples with the lowest activities; that is, the error bars overlapped. In such situations, one must ask whether the difference in the average value is statistically significant; is it a real difference, or did it just occur by chance? Using fancy statistics software and providing a table full of statistical minutiae, the authors claimed that the difference in activity between the knockouts and controls was *not* significant.

The problem with their claim is that for every time point in the graph, nine of them, the average activity for knockout neurons was always lower than for controls. If in reality there was no significant difference between the knockouts and controls, for each time point, there should a 50:50 chance that the average value for the knockouts would happen to be lower. This means the chance that all nine time points for the knockouts have lower average values would be one-half to the ninth power: that is, 50% of 50% of 50% of 50% of 50% of 50% of 50% of 50% of 50%, which equals 0.2%, or a 1 in 512 chance, to be exact. It is the same probability that, tossing a coin, you get heads nine times in a row. Clearly, I think the authors got it wrong, that in fact the average activity of the mitochondria in the knockout dopaminergic neurons really was lower than in controls. The moral of this story is to be wary of staking your reputation on software you do not understand.

In the nearly 40 years since its discovery, what has the parkinsonism-inducing drug MPTP taught us about the origins of Parkinson's disease? Much less than initially hoped (in my opinion), but still MPTP research has provided important clues. The number one lesson is that mitochondria are the Achilles' heel of the dopaminergic neurons of the substantia nigra.

Nevertheless, there are major differences between what happens in Parkinson's disease and MPTP poisoning. Parkinson's disease typically develops over a decade or more before parkinsonian symptoms first manifest, whereas MPTP causes them in a few days. Also, in Parkinson's disease, many more regions of the brain are affected. Indeed, one would be justified to question whether the mechanism of MPTP toxicity has anything to do with Parkinson's disease. However, genetics studies of Parkinson's disease patients support the notion that mitochondria might be the key to understanding the disease.

Two genes for mitochondrial proteins cause Parkinson's disease when mutated. These are the genes for parkin and PTEN-induced kinase 1 (PINK1). Neither protein is involved with complex I or other components of the respiratory chain. Instead, they both seem to be involved in recycling damaged mitochondria [41]. Normally in the cell, mitochondria are dynamic organelles that can merge and split into two. They have their own DNA and messenger RNA-translating ribosomal machinery, although they make only 13 proteins, a small but core subset of the proteins making up the respiratory chain complexes. Mitochondria are thought to have once been free-roaming bacteria. Over a billion years ago, the precursor of all eukaryotes ate the precursor of all mitochondria, but instead of being digested, it continued to live on inside the cell, taking advantage of the food molecules inside the cell and providing chemical energy ATP molecules in return.[1] Eukaryotes are organisms whose cells have nuclei, including all animals, plants, and fungi.

The processes of mitochondrial merging and splitting are called fusion and fission. During fusion, the DNA in the combined mitochondria can undergo recombination, which can repair damage and fix errors that have accumulated in each individual mitochondrion. Fission is used to make new mitochondria, and asymmetrical fission can occur, too, in which damaged components of the parent mitochondrion can be passed more to one daughter than the other. Damaged mitochondria are recycled in the cell in a process called mitophagy, a specialized form of lysosomal degradation.

Parkin and PINK1 are part of the mitophagy machinery on the outer membrane of mitochondria. When someone inherits two defective copies of the gene for one of these proteins, he or she has a high probability of

[1] Single-celled eukaryotes have been discovered that lack mitochondria, the genus *Monocercomonoides*, for instance. Also, certain cells in multicellular eukaryotes lack mitochondria: red blood cells, for example.

developing parkinsonism; that is, the mutations are Mendelian and recessive, and have high penetrance [42,43]. Impaired recycling of damaged mitochondria causes dopaminergic neuron loss in the substantia nigra, as with MPP+, although neural losses are seen elsewhere, too, in the locus coeruleus, for instance. In many cases, especially for those with parkin mutations, there is no Lewy body pathology [44], so technically, one can argue that these cases cannot be called Parkinson's disease, not that such distinctions are much solace to the patients.

So where does alpha-synuclein pathology fit into this picture? Is its mechanism of toxicity more like MPP+, inhibiting the respiratory chain, or like parkin and PINK1, interfering with recycling of damaged mitochondria, or does it involve a different mechanism altogether? To explore this question, we need to learn about amyloid oligomers.

References

[1] National Institute on Drug Abuse. Overdose Death Rates. https://www.drugabuse.gov/related-topics/trends-statistics/overdose-death-rates.
[2] Davis GC, et al. Chronic Parkinsonism secondary to intravenous injection of meperidine analogues. Psychiatry Research December 1979;1(3):249–54.
[3] Chiueh CC, et al. Neurochemical and behavioral effects of systemic and intranigral administration of N-methyl-4-phenyl-1,2,3,6-tetrahydropyridine in the rat. European Journal of Pharmacology April 20, 1984;100(2):189–94.
[4] Johannessen JN, et al. Differences in the metabolism of MPTP in the rodent and primate parallel differences in sensitivity to its neurotoxic effects. Life Sciences January 21, 1985;36(3):219–24.
[5] Schneider JS, Markham CH. Neurotoxic effects of N-methyl-4-phenyl-1,2,3,6-tetrahydropyridine (MPTP) in the cat. Tyrosine hydroxylase immunohistochemistry. Brain Research May 14, 1986;373(1–2):258–67.
[6] Barbeau A, et al. New amphibian models for the study of 1-methyl-4-phenyl-1,2,3,6-tetrahydropyridine (MPTP). Life Sciences March 18, 1985;36(11):1125–34.
[7] Pollard HB, et al. A parkinsonian syndrome induced in the goldfish by the neurotoxin MPTP. Federation of American Societies for Experimental Biology Journal September 1992;6(12):3108–16.
[8] Muñoz-Soriano V, Paricio N. Drosophila models of Parkinson's disease: discovering relevant pathways and novel therapeutic strategies. Parkinson's Disease March 3, 2011;520640.
[9] Meredith GE, Rademacher DJ. MPTP mouse models of Parkinson's disease: an update. Journal of Parkinson's Disease 2011;1(1):19–33.
[10] Singer TP, et al. Mechanism of the neurotoxicity of 1-methyl-4-phenylpyridinium (MPP+), the toxic bioactivation product of 1-methyl-4-phenyl-1,2,3,6-tetrahydropyridine (MPTP). Toxicology April 1988;49(1):17–23.
[11] Kitayama S, et al. MPP+ toxicity and plasma membrane dopamine transporter: study using cell lines expressing the wild-type and mutant rat dopamine transporters. Biochemica et Biophysica Acta September 16, 1998;1404(3):305–13.
[12] Calipari ES, Ferris MJ. Amphetamine mechanisms and actions at the dopamine terminal revisited. Journal of Neuroscience 2013;33(21):8923–5.

[13] Curtin K, et al. Methamphetamine/amphetamine abuse and risk of Parkinson's disease in Utah: a population-based assessment. Drug and Alcohol Dependence January 1, 2015;146:30–8.
[14] Takahashi N, et al. VMAT2 knockout mice: heterozygotes display reduced amphetamine-conditioned reward, enhanced amphetamine locomotion, and enhanced MPTP toxicity. Neurobiology September 2, 1997;94(18):9938–43.
[15] Meiser J, Weindl D, Hiller K. Complexity of dopamine metabolism. Cell Communication and Signaling May 17, 2013;11:1.
[16] Rodriguez-Menchaca AA, et al. S(+)amphetamine induces a persistent leak in the human dopamine transporter: molecular stent hypothesis. British Journal of Pharmacology April 2011;165(8):2749–57.
[17] Murphy MP, et al. Inhibition of complex I by hydrophobic analogues of N-methyl-4-phenylpyridinium (MPP+) and the use of an ion-selective electrode to measure their accumulation by mitochondria and electron-transport particles. Biochemical Journal March 1, 1995;306(Pt 2):359–65.
[18] Murphy MP. How mitochondria produce reactive oxygen species. Biochemical Journal January 1, 2009;417(Pt 1):1–13.
[19] Guo JD, et al. Damage to dopaminergic neurons by oxidative stress in Parkinson's disease. International Journal of Molecular Medicine January 19, 2018:1817–25.
[20] Redza-Dutordoir M, Averill-Bates DA. Activation of apoptosis signalling pathways by reactive oxygen species. Biochimica et Biophysica Acta December 2016;1863(12):2977–92.
[21] Streit WJ, Walter SA, Pennell NA. Reactive microgliosis. Progress in Neurobiology April 1999;57(6):563–81.
[22] Joshi AU, Paras S, Minhas PS, et al. Fragmented mitochondria released from microglia trigger A1 astrocytic response and propagate inflammatory neurodegeneration. Nature Neuroscience September 2019;22:1635–48.
[23] Callizot N, et al. Necrosis, apoptosis, necroptosis, three modes of action of dopaminergic neuron neurotoxins. PLoS One April 25, 2019;14(4):e0215277.
[24] Tanner CM, et al. Rotenone, paraquat, and Parkinson's disease. Evironmental Helath Perspectives June 2011;119(6):866–72.
[25] Corasaniti MT, Strongoli MC, Nisticà G. Determination of paraquat in rat brain using ion-pair solid-phase extraction and reversed-phase high-performance liquid chromatography with ultraviolet detection. Journal of Chromatography April 27, 1990;527(1):189–95.
[26] Wu B, et al. Central nervous system damage due to acute paraquat poisoning: an experimental study with rat model. Neurotoxicology March 2013;35:62–70.
[27] Bartlett RM, et al. Paraquat is excluded by the blood brain barrier in rhesus macaque: an in vivo PET study. Brain Research March 9, 2009;1259:74–9.
[28] Sherera TB, et al. Subcutaneous rotenone exposure causes highly selective dopaminergic degeneration and α-synuclein aggregation. Experimental Neurology January 2003;179(1):9–16.
[29] Lannuzel A, et al. Atypical parkinsonism in Guadeloupe: a common risk factor for two closely related phenotypes? Brain March 2007;130(3):816–27.
[30] Kuhn W, et al. Plasma levels of the beta-carbolines harman and norharman in Parkinson's disease. Acta Neurologica Scandanavica December 1995;92(6):451–4.
[31] Dauer W, et al. Resistance of alpha -synuclein null mice to the parkinsonian neurotoxin MPTP. Proceedings of the National Academy of Sciences, United States of America October 29, 2002;99(22):14524–9.
[32] Lee FJ, et al. Direct binding and functional coupling of alpha-synuclein to the dopamine transporters accelerate dopamine-induced apoptosis. Federation of American Societies for Experimental Biology Journal April 2001;15(6):916–26.
[33] Abeliovich A, et al. Mice lacking alpha-synuclein display functional deficits in the nigrostriatal dopamine system. Neuron January 2000;25(1):239–52.

[34] Greten-Harrison B, et al. αβγ-Synuclein triple knockout mice reveal age-dependent neuronal dysfunction. Proceedings of the National Academy of Sciences, United States of America November 9, 2010;107(45):19573–8.
[35] Devi L, et al. Mitochondrial import and accumulation of alpha-synuclein impair complex I in human dopaminergic neuronal cultures and Parkinson disease brain. Journal of Biological Chemistry April 4, 2008;283(14):9089–100.
[36] Nakamura K, et al. Direct membrane association drives mitochondrial fission by the Parkinson disease-associated protein α-synuclein. Journal of Biological Chemistry June 10, 2011;286(23):20710–26.
[37] Guardia-Laguarta C, et al. Novel subcellular localization for α-synuclein: possible functional consequences. Frontiers in Neuroanatomy February 23, 2015;9:17.
[38] Ellis CE, et al. Mitochondrial lipid abnormality and electron transport chain impairment in mice lacking alpha-synuclein. Molecular and Cellular Biology November 2005;25(22):10190–201.
[39] Beltrán-Heredia E, et al. Membrane curvature induces cardiolipin sorting. Communications Biology June 20, 2019;2:255.
[40] Harris JJ, Jolivet R, Attwel D. Synaptic energy use and supply. Neuron September 6, 2012;75(6):762–77.
[41] Pickrell AM, Youle RJ. The roles of PINK1, parkin, and mitochondrial fidelity in Parkinson's disease. Neuron January 21, 2014;85(2):257–73.
[42] Kitada T, et al. Mutations in the parkin gene cause autosomal recessive juvenile parkinsonism. Nature April 1998;392(6676):605–8.
[43] Valente EM, et al. Hereditary early-onset Parkinson's disease caused by mutations in PINK1. Science May 21, 2004;304(5674):1158–60.
[44] Farrer M, et al. Lewy bodies and parkinsonism in families with parkin mutations. Annals of Neurology September 2001;50(3):293–300.

CHAPTER 8

Attack of the Oligomers – Alpha-synuclein amyloid oligomers and their pathogenic roles

Part 1. Unmasking the villain

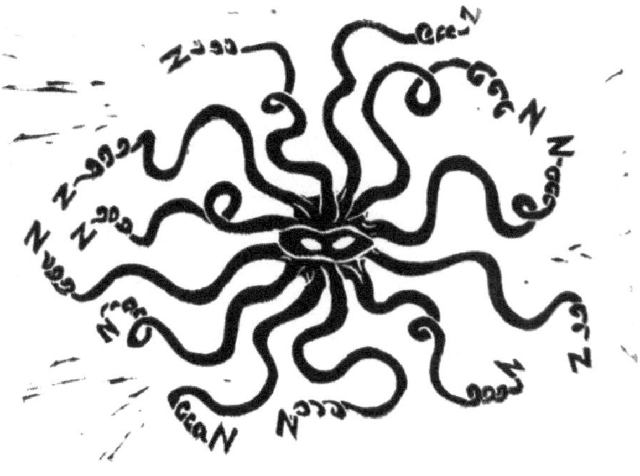

By the early 1990s, Alzheimer's disease researchers suspected they were missing something—yes, Alzheimer's, not Parkinson's, but I'm getting there. Alzheimer's disease was clearly associated with senile plaques, dense, extracellular protein deposits composed primarily of fibrils of the peptide amyloid-beta (scientists call a protein with a short sequence a peptide, and amyloid-beta has only 40 or 42 amino acids, which is about 10 times shorter than the average protein) [1]. However, the severity of the disease did not always correlate with the amount of plaque present in the brain. The plaques often interfered with blood flow and were associated with

Synuclein and the Coelacanth
ISBN 978-0-323-85707-9
https://doi.org/10.1016/B978-0-323-85707-9.00001-0

inflammation that could conceivably induce apoptosis in nearby neurons, but it was unclear how the protein deposits outside neurons were related to the pathology seen inside the cells, such as the neurofibrillary tangles composed of filaments of tau protein. Furthermore, it was unclear whether senile plaques were necessarily present during the mild cognitive impairment that typically precedes the full-blown dementia of Alzheimer's disease [2].

As soon as the amyloid-beta peptide sequence was determined, scientists started researching its behavior in test tubes. Sure enough, the peptide could form amyloid fibrils similar to those seen in the plaques, but the fibrils did not form right away; they took a few days to start forming on average. Before they formed fibrils, the individual peptides aggregated into multipeptide clumps (i.e., oligomers) consisting of two to 20 or more individual peptides.

In 1991, a team of researchers in the laboratory of Carl W. Cotman at the University of California, Irvine, discovered that amyloid-beta oligomers formed in test tubes were toxic to neuronal cell cultures, whereas non-clumped amyloid-beta peptides (i.e., monomers) were nontoxic [3]. By 1994, the laboratory of Caleb E. Finch at the University of Southern California reported additional evidence that these amyloid-beta oligomers were likely the neurotoxic culprits in the brains of Alzheimer's patients [4]. The Finch laboratory teamed up with the laboratory of William L. Klein at Northwestern University, and in 1998 they presented evidence that the oligomers were not only toxic, they interfered with the long-term potentiation of neurons, something required to form long-term memories [5]. These oligomers were termed amyloid-beta–derived diffusible ligands (ADDLs), which I presume was a tongue-in-cheek nod to the addled mental state of early dementia.

News of the discoveries about amyloid-beta oligomers spread through the neurological science fields, and Parkinson's disease researchers wondered whether something similar might be happening with alpha-synuclein. In 1998, two studies came out: one from the laboratory of Peter T. Lansbury at Harvard and another from a multinational team including Omar M.A. El-Agnaf at Queen's University in Belfast. They both obtained results similar to what had been seen with amyloid-beta: that after several days, solutions of monomeric alpha-synuclein started to aggregate into oligomers and fibrils [6]. The Belfast team also showed the aggregated alpha-synuclein became toxic to neuronal cells in vitro [7]. At first, it was unclear whether oligomers or newly matured fibrils were the toxic agent, but as more research was done, it became clear that alpha-synuclein oligomers were more toxic, although fibrils were toxic, too.

Soon, researchers were finding toxic amyloid oligomers for a wide variety of diseases, such as Creutzfeldt-Jakob disease and Huntington disease, tauopathies such as progressive supranuclear palsy, and even non-neurodegenerative diseases such as type 2 diabetes [8–10]. When exposed to the oligomers in vitro, the death rate of cultured cells increased in a dose-dependent manner. Using centrifugation and chromatographic techniques, researchers were able to purify amyloid oligomers from tissue samples. Some argued that the purified oligomers might be a side effect, perhaps forming during the harsher steps of the purification protocol rather than being present in the tissue samples. However, tissue from diseased patients contained much more oligomers than controls on average, and purified oligomers could be obtained by multiple techniques and protocols, which suggested that they really were present in the samples, not just a spurious by-product of the purification process.

Where there was amyloid disease, there were amyloid oligomers. Researchers started to speculate that there might be a general, shared mechanism of toxicity for the oligomers. In 2003, this speculation was bolstered by a surprising find. An antibody was discovered that bound to amyloid-beta oligomers consisting of at least eight monomers [11]. It did not bind to fibrils, smaller oligomers, or monomers. The paper was from the laboratory of Charles G. Glabe at the University of California Irvine, with Rakez Kayed as first author. Glabe had also been on the Irvine team with Cotman that published the 1991 paper first showing the toxicity of amyloid-beta oligomers. The most surprising aspect of the oligomer-specific antibody was that it could bind to amyloid oligomers from other diseases as well, including alpha-synuclein and prion protein (PrP) oligomers.

Antibodies recognize and bind to specific regions on biomolecules. The region to which the antibody binds is called the epitope. For a peptide with disordered or extended structure, the epitope is typically a specific sequence of amino acids. For a protein with well-defined structure, the epitope consists of the amino acids within the specific region of the protein surface to which the antibody binds. The astonishing thing about the oligomer-specific antibody is that because the amino acid sequences of the various amyloid oligomers are different, it seemed to be recognizing different sets of amino acids. In other words, some part of the oligomer surface was nearly identical for the different oligomers, with the same physical properties, shape, charge, and hydrophobicity despite having different amino acid sequences. Kayed and Glabe took the logic one step further: if the oligomers had similar toxicity and similar epitopes, the toxicity was probably related to the physical properties of the epitope [12].

When I first heard about the work of Kayed and Glabe in the mid-2000s, I was intrigued, but it had little to do with my own work at the time, which was about a component of the internal scaffolding of the cell, the tail domain of a protein called myosin [13]. The project was a collaboration with my lab mate Kae-Jung Hwang and the laboratory of Edward Korn at the National Institutes of Health (NIH). I had run into a roadblock: I could not see all the nuclear magnetic resonance signals from protein backbone. I already had a predicted model structure for the domain using a technique called protein threading, in which every protein structure that has ever been solved gets a profile consisting of the amino acid sequence plus the environment of each amino acid, that is, whether it is buried or exposed to water, and to what secondary structure each amino belongs, because some amino acids prefer certain secondary structures over others.

The website for the threading program was called Protein Homology/Analogy Recognition Engine (Phyre). (Now there is a newer version called Phyre2.) It was provided free to researchers by the Imperial College in London [14]. The Phyre server predicted that the profile that fit my protein sequence best was a pleckstrin homology domain structure, one of the most common protein folds [15].

The pleckstrin protein consists of two stacked beta sheets with a long alpha helix nestled in the loops along the top edge. The missing signals belonged to amino acids in the predicted beta sheet regions. We tried increasing the concentration of the protein to increase the signal strength, but that made matters worse. It was a sure sign that my protein was aggregating (that is, forming clumps), and the higher concentrations increased the fraction of protein in the clumped state. Curiously, only the beta strand signals were significantly affected; the loop and alpha-helical regions were strong, even at the higher concentrations.

How might the beta sheets be aggregating without involving the loop regions and alpha helix? Perhaps the sheet faces were stacking on top of each other, or the sheets were interacting edge to edge. In the stacking arrangement, the side chains of the amino acids forming the sheets would interact, whereas in the edge-to-edge arrangement, the backbones would interact via the same sort of hydrogen bonds that form the beta sheet structure.

I could not tell which situation was happening for my protein, but the possibility of an intermolecular beta sheet edge interaction reminded me of the amyloid oligomers I had just learned about [12]. One of the places in a protein structure where the backbone is most exposed is the edge of a beta sheet. If the oligomer-specific antibody were recognizing an exposed edge of a beta sheet edge, that could explain how it could recognize oligomers with different sequences; it would be recognizing the peptide backbone, which is the same for all amino acids (except proline, which is typically not found in beta strands).

This realization inspired my first foray into the amyloid field, "Do Amyloid Oligomers Act as Traps for Misfolded Proteins? A Hypothesis" [16]. In the paper, I reviewed the scientific evidence that amyloid oligomers might be interacting with the exposed beta sheet edges of transiently misfolded proteins, as well as nascent proteins being made at ribosomes translating messenger RNA (mRNA).

One of the pathologies observed in amyloid diseases is overloading of the ubiquitin-proteasome system (UPS) [17]. The UPS involves cellular machinery that detects damaged and misfolded proteins. When a faulty protein is detected, it attaches a string of small proteins, called ubiquitin, which signals the cell to transport the damaged or misfolded protein to the proteasome. The proteasome is a large complex of proteins that snips off the attached string of ubiquitin and degrades the damaged or misfolded protein into its constituent amino acids. The UPS is like a trash collection and recycling service for the cell.

The cell has proteins called chaperones that help newly formed and transiently misfolded proteins achieve their correctly folded shapes. The chaperones shield the proteins from the UPS, giving the proteins extra time to fold correctly. Basically, my hypothesis was that amyloid oligomers might be acting as antichaperones, to borrow the expression from my NIH colleague Alan Peterkofsky, preventing them from achieving their correct fold.

For fully folded proteins, their exposed beta sheet edges evolved to have amino acids that make them less prone to aggregating interactions [18]. For misfolded and newly forming proteins, the beta strands that normally would be buried can be exposed, allowing beta sheet edge interactions with oligomers that could temporarily trap them in their misfolded state. I hypothesized that amyloid oligomers would not shield the misfolded protein the way chaperones do, and UPS enzymes could then add a chain of ubiquitin proteins, dooming the protein to proteasomal degradation. If the oligomer dissociated before degradation, it could repeat the process with additional misfolded proteins, adding a catalytic aspect to the mechanism.

My misfolding trap hypothesis paper came out in 2008. That same month, another paper in a similar vein was published by Yuji Yoshiike in the laboratory of Akihiko Takashima at the RIKEN Brain Science Institute in Japan [19]. It turns out that the amyloid oligomer-specific antibody recognized not only amyloid oligomers but also a few other select proteins. Most of these proteins were chaperones! When the researchers compared the structures of the chaperones, they found that their shared structural feature was (you guessed it) exposed beta sheet edges. Some ideas are proposed by visionaries years before their time; the misfolding trap hypothesis was not one of them.

Since the 1990s, as soon as researchers started speculating that amyloid oligomers were the main neurotoxic agents in amyloid-related diseases, others sought to determine their structures. The oligomers did not yield their secrets easily. The biggest issue was the heterogeneity of their structures. Starting from a homogeneous solution of amyloid peptide monomers, a veritable zoo of oligomers could form, some in minutes, others appearing later, in hours or even days [20]. They had sizes ranging from a few monomers to several dozen, some more loose, others more compact. Using a technique called circular dichroism (CD) spectroscopy (see Twisting Light) they discovered the oligomers had beta sheet content intermediate between the monomers, which typically had none, and the corresponding amyloid fibrils [21]. Using another technique call infrared spectroscopy, which measures chemical bond vibrations and how they are affected by hydrogen bonds, the researchers also learned that the beta sheets in oligomers were predominantly antiparallel, in contrast to fibrils [22].

Twisting light – CD spectroscopy

Light is a curious thing: a wave, a particle; both are equally correct descriptions, with interwoven electric and magnetic fields oscillating perpendicular to the light's direction. Light can be linearly polarized; that is, its fields can be confined to oscillate in a single plane. Shine light through a polarized sunglass lens, and the fields of the light that emerges will be vertically polarized. Light can also be circularly polarized, with the fields oscillating so that they trace a circle spiraling perpendicular to the light's direction. Owing to light's curious nature, a linearly polarized light beam is equivalent to the sum of two circularly light beams, with the fields of one beam oscillating clockwise and the other counterclockwise.

Light interacts with matter through which it passes, with the negatively charged electrons and positively charged nuclei feeling the force of the oscillating electric field. This interaction slows down the light, and this phenomenon is responsible for why light bends at the interface between two media with different densities, such as air and water. Many molecules have a property called chirality, a fancy word for handedness, meaning their structure is not the same when reflected in a mirror. Amino acids are chiral, and so are the secondary structures they form in proteins. For instance, alpha-helices are right-handed; point your right thumb along the axis of the helix and the backbone will spiral upward in the direction the fingers are curling, counterclockwise.

When shone through a medium containing chiral molecules, light circularly polarized in one direction, say clockwise, can interact more strongly with the medium than light circularly polarized the other way. Chiral media with this property are said to be circularly dichroic. Something really cool happens when you shine a linearly polarized light beam through such a medium: it emerges linearly polarized, but in a different direction. In effect, the light beam gets twisted. Why does this happen? Remember the linearly polarized beam is equivalent to the sum of two circularly polarized beams rotating in opposite directions. Because one of the two circularly polarized beams will slow down more than the other, when you add them together after leaving the medium, their circularly polarized fields will meet up in a different direction.

The amount that circularly polarized light interacts with chiral media also depends on the wavelength of light. The secondary structural elements of proteins interact most strongly with near ultraviolet light, and circular dichroism spectrometers are designed to measure the angle that linearly polarized light gets twisted in this wavelength range. Because beta sheets and alpha-helices interact differently with circularly polarized light as one varies the wavelength, the measured spectrum can be used to determine the amount of each type of secondary structure in the protein.

To learn about the different oligomers, the researchers needed to find a way to separate the various species. They tried every trick in the proverbial book: size exclusion chromatography, centrifugation, varying the solution temperature, acidity, salt concentrations, agitating the samples, freezing, drying, and redissolving them [20]. Slowly, the oligomers divulged more secrets. To isolate the most stable species, time was the best tool, because over time more ephemeral species would morph into more stable ones or into fibrils, which could easily be removed by centrifugation. These stable oligomers were compact and toxic to cells. Cryoelectron microscopy revealed that these stable oligomers had short cylindrical structures that were hollow with thick walls, unlike the more rodlike, nonhollow filaments of amyloid fibrils [23]. The oligomer specific antibody was able to recognize the stable oligomers [11], but just where in their cylindrical structure the antibody was binding (that is, where the putative beta sheet edge might be) was unclear.

The researchers discovered more classes among the oligomers. Most were on-pathway; that is, they would ultimately morph into amyloid fibrils, but some were off-pathway, incapable of forming fibrils [24]. For alpha-synuclein, the on-pathway oligomers had core structures consisting of subsets of the amino acids that formed the core structure of the mature fibrils, whereas off-pathway oligomers had cores consisting of the N-terminal amino acids, tucking them away from the oligomer surface [25]. Intriguingly, the off-pathway oligomers were not toxic to cells, even though the oligomer specific antibody could still recognize them. Were the on-pathway oligomers toxic because of exposed beta sheet edges, or was having an exposed, membrane-binding N-terminal region somehow necessary for alpha-synuclein oligomer toxicity, or was it maybe a combination of both?

The researchers made another intriguing discovery, that dopamine could alter the types of oligomers created in test tubes, leading to additional on-pathway, toxic species [24]. Perhaps this could explain why the dopaminergic neurons of the substantia nigra seem especially vulnerable to Parkinson's disease pathology. Numerous types of neurons express alpha-synuclein and develop Lewy bodies and neurites, including many nondopaminergic types, but they usually have slower rates of neural loss than the substantia nigra.

Before we move on to explore the possible mechanisms of amyloid oligomer toxicity, let us imagine how oligomers transform into fibrils. In my mind's eye, I imagine each of the oligomer peptides adopting one or more intrapeptide, antiparallel beta hairpins, as suggested from experimental observations, but with perhaps a short region of interpeptide, parallel beta sheet, contributing to the various interpeptide interactions holding the oligomer together. In other words, as I picture it, much of the structure of the individual monomers that make up the oligomer is still involved in typical intramolecular interactions, but with just a few amino acids involved in the cross-beta intermolecular interactions found in amyloid fibrils. To fill out the picture in my mind's eye further, perhaps at the ends of the oligomer, the strands at the edges of the nascent interpeptide beta sheet are exposed, providing the epitope for the oligomer-specific antibody.

For the peptide at the tip of the growing amyloid fibril, I imagine the same short interpeptide beta sheet region of the oligomers becomes more expanded, with more and more amino acids participating as one progresses to the peptides farther from the tip. Eventually, the amino acids that formed intrapeptide, antiparallel beta hairpins in the oligomers begin to participate in the growing parallel beta sheet interactions, rotating their backbones 90 degrees, breaking the intrapeptide hydrogen bonds so that they, too, can form interpeptide hydrogen bonds pointing along the axis of the growing fibril. Basically, I envision the fibril tip as a transition zone, going from a mostly oligomer type structure at the very tip to a fully mature fibril cross-beta structure farther down the fibril. How many peptides lie in the transition zone? This is pure speculation, but I would guess two to three times the average number of peptides per oligomer, enough to allow a gradual transition in structure.

There is no direct experimental evidence regarding the structure at the tip of amyloid fibrils or the precise structural transitions undergone by oligomers transforming into fibrils. Peptide monomers can also attach to the ends of fibrils, and even away from the tips, where fibrils can serve as templates for monomers to aggregate into new oligomers [26]. Fibril growth is not necessarily a one-way process. Fibrils can break under shear forces, and probably their tips may sometimes detach, releasing oligomers. Different strains of amyloid fibrils can be more prone to breakage; there is even a special word for the propensity to break: "frangibility." [27].

Remember how alpha-synuclein fibrils injected into mice acted as prions, triggering pathology that spreads from the site of injection, and that various strains of fibrils could lead to varying degrees of pathology? In similar experiments, comparable amounts of alpha-synuclein oligomers injected into the brain produced less pathology than fibrils [28,29], despite evidence that oligomers are the more neurotoxic form [30]. The fibrils can act as templates for the seeding, growth, and shedding of toxic oligomers, recruiting the mouse's own peptides, transforming them into their oligomer forms. Although the oligomers are neurotoxic, it would seem they are ineffective at recruiting the native alpha-synuclein to form more oligomers; thus, they cannot amplify the pathology. Because fibrils can act as reservoirs and templates for oligomers, fibrils and oligomers are partners in pathology.

Part 2. Mitochondria, the scene of the crime

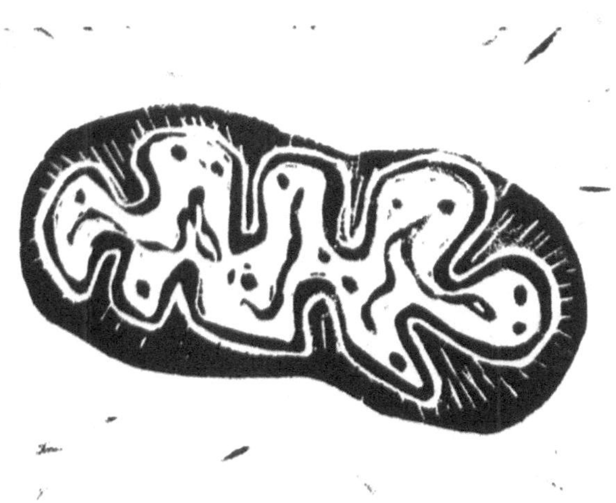

Overloading of the UPS machinery of the cell was not the only cellular pathology caused by amyloid oligomers. They also caused ion dyshomeostasis, a fancy word for imbalance. The levels of particular ions in the cytosol would increase dramatically upon exposure to the oligomers, especially calcium [31]. As described in earlier chapters, cells use the influx of certain ions such as calcium to trigger cellular machinery. When calcium levels stay elevated, the machinery cannot reset after being triggered and remains in the posttrigger state. The oligomers were letting the ions in somehow, and not just through the cell membrane, but also through membranes of other organelles such as mitochondria, lysosomes, and the endoplasmic reticulum, which also hold high concentrations of calcium in their interiors.

Upon closer inspection of the oligomer—membrane interactions, a difference was discovered between amyloid-beta and alpha-synuclein oligomers. At sufficiently high concentrations, amyloid-beta oligomers appeared to form pores at the membrane, allowing the ions to flow through [32]. For alpha-synuclein oligomers, however, ion leakage occurred even though no pores could be observed by microscopic techniques. Perhaps the pores were forming and dissociating too fast to be observable, but the consensus among alpha-synuclein researchers seems to favor a different picture: that alpha-synuclein oligomers thin and distort the membrane, introducing flaws in the lipid bilayer that help ions sneak through [33].

A major problem with these studies is that oligomers were typically added to the cell culture medium, or injected in the cerebrospinal fluid of test animals, so they interacted first with the outer cell membrane of the neurons. However, many of the disease-related amyloid peptides, such as alpha-synuclein, are primarily intracellular. Thus, the toxicity observed in the experiments, caused by oligomers introduced in an extracellular manner, might not be representative of what was actually happening to neurons in diseased tissue.

Another issue with these studies was that the toxic effects happened too quickly; dysfunction and an increased cell death rate started immediately upon exposure to the oligomers. In neurodegenerative amyloid-related diseases such as Parkinson's, the pathology is much slower, taking years to manifest. Was the neurotoxicity seen in cell cultures truly related to disease, or was it an artifact of the artificial in vitro system? To address this question, researchers turned to neural cell cultures and animal models in which alpha-synuclein was being overexpressed inside the neurons.

Supporting the previous work, increased intracellular expression produced much of the same pathology as the experiments in which aggregated alpha-synuclein was added extracellularly. In addition, aggregated alpha-synuclein was observed inside the cells, although the aggregations were much smaller than Lewy bodies and neurites. Similar ion dyshomeostasis and UPS disruption were observed [34,35]. There were alterations in vesicular trafficking, that is, aberrant sorting and delivery of vesicles to their proper locations in cells [36], and there were changes in gene expression [37], although it was not always clear whether the effects observed were caused by aggregated alpha-synuclein or by increased levels of the monomer form of the protein. Finally, similar, but more pronounced, mitochondrial dysfunction was observed, with reduced adenosine triphosphate (ATP) production and increased levels of reactive oxygen species [38,39].

What was causing the mitochondrial dysfunction? Perhaps alpha-synuclein oligomers increased the permeability of the mitochondrial membrane, not just causing calcium to leak out, but letting hydrogen ions in and reducing the ion and electric potential gradients needed to produce ATP molecules. To do so, oligomers would have to reach the inner membrane of the mitochondrion, which holds back the excess of calcium and hydrogen ions stored within the mitochondrial matrix. The inner membrane is also where the respiratory chain and ATP-producing proteins are located.

However, to get to the inner mitochondrial membrane, proteins must first pass through the mitochondrial outer membrane via protein transport complexes. Alpha-synuclein oligomers have been shown to bind two of the protein transport components: the proteins TOM40 and TOM20 [40]. Proteins destined for transport to mitochondria contain a conserved sequence at the N-termini that is recognized by the protein transport machinery. The N-terminal sequence of alpha-synuclein resembles this mitochondrial targeting sequence. The binding of alpha-synuclein oligomers to the protein transport components interferes with the normal transport of proteins to the inner membrane. Could it be that the outer membrane is where the oligomers do the bulk of their molecular dirty work?

Mitochondria are special organelles that divide to create new mitochondria to meet the energy demands of the cell, a process called mitochondrial fission. They can fuse together, too, combining their mitochondrial chromosomes to help repair the DNA damage that occurs due to exposure to reactive oxygen species and other free radicals produced as by-products of energy production [41]. When a mitochondrion becomes too damaged to repair, it signals the cell to engulf it, digesting it inside lysosomes in a process called mitophagy.

In transgenic mice and in neural cell cultures overexpressing alpha-synuclein, fission is stimulated, and the number of smaller-sized mitochondria increases [42]. In molecular biology jargon, the situation is called mitochondrial fragmentation. Many researchers have shown alpha-synuclein overexpression promotes fission of mitochondria and stimulates mitophagy [43].

In damaged mitochondria, the lipid cardiolipin, which is normally concentrated on the inner surface of the inner membrane, becomes translocated to the outer membrane, probably one of the molecular cues that tag a mitochondrion for mitophagy [44]. Unfortunately, alpha-synuclein likes to bind to cardiolipin, too. For the toxic, on-pathway alpha-synuclein oligomers, the N-terminal regions that bind cardiolipin are exposed. Excess cardiolipin on the outer membrane of damaged mitochondria is an invitation for alpha-synuclein oligomers to bind.

Even in healthy mitochondria, small amounts of cardiolipin can be found at the outer membrane. The high affinity of alpha-synuclein oligomers for cardiolipin could help to stabilize and accumulate more cardiolipin at the outer membrane. To function properly, the mitochondrion needs over 1000 proteins: for instance, hemoglobin mentioned in Chapter 5. However, as mentioned earlier, it makes only 13; all other proteins are expressed outside the mitochondrion and are transported there by the cell, to be delivered to their ultimate destinations by the mitochondrial protein transport machinery. The combination of oligomers binding the protein transport machinery, impeding transport of needed proteins across the outer membrane, and binding cardiolipin at the outer membrane could work together to push the mitochondrion down the mitophagy pathway.

Because experiments have shown that alpha-synuclein overexpression stimulates clearance of mitochondria by mitophagy, could it be that Parkinson's disease is caused by reduced numbers of mitochondria? That is, could the disease be caused by what would amount to a cellular energy crisis? Animal disease models might be the workhorses of medical research, but mice are not humans. To answer the question definitively, data from human brain tissue were required.

Brain tissue from Parkinson's disease patients was autopsied to determine whether mitochondria numbers were depressed. In particular, the substantia nigra, where the most severe neuron loss occurs, was microscopically examined [45]. Surprisingly, the numbers of mitochondria the researchers observed in the dopaminergic neuron cell bodies were no different from those in controls, although the researchers saw impaired mitophagy and depletion of lysosomes, confirming earlier research [46]. Using the neurotoxin MPTP, they also discovered evidence linking the depletion of lysosomes to increased levels of reactive oxygen species from defective mitochondria.

In Parkinson's disease, loss of presynaptic boutons of the dopaminergic neurons of the substantia nigra precedes loss of the neurons themselves. Could it be that mitochondria were first depleted in the boutons and in the axonal branches connecting them to the cell body? A team of researchers from Newcastle University in the United Kingdom set out to find the answer [47]. They examined brain tissue from 40 patients, 10 with Parkinson's disease, 10 with dementia with Lewy bodies, 10 with Alzheimer's disease, and 10 with no neurodegenerative diseases.

Using four different fluorescently labeled antibodies, one specific to dopaminergic neurons and three specific to different mitochondrial proteins, the Newcastle researchers stained the striatum tissue from the brains of patients in the study. Using microscopes equipped with fluorescence detectors, they were able to locate the axons and presynaptic boutons in the striatum coming from the dopaminergic neurons of the substantia nigra, and they could count the number and measure the volume of the mitochondria inside. Yet another surprising result: the axons and presynaptic boutons of Parkinson's patients had *more* mitochondria than in controls and Alzheimer's patients!

The Newcastle team found that the average volume of the individual mitochondria in Parkinson's patients was the same as for the controls; that is, the volume was not smaller, in contrast to earlier in vitro experiments of alpha-synuclein that induced mitochondrial fragmentation. The same results held true for patients with dementia with Lewy bodies, although to a lesser degree. They also examined substantia nigra tissue from the same set of patients and confirmed that, unlike the axons and boutons, the number of mitochondria in the cell bodies of the dopaminergic neurons was the same in Parkinson's patients and controls, as seen in previous studies [45].

The paper described additional fascinating details. The size of presynaptic boutons in Parkinson's patients was larger. In addition, a significant number of the presynaptic boutons were empty of mitochondria, and the highest empty fraction was seen for controls and the smallest for Parkinson's patients. Without mitochondria, boutons are nonfunctional. Why would our brains have evolved to have nonfunctional synapses? And why did Parkinson's patients have fewer nonfunctional synapses than did controls?

The Newcastle researchers suggested that the increased number of mitochondria and larger presynaptic bouton size observed for Parkinson's patients were to compensate for other presynaptic boutons that had already been lost, to maintain sufficient dopamine levels in the striatum. The lower fraction of boutons empty of mitochondria might also be part of this compensation, with neurons somehow activating these boutons as others become lost. Earlier studies employing positron emission tomography using radioactive fluorodopa showed that the annual rate of decline in striatal dopamine levels in Parkinson's patients is more rapid in the earlier stages of the disease [48]. In addition, all primates, including humans without neurodegenerative disease, have some loss of dopaminergic substantia nigra neurons as they age [49], so it is plausible that a reserve of nonfunctioning boutons evolved, becoming activated as needed to compensate for age-related neural loss.

I mentioned that the Newcastle team used three antibodies to visualize mitochondria. One of the antibodies was for a protein universally present in mitochondria, but the other two were for respiratory chain proteins encoded by mitochondrial DNA. Mitochondria with less of these respiratory chain proteins presumably have reduced or defective energy production. In Parkinson's patients, more dopaminergic neurons in the substantia nigra displayed reduced amounts of these two proteins, although the researchers cautioned that the difference was not statistically significant given the small number of patients examined (more brains needed!). Curiously, they saw this difference only in dopaminergic cell bodies, not in their axons and presynaptic boutons.

The observation of increased numbers of substantia nigra neurons with mitochondria deficient in respiratory proteins resonated with earlier studies of Parkinson's patients, suggesting that their mitochondrial DNA accumulated more damage with age compared with controls [50]. The two respiratory proteins for which the Newcastle team tested are encoded by mitochondrial DNA genes; thus, their deficiency might reflect increased mitochondrial DNA damage. In fact, the dopaminergic neurons showing cell body mitochondria to be more deficient in these proteins might be the ones next in line to die, but as of yet, no one has devised a way to test this hypothesis experimentally.

Another caveat that the Newcastle team pointed out is that they had no way of pairing the specific dopaminergic neurons in the substantia nigra with their corresponding axons in the striatum. Although they saw fewer dopaminergic synapses in the striatum tissue from Parkinson's patients, their synapses appeared healthy, with no respiratory chain protein deficiency compared with controls. Did the neurons with deficient mitochondria in their cell bodies have fewer surviving synapses? Again, no experiment has yet been devised to answer this question.

The neurotoxin MPTP works by first being taken up and converted to MPP+ (1-methyl-4-phenylpyridinium) in dopaminergic neurons, which then short-circuits the respiratory chain protein machinery, causing the production of free radicals and other reactive oxygen molecules. These toxic chemicals not only damage other parts of the cell, they trigger inflammation, stimulating surrounding glial cells to speed the destruction and removal of the afflicted neurons. MPTP causes dopaminergic neuron death in the substantia nigra in a few days, whereas in Parkinson's disease, the loss of the same neurons takes decades. What could explain this huge discrepancy in time frames?

One possible scenario is that by stimulating mitochondrial fission over fusion, the alpha-synuclein oligomers slightly tip the balance toward a greater number of mitochondria with less of the mitochondrial DNA repair that occurs owing to fusion. Only after years of this subtle imbalance does DNA damage accumulate to the point at which the levels of reactive oxygen species produced are sufficient to stimulate inflammation and synapse loss. The afflicted neurons fight back, seeking to boost the percentage of healthy mitochondria in their synapses by shuttling the most defective mitochondria back to the cell body where they can be more efficiently degraded. Eventually, the defective energy production and increased reactive oxygen species levels in the cell body lead to reduced production of lysosomes, slowing mitophagy and hastening cell death [45].

More evidence linking Parkinson's disease to disrupted mitochondrial maintenance comes from recessive mutations in the *PINK1* and *parkin* genes, causing rare, familial forms of the disease. Normally, the PINK1 protein arrives at the outer mitochondrial membrane, and in mitochondria healthy enough to maintain the normal gradient of electric potential across their inner membrane, PINK1 gets translocated to the inner membrane where protease enzymes cut the protein into pieces [51].

However, in mitochondria damaged to the point at which the inner membrane electric potential gradient is lost, PINK1 accumulates at the outer membrane. There, it phosphorylates proteins, including ubiquitin, and recruits the parkin protein from the cytosol, and together they stimulate fission, axonal transport of the mitochondrion, and mitophagy [52]. As noted in Chapter 6, for many patients with recessive *parkin* mutations, there is little or no evidence of Lewy pathology, which suggests that disruption of mitochondrial maintenance alone, even in the absence of alpha-synuclein oligomers, can cause the pathology of Parkinson's disease.

Knockout mice for the *PINK1* and *parkin* genes appear normal, with no obvious evidence of impaired mitophagy. However, when subjected to prolonged exercise, which taxes the mitochondrial machinery, the knockout mice produced more inflammatory signals than control mice, releasing cytokines such as interferon [53]. If the neuron produces too many inflammatory signals, it can be triggered to undergo apoptosis. It seems that one of the roles of parkin is to suppress inflammatory signals in the normal mitophagy process. Do *parkin* mutations trigger parkinsonian pathology in the same manner as alpha-synuclein oligomers? Not necessarily, but it could be that alpha-synuclein oligomers bound to the mitochondrial outer membrane somehow interfere with the normal function of parkin, allowing the inflammatory signals to escape suppression, helping to cause to synapse loss and ultimately, cell death.

Mutations in other Parkinson's related genes, such as *LRRK2* and *DJ-1*, also seem to hinder mitophagy, although the actual roles of the two proteins in normal cell functioning and disease pathology are perhaps more complex than PINK1 and parkin [54,55]. I suspect mutations in the gene for the tau protein might also contribute by interfering with axonal transport of mitochondria [56]. Tau is an axon-specific protein associated with tubulin, a protein that polymerizes into long chains throughout the cell called microtubules, which serve as molecular highways for transporting cellular cargo from one end of the cell to the other. Tau modulates protein trafficking, like speed bumps on the microtubule highway, to paraphrase my NIH colleagues. In the tauopathy diseases, tau dissociates from microtubules, which become destabilized, and the dissociated tau becomes hyperphosphorylated and forms neurofibrillary tangles.

Lewy neurites, deposits of alpha-synuclein fibrils in axonal branches, interfere with transport of large vesicle cargoes such as autophagosomes [57], which are double-membrane vesicles involved in macroautophagy. Here is another instance of fibrils acting as partners in crime with the oligomers. The transport machinery does its best to go around the deposits, avoiding complete blockage of the axonal branches, at least in rat neuronal cell cultures overexpressing alpha-synuclein. Tau is thought to be important in the flexible remodeling of microtubules. Whether impeded transport caused by neurites, tau-related dysfunction, or other oligomer-related disruption of axonal transport ends up causing the most damage is not known. Nevertheless, the striatum in the brains of affected individuals becomes littered with Lewy neurites as dopaminergic synaptic boutons die off in the middle stages of Parkinson's disease [58].

One of the curious things about Lewy bodies is that the proportion of substantia nigra dopaminergic neurons that have them stays relatively constant, regardless of how many neurons have already been lost [59]. There are at least two possible ways to explain this. One is that the neuron loss rate is relatively constant and Lewy bodies form toward the end of the neuron's life. The other is that Lewy bodies form during only a relatively brief period shortly after onset of the disease. Because of these different interpretations, estimates in the scientific literature for the age of Lewy bodies observed upon autopsy range from months to decades. Oddly, Lewy bodies seem to be degraded once the neuron dies, either by extracellular enzymes or by glial cells, because few extracellular Lewy bodies are seen in autopsied brain tissue [60].

The consensus among my colleagues has been that Lewy bodies are formed in some sort of macroautophagy process, possibly nucleated by an aggresome, mentioned earlier. In macroautophagy, organelles and cytosol are enveloped in autophagosomes, which then merge with lysosomes, whose enzymes break down the enveloped contents for reuse in the cell. For years, no actual Lewy body formation had been observed in cell culture, despite accumulating circumstantial evidence supporting this picture. Specifically, among the components found in significant amounts in Lewy bodies are mitochondrial and lysosomal proteins and lipids [61,62]. The alpha-synuclein fibrils, which are resistant to proteolytic degradation, presumably survive the macroautophagy process with their fibril core intact, although often with part of the C-terminus missing.

Finally, in 2020, Lewy body formation in a neuronal cell culture was reported in a study from the laboratory of Hilal A. Lashuel at École Polytechnique Fédérale, Lausanne, Switzerland [63]. They added preformed alpha-synuclein fibrils to the culture, and after 3 weeks, large aggregations with the distinctive features of Lewy bodies appeared. Did initial fibrillar clumps grow into Lewy bodies by directly adding oligomers to their fibrils, or were their large spherical shapes the result of layers of new fibrils deposited on the Lewy body surface in repeated rounds of autophagy, like the layers of a pearl? It is likely both processes contribute. Curiously, though, the inner layers show higher amounts of C-terminally truncated alpha-synuclein, whereas the outer layers of Lewy bodies show more phosphorylation of alpha-synuclein [62]. This suggests that truncated synuclein has a role in initiating formation of the Lewy body, and that phosphorylation primarily occurs at a later stage in the pathology.

It would be wrong for me to imply there is any consensus among Parkinson's disease researchers regarding the proximal cause of neurodegeneration. Although the experimental and genetic evidence for mitochondrial fragmentation is robust, there is not complete agreement regarding how it is caused and, in particular, whether altered mitophagy, especially that observed in cell cultures in vitro actually has a significant role in the cell in vivo. Some argue that interference with the respiratory chain, either by direct interaction with oligomers with respiratory chain proteins or indirectly via disruption of the inner mitochondrial membrane, and the reactive oxygen species thus generated, is the more proximal cause of the disease.

Others might not even be convinced that mitochondrial dysfunction is the main cause of neurodegeneration, although probably all agree that it has at least a minor role. Oligomers cause dysfunction in many components of the cell, so neurodegeneration might result from the systemic failing of many cellular components together, rather than be dominated by mitochondrial dysfunction. I do not necessarily disagree with this more nuanced viewpoint, but in my opinion, impaired mitochondrial maintenance is the proverbial straw that broke the camel's back. Specifically, only when the mitochondria become damaged enough to produce sufficiently high levels of reactive oxygen species do inflammatory signals reach the threshold that finally triggers surrounding glial cells to destroy the diseased presynaptic boutons and ultimately, the cell body of the neuron.

Part 3. Alpha-synuclein mutants, the final mystery

We will turn our attention to these other areas of alpha-synuclein—induced dysfunction in the next chapters, but for now let us return our focus to mutations in alpha-synuclein. Why do the familial alpha-synuclein mutants cause the disease? Recall that the mutations are dominant with high penetrance. So far, six mutations, all very rare, have been found [64]. They are designated A30P, E46K, H50Q, G51D, A53T, and A53E, in which the first letter corresponds to the normal amino acid, the number its position in the protein sequence, and the second letter is the mutant amino acid that replaces the normal one. Such mutations, in which one amino acid is replaced by another, are called point mutations. The normal version of the gene is called the wild type.

There is no shared pattern to the mutations; that is, they alter the physical properties of alpha-synuclein in different ways. Two of the amino acids have negative charges, E (for glutamate) and D (for aspartate), and two have positive charges, K (for lysine) and H (for histidine) (which is positively charged only under acidic conditions). The other amino acids involved are neutral. A (for alanine) has a small side chain that is mildly hydrophobic, whereas G (for glycine) has no side chain at all. Q (for glutamine) and T (for threonine) are hydrophilic amino acids, preferring to form hydrogen bonds with water and all other hydrophilic amino acids, including the charged ones. Finally, P (for proline) is mildly hydrophobic and has a side chain connected to the peptide backbone at two places, which makes it unable to form one of the hydrogen bonds needed for the helix and beta sheet secondary structure.

Many hypotheses have been proposed, most commonly, that the mutant forms of alpha-synuclein might be neurotoxic in some way, the so-called toxic gain-of-function hypothesis [65]. Dominant mutations naturally lend themselves to the gain-of-function hypothesis: all that is needed is for one copy of the gene to produce the toxic form of the protein. In contrast, the alternative loss-of-function hypothesis is typically proposed for recessive mutations, in which both genes are mutated in a way that inactivates the function of the protein. It was because of the presumption of the gain-of-function hypothesis that studies of alpha-synuclein mutant transgenic mice were pursued with such gusto. However, as we saw earlier, the evidence was questionable whether overexpression of mutant alpha-synuclein was worse than overexpression of wild type; both were toxic.

Of course, as soon as the various mutant forms of alpha-synuclein were discovered, researchers rushed to characterize their physical properties. They seem to fall into two classes. One class (E46K, H50Q, and A53T) was capable of forming fibrils more quickly and binding negatively charged lipid membranes more tightly, and the other (A30P, G51D, and A53E) showed the opposite trends [64,66,67]. Wild-type alpha-synuclein fell somewhere in the middle. That G51D and A53E bound negatively charged membranes more weakly was unsurprising because the mutations made alpha-synuclein more negative, and like charges repel. The A30P mutation introduced a proline into a region that is normally alpha-helical upon binding membrane, and because proline disrupts helices, its weaker binding was also unsurprising. The presence of lipids altered the various rates of fibril formation, too.

The properties of the mutant oligomers were also characterized. A30P, E46K, H50Q, and A53T formed oligomers more quickly, and G51D and A53E formed them more slowly than wild type [66,68–70]. All of them were neurotoxic in cell cultures, but surprisingly, G51D was unable to permeabilize membranes, unlike the wild type and the other mutants [71]. This poked a hole (sorry, I couldn't help it) in the theory that membrane permeabilization was important for disease pathology; G51D mutation carriers definitely develop Parkinson's disease. The reason for the differences in the oligomer and fibril formation speeds was not clear. Nevertheless, the more important point was that there was no obvious physical attribute shared by the mutants, but lacking in the wild type, that could explain their putative toxicity.

Have you noticed that I have not yet mentioned the two other familial Parkinson's alpha-synuclein mutants? These are the cases in which one of the two synuclein genes has been duplicated or triplicated [72,73]. These mutations also cause the disease with high penetrance, especially for cases with gene triplication. The obvious implication is that an increased level of alpha-synuclein alone is sufficient to cause the disease, as mentioned in the second chapter. We will call this the overexpression model, to contrast with the toxic gain-of-function model. A single duplication suggests a 50% increase in gene expression, because the patient would have the extra, duplicate gene plus the two normal genes everyone has. Similarly, gene triplication suggests a 100% increase: that is, double the gene expression. Gene regulation is complex, however, with many checks and balances, so it could be that gene expression and resulting alpha-synuclein protein levels are not increased the full 50% or 100%, respectively.

Even a modest increase in levels has major implications for the aggregation of alpha-synuclein into oligomers. In its simplest form, aggregation is what chemists call a second-order process, which means it depends on the square of the concentration. If you double the concentration of alpha-synuclein, the rate at which monomers form oligomers quadruples. The actual picture is slightly more complicated, because one needs also consider oligomer—monomer and oligomer—oligomer encounters, but the process is still strongly promoted by higher concentrations of the constituents. In light of the strong concentration dependence, it is unsurprising that the familial alpha-synuclein duplication and triplication mutations cause Parkinson's disease.

Because elevated alpha-synuclein expression levels seemed to cause familial Parkinson's disease in the duplication and triplication patients, researchers wondered whether idiopathic Parkinson's patients also had elevated expression. They used a technique called real-time polymerase chain reaction to quantify the amount of alpha-synuclein mRNA expressed. The results were all over the map. Some saw increased expression [74], whereas others saw a reduction [75]. Part of the problem was that in the brains of deceased Parkinson's patients, there was already substantial loss of dopaminergic neurons, especially in the substantia nigra. As a consequence, brain tissue from the patients had a lower percentage of dopaminergic neurons compared with controls; thus, it could appear that brain tissue from the patients had lower alpha-synuclein expression.

Still, the mRNA expression studies provided evidence of other changes in dopaminergic neurons during the disease. In a 2008 study from the laboratory of Birgit Liss at the University of Ulm, Germany, they used microscopes and lasers to dissect individual neuron cell bodies from tissue slices of the substantia nigra. Analyzing the RNA of each cell body individually, the researchers were able to identify the dopaminergic ones, and showed that in addition to increased alpha-synuclein mRNA levels, two other genes were expressed at higher levels in Parkinson's patients [76]. One was the gene for tyrosine hydroxylase, which is needed to make dopamine. This hinted that the remaining dopaminergic neurons in Parkinson's patients appeared to be compensating for the loss of their neighbors by upping dopamine production (a scenario confirmed by the Newcastle team [47]), and perhaps alpha-synuclein expression was also increased as a consequence.

In idiopathic patients, did alpha-synuclein expression increase only as the disease progressed, or was it already elevated before the pathology first appeared? If the pathology was triggered by alpha-synuclein aggregation, higher prepathology alpha-synuclein levels should increase the odds of triggering the pathology. But how could one determine prepathology expression levels? One was faced with difficult requirements: obtaining prepathology brain tissue from the patient, and knowing that the patient would go on (for a live patient) or would have gone on (for a deceased patient) to develop idiopathic Parkinson's disease.

While pondering this conundrum, I was trying to get my mind around another even more puzzling claim regarding the familial alpha-synuclein point mutations. In 2003, a multinational team headed by Nobutaka Hattori at the Juntendo University School of Medicine, Tokyo, came out with the startlingly titled (to me) report, "Haploinsufficiency at the Alpha-Synuclein Gene Underlies Phenotypic Severity in Familial Parkinson's Disease" [77].

Allow me to translate. Haploinsufficiency means that one copy of a gene is defective, and not enough protein is made to support normal function in the cell. The study looked at two familial alpha-synuclein point mutants, A30P and A53T. The researchers claimed that the mutant gene copy was being expressed at much lower levels than the wild-type copy, and that this was causing, or at least correlated with, a more severe phenotype, which in this case meant more severe symptoms. In general, phenotype means the trait associated with the two copies of a gene (which might be different variants), which is called the genotype.

If their claim were true, this would seem to invalidate both the toxic gain-of-function model and the elevated expression model for Parkinson's disease pathology. The study was performed on members of three families with mild to severe symptoms, five with the A30P mutant and two with the A53T mutant. The results showed a clear correlation between symptom severity and the ratio of wild type to mutant mRNA expression. Focusing on the ratio rather than absolute mRNA amounts strengthened their claim, because systematic errors in absolute amounts in each individual sample should be correlated and become divided out, that is, removed, by taking the ratio.

However, two tables in the report included the absolute numbers for the mRNA amounts, and the amounts caught my attention. The mRNA numbers for the wild type for those with the severe symptoms were much closer to the values reported for patients with idiopathic Parkinson's disease mentioned earlier, and the mutant mRNA amounts were closer to the control values. Had the researchers misinterpreted the situation? Maybe the changes they saw actually resulted from an increase in wild-type mRNA rather than a decrease in mutant mRNA? It could be that what they were seeing was a haplosurplus of wild-type mRNA rather than an insufficiency of mutant mRNA. They even acknowledged that their assumption of haploinsufficiency might be wrong: "Disease progression and the presence of dementia appear to be associated with either increased expression of the wild-type allele or reduced expression of the mutant allele" (allele means the version of the gene copy).

But wait, there's more! The samples they used were not brain tissue. Instead, they were derived from white blood cells taken from the family members and transformed into cancer-like lymphoblastoid cells, capable of living in cell culture in vitro. White blood cells also express alpha-synuclein, which allowed the researchers to study its mRNA expression without needing brain tissue, a major advantage. The white blood cells were normal, so the expression differences were likely due to the point mutation itself, unlike the other studies, which might have been skewed by the diseased state of the brain tissue. There were drawbacks, too, of course; white blood cells are not neurons, and one should not assume their alpha-synuclein expression levels are necessarily similar. Still, it is hard to explain otherwise the correlation of more severe symptoms with the greater alteration of synuclein expression.

Since 2003, I have found only one new lymphoblastoid study, by a team at the University of Patras, Greece [78]. That report examined a single patient and confirmed the results for the A53T mutation, showing practically no expression of mutant mRNA, and more wild-type mRNA from the patient's single wild-type copy than in controls with two wild-type copies. The lymphoblastoid studies looked at a total of only eight people, so the chance that results were simply a statistical fluke cannot be dismissed.

Unfortunately, point mutations of alpha-synuclein are rare, and large cohort studies with the statistical power to rule out spurious results have not been an option. However, the options are changing. With pluripotent stem cells derived from patients, and with gene-editing technology, these types of studies can be now redone on a much larger scale. Perhaps it would be wise to await more definitive studies using neural cell lines rather than to formulate grand hypotheses based on two studies of eight patients using techniques with which I have no personal experience ... Who am I kidding? Let's speculate!

Recall from the first chapter that alpha- and beta-synuclein have highly conserved sequences across species, marking them as hub proteins; that is, proteins with many important interactions in the cell. These interactions include protein—protein interactions, and likely interactions at the DNA/RNA level as well. For the synucleins, lipid interactions are also important. Chances are good that a particular mutation will result in perturbation of one or more interactions.

One way for the cell to counter perturbed interactions is to increase wild type and decrease mutant expression. This could explain the imbalanced expression seen in the lymphoblastoid studies. What physical interactions might be perturbed in the familial alpha-synuclein mutants? So far, research has focused mostly on varying interactions of mutants with lipid membranes and their propensity to interact with themselves to form oligomers. Chances are that these two interactions represent just a tiny fraction of all interactions that can potentially be perturbed by the alpha-synuclein point mutants.

How could wild-type expression in those with familial point mutations be even higher than the total alpha-synuclein expression in controls? I suspect the answer lies, at least in part, with the toxic gain-of-function model. One type of toxic interaction is when the mutant interacts with an interaction partner, but fails to do so correctly, interfering with the function of the interaction. To correct the interaction, the expression of the wild type needs to be more than doubled. Twice as much wild-type expression is needed to restore function back to the normal level to supply normal function, plus an extra amount is needed to crowd out incorrect interactions with the mutant.

Why not just turn off the mutant alpha-synuclein gene? Perhaps sometimes the cell can suppress expression only to a point, and the remaining compensation must rely on increasing wild-type expression. The resulting situation is basically the overexpression model; the greater concentration of wild type plus mutant alpha-synuclein increases the odds of oligomer formation. Thus, if this line of speculation is correct, the toxic gain-of-function model leads to the overexpression model of Parkinson's disease for those with familial alpha-synuclein point mutations. Alas, there is a caveat to this picture; in the study of rat G51D knock-in rats mentioned in Chapter 4 [79], alpha-synuclein levels were the same in the knock-in rats and controls, although as far as I could tell, only rats homozygous for the G51D mutations (that is, with both alpha-synuclein genes mutated) were tested, not those with one wild type and one G51D mutation.

Amyloid oligomers of alpha-synuclein appear to be the primary toxic agent in Parkinson's disease. In my opinion, the proximal cause of synapse loss and neuronal death is their interference with protein transport and signaling at the mitochondrial outer membrane, stimulating fragmentation and mitophagy. By favoring mitochondrial fission over fusion, DNA damage accumulates at an accelerated rate, ultimately leading to deficits in the respiratory chain machinery and increased production of reactive oxygen species.

Oligomers alone do not appear to be able to cause the disease to spread through the brain, whereas injected fibrils can cause pathology that spreads from the site of injection. Apparently, fibrils are required, serving as reservoirs and templates for the generation of new oligomers. In addition, Lewy neurites disrupt axonal transport of large organelles, such as autophagosomes containing damaged mitochondria, hampering mitochondrial turnover.

Studies of familial alpha-synuclein mutants imply that an increase in the total amount of alpha-synuclein expression is ultimately to blame for triggering Parkinson's disease. The increased expression is either direct, in the cases of gene duplication or triplication, or indirect, arising from wild-type gene overexpression to compensate for the toxic gain-of-function in the familial point mutations. The increased alpha-synuclein expression increases the odds that oligomers will form. The more oligomers, the greater the production of fibrils, which can then spread the disease to other areas of the brain.

I suspect that alpha-synuclein oligomers owe most of their toxicity to their multiple N-terminal membrane-interacting regions being in close proximity. Alpha-synuclein interacts with two proteins at the outer mitochondrial membrane, TOM20 and TOM40, which are part of the machinery that governs entry of proteins through the membrane. TOM20 and TOM40 recognize proteins that have a mitochondrial targeting sequence, which consists of an N-terminal membrane binding helix, similar to the one at the N-terminus of alpha-synuclein.

In fact, this N-terminus helix interaction could be how alpha-synuclein oligomers gum up the works. One peptide of the oligomer could have its N-terminal helix sequence bound to TOM20 or TOM40, while the N-termini of the other oligomer peptides interact with cardiolipin and other acidic lipids of the mitochondrial membrane, anchoring the oligomer to the membrane surface. Basically, alpha-synuclein oligomers could be getting stuck in the turnstiles, not letting other proteins, such as PINK1, get through. Experiments in rats overexpressing alpha-synuclein support this picture [80]. When TOM20 is also overexpressed, dopaminergic neurodegeneration is reduced.

You might be wondering, "What happened to the common mechanism of oligomer toxicity and that exposed beta sheet business?" I admit that for me, this idea has fallen by the wayside. It could be that exposed beta sheet edge interactions with oligomers cause at least some of the overloading of the UPS seen in various amyloid-related diseases, but I doubt UPS overloading is the main cause of Parkinson's disease. My misfolding trap hypothesis rode the trending wave of the times, but a paradigm-shifting tsunami it was not.

References

[1] Glenner GG, Wong CW. Alzheimer's disease: initial report of the purification and characterization of a novel cerebrovascular amyloid protein. Biochemical and Biophysical Research Communications May 16, 1984;120(3):885–90.

[2] Nelson PT, Braak H, Markesbery WR. Neuropathology and cognitive impairment in Alzheimer disease: a complex but coherent relationship. Journal of Neuropathology and Experimental Neurology January 2009;68(1):1–14.

[3] Pike CJ, et al. In vitro aging of beta-amyloid protein causes peptide aggregation and neurotoxicity. Brain Research November 1, 1991;563(1–2):311–4.

[4] Oda T, et al. Clusterin (apoJ) alters the aggregation of amyloid β-peptide (Aβ1-42) and forms slowly sedimenting Aβ complexes that cause oxidative stress. Experimental Neurology November 1995;136(1):22–31.

[5] Lambert MP, et al. Diffusible, nonfibrillar ligands derived from Abeta1-42 are potent central nervous system neurotoxins. Proceedings of the National Academy of Sciences, United States of America May 26, 1998;95(11):6448–53.

[6] Conway KA, Harper JD, Lansbury PT. Accelerated in vitro fibril formation by a mutant α-synuclein linked to early-onset Parkinson disease. Nature Medicine 1998;4:1318–20.

[7] El-Agnaf OM, et al. Aggregates from mutant and wild-type alpha-synuclein proteins and NAC peptide induce apoptotic cell death in human neuroblastoma cells by formation of beta-sheet and amyloid-like filaments. FEBS Letters November 27, 1998;440(1–2):71–5.

[8] Haass C, Selkoe DJ. Soluble protein oligomers in neurodegeneration: lessons from the Alzheimer's amyloid beta-peptide. Nature Reviews Molecular Cell Biology February 2007;8:101–12.

[9] Lasagna-Reeves CA, et al. Preparation and characterization of neurotoxic tau oligomers. Biochemistry November 30, 2010;49(47):10039–41.

[10] Haataja L, et al. Islet amyloid in type 2 diabetes, and the toxic oligomer hypothesis. Endocrine Reviews May 2008;29(3):303–16.

[11] Kayed R, Head E, et al. Common structure of soluble amyloid oligomers implies common mechanism of pathogenesis. Science April 18, 2003;300(5618):486–9.

[12] Kayed R, Glabe CG. Conformation-dependent anti-amyloid oligomer antibodies. Methods in Enzymology 2006;413:326–44.

[13] Hwang KJ, et al. Intramolecular interaction in the tail of Acanthamoeba myosin IC between the SH3 domain and a putative pleckstrin homology domain. Proceedings of the National Academy of Sciences, United States of America January 16, 2007;104(3):784–9.

[14] Kelley LA, et al. The Phyre2 web portal for protein modeling, prediction and analysis. Nature Protocols June 2015;10(6):845–58.

[15] Müller A, MacCallum RM, Sternberg MJ. Structural characterization of the human proteome. Genome Research November 2002;12(11):1625–41.

[16] Gruschus JM. Do amyloid oligomers act as traps for misfolded proteins? A hypothesis. Amyloid September 2008;15(3):160–5.

[17] Bence NF, Sampat RM, Kopito RR. Impairment of the ubiquitin-proteasome system by protein aggregation. Science May 25, 2001;292(5521):1552–5.

[18] Richardson JS, Richardson DC. Natural beta-sheet proteins use negative design to avoid edge-to-edge aggregation. Proceedings of the National Academy of Sciences, United States of America March 5, 2002;99(5):2754−9.
[19] Yoshiike Y, et al. Amyloid oligomer conformation in a group of natively folded proteins. PLoS One September 18, 2008;3(9):e3235.
[20] Chen SW, et al. Structural characterization of toxic oligomers that are kinetically trapped during α-synuclein fibril formation. Proceedings of the National Academy of Sceinces, United States of America April 21, 2015;112(16):E1994−2003.
[21] Uversky VN, Li J, Fink AL. Evidence for a partially folded intermediate in α-synuclein fibril formation. Journal of Biological Chemistry April 6, 2001;276:10737−44.
[22] Celej MS, et al. Toxic prefibrillar α-synuclein amyloid oligomers adopt a distinctive antiparallel β-sheet structure. Biochemical Journal May 1, 2012;443(3):719−26.
[23] Zhang H, et al. In vitro study of α-synuclein protofibrils by cryo-EM suggests a Cu^{2+}-dependent aggregation pathway. Biophysical Journal June 18, 2013;104(12):2706−13.
[24] Pieri L, Madiona K, Melki R. Structural and functional properties of prefibrillar α-synuclein oligomers. Scientific Reports 2016;6:24526.
[25] Fusco G, et al. Structural basis of membrane disruption and cellular toxicity by α-synuclein oligomers. Science December 15, 2017;358(6369):1440−3.
[26] Gaspar R, et al. Secondary nucleation of monomers on fibril surface dominates α-synuclein aggregation and provides autocatalytic amyloid amplification. Quarterly Reviews of Biophysics January 2017;50:e6.
[27] Chuang E, et al. Amyloid assembly and disassembly. Journal of Cell Science April 13, 2018;131(8):jcs189928.
[28] Peelaerts W, et al. α-Synuclein strains cause distinct synucleinopathies after local and systemic administration. Nature June 18, 2015;522(7556):340−4.
[29] Froula JM, et al. Defining α-synuclein species responsible for Parkinson's disease phenotypes in mice. Journal of Biological Chemistry July 5, 2019;294(27):10392−406.
[30] Winner B, et al. In vivo demonstration that alpha-synuclein oligomers are toxic. Proceedings of the National Academy of Sceinces, United States of America March 8, 2011;108(10):4194−9.
[31] Demuro A, et al. Calcium dysregulation and membrane disruption as a ubiquitous neurotoxic mechanism of soluble amyloid oligomers. Journal of Biological Chemistry April 29, 2005;280(17):17294−300.
[32] Quist A, et al. Amyloid ion channels: a common structural link for protein-misfolding disease. Proceedings of the National Academy of Sciences, United States of America July 26, 2005;102(30):10427−32.
[33] van Rooijen BD, Claessens MM, Subramaniam V. Membrane permeabilization by oligomeric α-synuclein: in search of the mechanism. PLoS One December 13, 2010;5(12):e14292.
[34] Shen YF, et al. Adaptive changes in autophagy after UPS impairment in Parkinson's disease. Acta Pharmacologica Sinica May 2013;34(5):667−73.
[35] Slone SR, et al. Increased 14-3-3 phosphorylation observed in Parkinson's disease reduces neuroprotective potential of 14-3-3 proteins. Neurobiolocial Disorders July 2015;79:1−13.
[36] Hunn BH, et al. Impaired intracellular trafficking defines early Parkinson's disease. Trends in Neurosciences March 2015;38(3):178−88.
[37] Garcia-Esparcia P, et al. Altered machinery of protein synthesis is region- and stage-dependent and is associated with α-synuclein oligomers in Parkinson's disease. Acta Neuropathologica Communications December 1, 2015;3:76.
[38] Chinta SJ, et al. Mitochondrial α-synuclein accumulation impairs complex I function in dopaminergic neurons and results in increased mitophagy in vivo. Neuroscience Letters December 17, 2010;486(3):235−9.

[39] Xie W, Chung KKK. Alpha-synuclein impairs normal dynamics of mitochondria in cell and animal models of Parkinson's disease. Journal of Neurochemistry 2012;122(2):404–14.
[40] Di Maio R, et al. α-Synuclein binds to TOM20 and inhibits mitochondrial protein import in Parkinson's disease. Science Translational Medicine June 8, 2016;8(342):342ra78.
[41] Youle RJ, van der Bliek AM. Mitochondrial fission, fusion, and stress. Science August 2012;337(6098):1062–5.
[42] Vekrellis K, et al. Pathological roles of α-synuclein in neurological disorders. The Lancet Neurology November 2011;10(11):1015–25.
[43] Nakamura K. α-Synuclein and mitochondria: partners in crime? Neurotherapeutics July 2013;10(3):391–9.
[44] Dudek J. Role of cardiolipin in mitochondrial signaling pathways. Frontiers in Cell and Developmental Biology September 29, 2017;5:90. https://doi.org/10.3389/fcell2017.00090.
[45] Dehay B, et al. Pathogenic lysosomal depletion in Parkinson's disease. Journal of Neuroscience September 15, 2010;30(37):12535–44.
[46] Chu Y, et al. Alterations in lysosomal and proteasomal markers in Parkinson's disease: relationship to alpha-synuclein inclusions. Neurobiology of Disease September 2009;35(3):385–98.
[47] Reeve AK, et al. Mitochondrial dysfunction within the synapses of substantia nigra neurons in Parkinson's disease. npj Parkinson's Disease March 26, 2018;4:9.
[48] Morrish PK, Sawle GV, Brooks DJ. An [18F]dopa-PET and clinical study of the rate of progression in Parkinson's disease. Brain April 1996;119(2):585–91.
[49] Emborg ME, et al. Age-related declines in nigral neuronal function correlate with motor impairments in rhesus monkeys. Journal of Comparitive Neurology November 16, 1998;401(2):253–65.
[50] Smigrodzki R, Parks J, Parker WD. High frequency of mitochondrial complex I mutations in Parkinson's disease and aging. Neurobiology of Aging November–December 2004;25(10):1273–81.
[51] Jin SM, et al. Mitochondrial membrane potential regulates PINK1 import and proteolytic destabilization by PARL. Journal of Cell Biology November 29, 2010;191(5):933–42.
[52] Gladkova C, et al. Mechanism of parkin activation by PINK1. Nature July 2018;559(7714):410–4.
[53] Sliter DA, et al. Parkin and PINK1 mitigate STING-induced inflammation. Nature September 2018;561(7722):258–62.
[54] Hsieh CH, et al. Functional impairment in miro degradation and mitophagy is a shared feature in familial and sporadic Parkinson's disease. Cell Stem Cell December 1, 2016;19(9):709–24.
[55] Thomas KJ, et al. DJ-1 acts in parallel to the PINK1/parkin pathway to control mitochondrial function and autophagy. Human Molecular Genetics January 1, 2011;20(1):40–50.
[56] Zhang X, et al. Tau pathology in Parkinson's disease. Frontiers in Neurology 2018;9:809.
[57] Volpicelli-Daley LA, et al. Formation of α-synuclein Lewy neurite-like aggregates in axons impedes the transport of distinct endosomes. Molecular Biology of the Cell December 15, 2014;25(25):4010–23.
[58] Braak H, et al. Extensive axonal Lewy neurites in Parkinson's disease: a novel pathological feature revealed by alpha-synuclein immunocytochemistry. Neuroscience Letters April 9, 1999;265(1):67–9.

[59] Parkkinen L, et al. Disentangling the relationship between Lewy bodies and nigral neuronal loss in Parkinson's disease. Journal of Parkinson's Disease 2011;1(3):277–86.
[60] Den Hartog Jager WA, Bethlem J. The distribution of Lewy bodies in the central and autonomic nervous systems in idiopathic paralysis agitans. Journal of Neurology, Neurosurgery and Psychiatry 1960;23:283–90.
[61] Wakabayashi K, et al. The Lewy body in Parkinson's disease and related neurodegenerative disorders. Molecular Neurobiology April 2013;47(2):495–508.
[62] Shahmoradian SH, et al. Lewy pathology in Parkinson's disease consists of crowded organelles and lipid membranes. Nature Neuroscience July 2019;22(7):1099–109.
[63] Mahul-Mellier A-L, Burtscher J, et al. The process of Lewy body formation, rather than simply α-synuclein fibrillization, is one of the major drivers of neurodegeneration. Proceedings of the National Academy of Sciences, United States of America February 19, 2020;117(9):4971–82.
[64] Petrucci S, Ginevrino M, Valente EM. Phenotypic spectrum of alpha-synuclein mutations: new insights from patients and cellular models. Parkinsonism and Related Disorders January 2016;Suppl 1:S16–20.
[65] Bertoncini CW, et al. Familial mutants of alpha-synuclein with increased neurotoxicity have a destabilized conformation. Journal of Biological Chemistry September 2, 2005;280(35):30649–52.
[66] Rutherford NJ, et al. Divergent effects of the H50Q and G51D SNCA mutations on the aggregation of α-synuclein. Journal of Neurochemistry December 2014;131(6):859–67.
[67] Greenbaum EA, et al. The E46K mutation in α-synuclein increases amyloid fibril formation. The E46K Mutation in Alpha-Synuclein Increases Amyloid Fibril Formation March 4, 2005;280(9):7800–7.
[68] Conway KA, et al. Acceleration of oligomerization, not fibrillization, is a shared property of both alpha-synuclein mutations linked to early-onset Parkinson's disease: implications for pathogenesis and therapy. Proceedings of the National Academy of Sciences, United States of America January 18, 2000;97(2):571–6.
[69] Ghosh D, et al. The newly discovered Parkinson's disease associated Finnish mutation (A53E) attenuates α-synuclein aggregation and membrane binding. Biochemistry October 21, 2014;53(41):6419–29.
[70] Li J, Uversky VN, Fink AL. Effect of familial Parkinson's disease point mutations A30P and A53T on the structural properties, aggregation, and fibrillation of human alpha-synuclein. Biochemistry September 25, 2001;40(38):11604–13.
[71] Stefanovic AN, et al. Oligomers of Parkinson's disease-related α-synuclein mutants have similar structures but distinctive membrane permeabilization properties. Biochemistry May 26, 2015;54(20):3142–50.
[72] Ibáñez P, et al. Causal relation between alpha-synuclein gene duplication and familial Parkinson's disease. Lancet September 2004;364(9440):1169–71.
[73] Singleton AB, et al. alpha-Synuclein locus triplication causes Parkinson's disease. Science October 31, 2003;302(5646):841.
[74] Chiba-Falek O, Lopez GJ, Nussbaum RL. Levels of alpha-synuclein mRNA in sporadic Parkinson disease patients. Movement Disorders October 2006;21(10):1703–8.
[75] Su X, et al. Alpha-synuclein mRNA is not increased in sporadic PD and alpha-synuclein accumulation does not block GDNF signaling in Parkinson's disease and disease models. Molecular Therapy October 4, 2017;25(10):2231–5.
[76] Gründemann J, et al. Elevated alpha-synuclein mRNA levels in individual UV-laser-microdissected dopaminergic substantia nigra neurons in idiopathic Parkinson's disease. Nucleic Acids Research April 2008;36(7):e38.
[77] Kobayashi H, et al. Haploinsufficiency at the alpha-synuclein gene underlies phenotypic severity in familial Parkinson's disease. Brain January 2003;126(1):32–42.

[78] Voutsinas GE, et al. Allelic imbalance of expression and epigenetic regulation within the alpha-synuclein wild-type and p.Ala53Thr alleles in Parkinson disease. Human Mutation June 2010;31(6):685–91.
[79] West SA. Phenotypic analysis of the CRISPR-engineered G51D alpha-synuclein rat. Edinburgh: University of Edinburgh; 2019.
[80] DeMiranda BR, et al. Protection from α-synuclein-induced dopaminergic neurodegeneration by overexpression of the mitochondrial import receptor TOM20 in the rat midbrain. bioRxiv May 13, 2020. https://doi.org/10.1101/2020.05.11.089078.

CHAPTER 9

An Evolutionary Affair – The connection between Gaucher disease and Parkinson's disease

In October 2013, I found myself at home for an unplanned vacation of unknown duration. A federal budget impasse between President Obama and the Republican-controlled Congress had resulted in a federal government shutdown. I was at a loss with what to do with my unexpected free time. I had been through a government shutdown before, in the winter of 1995–96. Back then, I was a postdoctoral fellow just starting my research in the laboratory of Jim Ferretti. Despite the shutdown, we quietly snuck into work at the Bethesda campus each day, keeping a low profile and continuing our various research projects. After all, I was a young scientist and had my future career to consider. Without a research project, I would

Synuclein and the Coelacanth
ISBN 978-0-323-85707-9
https://doi.org/10.1016/B978-0-323-85707-9.00015-0

have no results to publish, and without published papers, I had no future as a research scientist. "Publish or perish" was, and still is, our motto.

In 2013, it was different. We were informed by our superiors in no uncertain terms that unless we were considered essential, we were not to come to work. Essential employees were ones involved in patient care and facilities that required daily maintenance. I, like thousands of others at the National Institutes of Health (NIH), was deemed nonessential. A lot of egos were bruised. Still, I wanted to use my time at home constructively, something work-related, but novel, outside my usual area of research.

I had heard of a new bioinformatics technique, correlated mutation analysis, which could be used to predict protein structure and interactions. Because of my previous work in predicting the structure of protein complexes, I was interested to learn more. When I first glanced at papers on the subject, I saw it involved mathematics with which I was unfamiliar, definitely not something I could learn without some in-depth reading. So, I got comfortable at my home computer, put in one of my favorite music CDs, and dove in.

During the course of evolution, deleterious mutations can occur, ones that destabilize protein structure or otherwise interfere with function and/or interactions. The basic idea behind correlated mutation analysis is that if a mutation is mild enough that it does not immediately get selected out in a population, there is a chance that a compensating mutation can occur that counteracts the deleterious effect. It is also possible a neutral mutation can occur, which when paired with another chance mutation, actually confers some benefit. Over time, as more species evolve, additional correlated mutations can occur. By looking at the protein sequences across species, one can identify which pairs of amino acids tend to mutate together: in other words, which amino acids show correlated mutation patterns.

The idea of correlated mutations in protein evolution had been around for decades, but interest in the subject surged in the early 1990s. In 1994, several research groups published papers showing statistically robust evidence of correlated mutations across species in a large number of proteins, most prominently by Chris Sander of the European Molecular Biology Laboratory in Heidelberg, Germany and a multinational team of collaborators [1,2]. What prompted this surge? In that same year the first worldwide competition in protein structure prediction, Critical Assessment of Protein Structure Prediction (CASP) was held. Interest in the field of protein structure prediction was exploding.

The primary goal of CASP was to take new, experimentally determined protein structures not yet made publicly available and to challenge research groups to predict their structures knowing only their amino acid sequences. Each group would submit their best guess(es) for the structures, and the CASP crew would score them based on their similarity to the experimental structures. The research groups could use whatever prediction methods they wished, such as homology modeling and threading, methods that took advantage of other, known protein structures. However, some of the new proteins had structures unrelated to any known protein structure, dooming such methods to failure.

Other research groups tried different approaches that did not rely on known protein structures, known collectively as de novo structure prediction methods. Some of the de novo researchers took a brute force approach, using computers to mimic actual protein folding in a watery environment. These in silico approaches attempted to explore all of the different ways the protein could fold, hoping to find the ideal combination of secondary structure and optimum packing of the hydrophobic side chains into the core of the protein. Unfortunately, at the time, computers were not up to the task; the number of ways to fold a protein was astronomically large. Somehow the search needed to be narrowed, and correlated mutation analysis might do just the trick.

What the correlated mutation analysis papers of 1994 had shown was not just that correlated mutations existed, but that the correlated amino acids pairs were often in direct contact. This meant the pairs could be used to narrow the protein structure searches to focus on structures with the correlated pairs near each other, and many research groups incorporated such restraints into de novo structure prediction methods.

Soon, CASP broadened to include prediction of protein interactions, and in 2001 this branched off to become its own competition, Critical Assessment of Prediction of Interactions (CAPRI). Correlated mutation analysis could help with protein interactions, too. The same evolutionary processes that led to correlated mutations within a protein could also occur for amino acids in contact at the interface of interacting proteins. Interacting proteins with correlated mutations are said to have coevolved.

Did the methods that incorporated correlated mutation analysis sweep the CASP and CAPRI competitions? Um, no. Although its promise was clear, correlated mutation analysis had its own issues. The earliest demonstrations of the technique used proteins with 1000 or more known sequences from different species, nearly all from bacteria; the high number of sequences was needed to ensure accurate identification of correlated pairs. To see why this could be a problem, consider vertebrate species, for which fewer than 200 genomes have been sequenced to date. If the unknown protein happened to be specific to vertebrates, would less than 200 sequences be sufficient to predict enough correlated pairs to be useful for protein structure prediction? And how many of the predicted pairs could be wrong (i.e., false positives) before structure prediction was no longer viable?

There were other issues with the analysis, too, such as indirect correlations, in which amino acid A would be correlated with amino acid C via contacts with an intermediate amino acid B. Because of details such as varying mutation rates at different sites in the protein sequence, the correlation between amino acids A and C could sometimes look stronger than between A and B or B and C, even though the direct interactions were with B. Mathematical improvements were needed to correct for indirect correlations and to boost accuracy in cases with a limited number of sequences.

We will get into more of the gory mathematical details in a bit, but first a confession. I was not studying correlated mutation analysis just for my scientific edification; I had an ulterior motive. To explain, I need to back up 3 more years. In 2010, my NIH colleague Thai Leong Yap, in the laboratory of Jennifer C. Lee, discovered that alpha-synuclein could bind to the enzyme glucocerebrosidase [3]. Mutations in the gene for glucocerebrosidase, *GBA1*, cause Gaucher disease, a recessive genetic disorder, and, as mentioned in Chapter 2, *GBA1* mutations are a risk factor for developing Parkinson's, with carriers being five time more likely to develop the disease.

The genetic connection between *GBA1* and Parkinson's disease had been shown by NIH scientist Ellen Sidransky and collaborators several years before Yap's discovery of the interaction between alpha-synuclein and glucocerebrosidase [4]. In fact, the Sidransky laboratory provided the glucocerebrosidase that Yap used for his research. Why did it take several years for the interaction to be discovered? Most likely because other researchers were looking for the interaction under neutral pH conditions, which is standard for most biological experiments. However, the glucocerebrosidase enzyme does its work inside lysosomes, and lysosomes are acidic. Yap discovered that the interaction occurs only under acidic conditions; at neutral pH, the interaction disappears.

Working with the Lee laboratory, we looked at the interaction using nuclear magnetic resonance (NMR) spectroscopy. The spectra showed that a segment of the C-terminal region of alpha-synuclein was binding to glucocerebrosidase [3]. This segment was dominated by negatively charged amino acids and included three tyrosine amino acids, an amino acid with a bulky side chain with mixed hydrophobic and hydrophilic properties. The interaction apparently required this combination, negatively charged amino acids in combination with tyrosine, because other regions in alpha-synuclein with just negatively charged amino acids or just tyrosine did not interact with glucocerebrosidase.

When we wrote up and submitted a paper on our results, it was sent back with the criticism that we had not demonstrated that the interaction of glucocerebrosidase and alpha-synuclein was specific. Perhaps it could be that under acidic conditions, alpha-synuclein had a tendency to interact with other proteins in a similar manner? And how could we be certain the two proteins really interacted inside lysosomes? These were fair questions. Nearly all molecular biological techniques to detect protein interactions do so under artificial conditions, and many detected interactions end up being experimental artifacts.

So, we went back to work. Using NMR spectroscopy, I looked for an interaction between alpha-synuclein and the lysosomal protein saposin C under the same acidic conditions and saw no interaction. Using fluorescence spectroscopy, Yap looked for an interaction with another lysosomal enzyme, acid alpha-glucosidase, and saw nothing. Working with Arash Velayati in the Sidransky laboratory, using samples of human brain tissue, Yap demonstrated that alpha-synuclein and glucocerebrosidase could be coimmunoprecipitated under acidic conditions. Using fluorescence microscopy of cultured neuronal cells, Wendy Westbroek in the Sidransky laboratory showed that both alpha-synuclein and glucocerebrosidase could be seen inside the same lysosomes; that is, they were colocalized.

The critics were not completely silenced by our additional results, but the paper was finally accepted. Still, we had lingering questions regarding the significance of the interaction between alpha-synuclein and glucocerebrosidase. Was the interaction a normal aspect of their biological function, or was it a damaging interaction that promoted the development of Parkinson's disease? Or maybe the interaction really was a red herring and had no significant role inside the cell. Along these lines, I developed three competing hypotheses, each named for their strongest advocates.

First, there was the Wendy hypothesis, which proposed that the interaction was not biologically important. Perhaps once inside the lysosome, alpha-synuclein was degraded so quickly that it did not have time for a significant interaction with glucocerebrosidase. In this hypothesis, mutant glucocerebrosidase promoted Parkinson's disease by reducing lysosomal function, which reduced clearance of alpha-synuclein, raising its levels, helping to trigger the Parkinson's disease.

Next, there was the Jim hypothesis, named for myself, which proposed that the interaction was normal and important for the lysosomal clearance of alpha-synuclein. The specific interaction I had seen by NMR was typical of proteins known to interact, and it was hard for me to imagine such a specific interaction happening by accident. We already knew that alpha-synuclein bound more weakly to at least one disease-associated glucocerebrosidase mutation [3]; thus, the mutations might be interfering with the normal biological interaction, leading to less synuclein being degraded and causing its level to rise in neurons.

Finally, there was the Arash hypothesis, which proposed that the interaction was deleterious. Velayati and Yap had shown that the interaction with alpha-synuclein inhibited glucocerebrosidase when bound to vesicles [5]. Binding to vesicles is important, because the target of glucocerebrosidase, glucosylceramide, also known as glucocerebroside, is present only in vesicles, delivered to the lysosome via multivesicular bodies derived from cell membrane endosomes. Furthermore, it was known that lysosomal function was reduced in the brains of Parkinson's patients [6]. The Arash hypothesis proposed that the two experimental results were causally related, producing a vicious cycle: mutant glucocerebrosidase resulted in reduced lysosomal function, which reduced clearance of alpha-synuclein, causing its level in the cell to rise, further inhibiting glucocerebrosidase and reducing lysosomal function.

There was another hypothesis, the excess substrate hypothesis [7]. It suggested that mutant glucocerebrosidase led to an increase in its substrate glucosylceramide. Increased glucosylceramide in the cell could then promote oligomerization of alpha-synuclein, and the oligomers would then impede delivery of vesicles to lysosomes, reducing lysosomal function. For my NIH colleagues and me, the excess substrate hypothesis had potential problems. In carriers of *GBA1* mutations, most studies have observed neuronal glucosylceramide levels in the same range as that seen for controls (a curious result to which we will return later) [8,9]. However, in vitro experiments showing promotion of alpha-synuclein oligomerization [10] had glucosylceramide levels hundreds of times higher than actually observed in Gaucher patients, and tens of thousands of times higher than controls (and therefore, carriers with a single *GBA1* mutation) [11]. To be fair, physiological glucosylceramide levels were measured for brain tissue, and lysosomes constitute only a small fraction of the tissue volume. It is conceivable that within lysosomes, levels could be elevated without significantly affecting the overall glucosylceramide level for the entire cell. Despite these potential issues, the excess substrate hypothesis appears to have the most adherents among Parkinson's disease researchers.

My NIH colleague Ehud Goldin proposed a possible fix for the excess substrate hypothesis, what he dubbed the double-hit hypothesis. Perhaps in carriers of *GBA1* mutations, a second mutation might occur in the other, good copy of *GBA1* in a dividing cell: in particular, a white blood cell called a macrophage. In Gaucher disease patients, macrophages accumulate so much of the substrate that it changes the appearance of the cell, forming what are called Gaucher cells. In carriers in whom a second *GBA1* mutation occurs in a macrophage, this cell would become a Gaucher cell. Substrate levels in Gaucher cells should be high enough, in theory, to promote the alpha-synuclein oligomerization observed in experiments.

When the brain experiences trauma, white blood cells, including macrophages, can enter the damaged part of the brain [12]. If one or more of these cells became Gaucher cells owing to a second mutation, causing alpha-synuclein to aggregate in them, they could release aggregated alpha-synuclein into the brain, starting the cascade of events that ultimately leads to Parkinson's disease. I like this hypothesis for its outside-the-box thinking. However, playing devil's advocate, because in Gaucher patients all macrophages already have two mutant forms of *GBA1*, should not developing Parkinson's disease be inevitable? In actuality, only about 1 in 10 Gaucher disease patients develops Parkinson's disease [13]. However, mutations that cause more severe Gaucher disease are also correlated with a greater risk for developing Parkinson's disease in carriers [14]. Because Gaucher disease patients with the most severe forms of the disease typically do not reach the ages when Parkinson's disease manifests, the 1 in 10 result would apply only to those with milder mutations.

Back to my ulterior motive. In the Jim hypothesis, I claimed that the interaction seemed too specific to have occurred by chance. Perhaps using correlated mutation analysis, I could discover correlated pairs of amino acids in alpha-synuclein and glucocerebrosidase, in particular, ones that involved amino acids in the interacting region. If such pairs existed, this would support the idea that the interaction was not an accident and the proteins had coevolved. By the end of the government shutdown, I had already begun gathering and aligning the alpha-synuclein and glucocerebrosidase sequences from different species that I would need to perform the analysis.

After toying with the various approaches to correlated mutation analysis employed by research groups over the years, I decided to write my own program so I could tinker with and combine ideas from the most successful approaches. I ultimately settled on a type of correlation called mutual information combined with a statistical technique called Z-value analysis. These are explained in more detail in the section "Are we correlated?"

Are we related? – Correlated mutation analysis

There are many ways to measure how closely two data sets are related. The most straightforward type of correlation analysis is based on covariance, a mathematical measurement of how much two data sets vary together. If a variation occurs for a point in one data set, how often is that variation mirrored in the other data set? Covariance measures how much the frequencies of particular combinations data point pairs deviate from what one would expect simply by chance.

Let's say we have the amino acid sequences for two proteins for 150 species. Imagine for each protein, we load the sequences into a spreadsheet, with each row being the sequence for a particular species with one amino acid per cell. Each data set for the correlation analysis would consist of a column in the first spreadsheet. i.e., the list of amino acids that occur at a particular site in the first protein sequence. and a column in the other spreadsheet, corresponding to a particular site in the second protein. If the first protein is 100 amino acids long and the second is 200, there are 100 times 200 pairs of sequence positions: that is, 20,000 data sets of 150 amino acids pairs each.

To understand how to calculate how much the frequency of pairs of amino acids deviate from the expected result, consider the following example. Say amino acid *a* occurs in 30 species out of 150 at the first site and amino acid *b* occurs 15 times at the second site. The odds that we see the particular combination of *a* and *b* just by chance is the frequency at which *a* occurs, 20% (30 divided by 150), times the frequency at which *b* occurs, 10% (15 divided by 150), which equals 2% (20% of 10% is 2%). In this case, if the combination of *a* and *b* occurs in more than three species (2% of 150 is 3), it indicates that the pair occurs more often than predicted by chance. Covariance is just the sum of the observed pair frequencies minus the frequencies expected by chance. The more combinations showing higher than expected frequencies, the higher the covariance, meaning the pair of sites is more correlated.

The mutual information value for a particular pair of positions is similar to the covariance, except that it is formulated in a way that makes the result even larger if a greater variety of correlated amino acid pair combinations is seen for the two sites. Specifically, mutual information uses the frequency of a particular amino acid combination divided by the frequency one would expect from chance. If the number is greater than 1, this means the combination occurs more often than predicted. Then, the natural logarithm of the ratio is taken and multiplied by the frequency of the particular combination. This formulation, frequency times the natural log of the frequency ratio, is reminiscent of entropy and has a central role in information theory.

The rationale for using mutual information is that if multiple, different correlated mutations occur for a pair of sites, it is less likely that the correlation arose by accident. For example, consider two noninteracting amino acid sites in the protein complex, and in a particular species, just by chance, mutations have occurred at both of these sites. Let us say that the chance of this one pair of mutations happening is 10%. However, the chance for mutations in this same pair of amino acid positions happening twice by accident in two unrelated species is one-tenth the chance, 1%.

One of the challenges in correlated mutation analysis is correcting for indirect correlations, in which sequence site *X* appears to be correlated to *Y*, but only because they both happen to be correlated to an intervening sequence site *Z*. Another issue is called phylogenetic bias, in which mutations in two noninteracting sites occur by chance in the common ancestor of a family (or order, class, or phylum) of species, and the mutations are passed on to the daughter species, resulting in what looks like a *bona fide* correlation, but really is just a kind of error propagation. For both of these issues, one ends up with more pairs of sites that share the same pattern of correlated mutations across species. Z-value analysis looks at the patterns of correlated mutations for all of the pairs of sites and scores them based on their uniqueness. In statistics jargon, the Z-value is how many standard deviations the result varies from the mean.

For cases of indirect correlations, all of the pairs with the same pattern will have decreased Z-values. Patterns of correlated mutations that correspond to deep divisions in the evolutionary tree, between mammals and nonmammals, for instance, will also be more common and have reduced Z-values. The Z-value approach is not perfect; truly correlated sites have their values reduced along with the indirectly correlated sites to which they give rise, and many of the sites with correlated mutations mirroring divisions in the evolutionary tree could be genuinely interacting. The Z-value analysis helps identify the strongest correlations that are the least likely to be due to indirect correlation and phylogenetic bias.

Once I had written my mutual information Z-value analysis program, the next step would be to test it on proteins with known interactions, but I was impatient and went straight to trying it out on alpha-synuclein and glucocerebrosidase. Before I spent a bunch of time testing my program, I wanted to see whether there would be interesting results in the end. The results were actually startling. The top correlated pair of sequence sites included amino acid 53 of alpha-synuclein, the site of the first discovered familial Parkinson's disease mutation. Even more startling, the mutation seen across species was the same as that seen in the disease, from a threonine to an alanine amino acid. What did this mean? Well, at least I knew my end result would be interesting, so I took a step back and went to work vetting my program.

Earlier programs had been tested using mostly bacterial proteins, because so many more bacterial genomes had been sequenced than those of other organisms. Bacterial genomes are short, their genes are simple, and thousands of bacterial species have had their genomes sequenced. However, bacteria do not have alpha-synuclein or glucocerebrosidase, and I wanted to test my program using the same species I would be employing for my two subject proteins. Glucocerebrosidase occurs in all animals, but the synucleins are found only in vertebrates, so I wanted a vertebrate test case. I chose the vertebrate-specific protein hemoglobin.

Hemoglobin consists of four chains of amino acids, two alpha chains and two beta chains. To test my program, I would see how well it predicted correlated pairs at the interfaces between the chains. Because correlated mutation analysis is also used for protein structure prediction, I would also test how well my program predicted correlated pairs within each chain. I was not the first to choose hemoglobin as a test case for correlated mutation analysis, although I did not know it at the time [15]. Its variations across vertebrate species have been studied for decades [16]. Hemoglobin protein sequences are known for over 300 species, perhaps the most for any vertebrate-specific protein. In contrast, I had alpha-synuclein sequences from only 72 species with which to work. One of the things I needed to test was whether the results were still meaningful with so few sequences.

Testing of my mutual information/Z-value analysis program went well when using all 300-plus hemoglobin sequences, especially for predicting intrachain contacts; five of the top 10 Z-value pairs corresponded to amino acids in contact in the protein structure. This result was comparable to other published methods at the time. The number of correctly predicted interchain contacts was not as high, only 2 of 10, but this was expected. The reason was that for the interchain prediction, the total number of possible pairs was much greater, but at the same time, the fraction of pairs actually in contact was smaller.[1]

When I reduced the number of species to match the 72 corresponding to those with known alpha-synuclein sequences, the hemoglobin results were not as good, but not completely bad either. The number of correct intrachain contacts fell from five to three in the top 10 Z-values. The number of correct interchain contacts fell also, from two to one. This implied that on average, only one of the top 10 interchain mutual information/Z-values was a *bona fide* correlated mutation pair. Not an ideal situation, but still, one was

[1] To be explicit regarding why predicting interchain contacting amino acids is expected to be more difficult than for intrachain, consider the numbers. For the hemoglobin alpha-chain there are 537 pairs of amino acids in contact, out of 8515 pairs altogether, whereas the number of contacting amino acids between the alpha- and beta-chains is only 149, but the number of possible pairs is 17,685. The chance of predicting an amino acid pair in contact just by chance for the intrachain case is 6.3% (537 / 8515), whereas the corresponding chance for the interchain case is 0.84% (149 / 17,865). The bottom line is that the chance of predicting correct interchain contacts by chance, 0.84%, is much lower than the actual result, 20%, so not so bad after all.

better than nothing. Good enough for me! There was a significant chance my startling result for alpha-synuclein coevolving with glucocerebrosidase was correct. I wrote up my results and sent the manuscript for publication in a peer-reviewed journal.

Rejection. My peers did not agree with me, and the manuscript would not be published. The issue was not with the result, but with the method. There were already correlated mutation analyses that had been tested and proven successful. For my paper to be considered for publication, I would have to include results from an already vetted analysis program.

I could try submitting my paper to another journal, hoping for more sympathetic peer reviews, but I was likely to run into the same credibility issue. I resigned myself to finding a program with established credentials. I settled on an analysis called direct coupling, developed by a US and Italian team that included correlated mutation analysis pioneer Chris Sander and José N. Onuchic, a colleague I knew from my work decades earlier on electron transfer in proteins [17]. The analysis was also based on mutual information values, but instead of Z-values, it relied on much fancier mathematics to disentangle direct interactions from indirect ones.

Before I used the direct coupling analysis on alpha-synuclein and glucocerebrosidase, I wanted to make sure it worked as well as or better than my Z-value analysis. Using the 300-plus sequences of hemoglobin, the direct coupling analysis did perform better for intrachain contacts, correctly predicting 6 out of 10 contacting amino acids. However, for interchain contacts, it did worse, getting only 1 out of 10 correct. When the number of species was reduced to 72, its performance for interchain contacts went from poor to abysmal, with 0 predicted pairs in the top 10 and only 2 in the top 100, compared with 1 and 5 in the top 10 and 100 for the Z-value analysis. Sweet!

Why was I happy? Although peer reviewers are leery of publishing results based on untested methods, they love papers that compare new methods with old ones, especially when the newer method shows some advantage. I rewrote my paper to include the comparison of the Z-value and direct coupling analyses of hemoglobin. However, for alpha-synuclein and glucocerebrosidase, I focused on the Z-value analysis results because of the method's superior performance. Why did the Z-value analysis perform better for interchain contacts? I suspect it might have something to do with the simpler mathematics of the Z-value analysis making it more robust when the species number was reduced, but to be honest, I do not really know, it could have just been luck.

There was something else especially interesting about the top correlated pair of amino acids in alpha-synuclein and glucocerebrosidase. The change from a threonine amino acid to alanine in position 53 of alpha-synuclein was seen for apes and Old World monkeys plus squirrel and capuchin monkeys of the New World. I was not the first to notice the connection between this mutation and Old World monkeys; Bruce A. Hamilton at University of California, San Diego, had published a paper on the connection in 2004 [18]. For glucocerebrosidase, the amino acid change, in position 115, was seen only in Old World monkeys and apes and a single frog species, not in any of the other vertebrates. Could the correlated pair of mutations have had a part of the evolution of the Old World monkey and ape brain? Unfortunately, there were several issues with this correlated pair that suggested caution before jumping to any conclusions.

The first issue has to do with phylogenetic bias. The mutation pattern seen for the top correlated pair could easily have happened by chance. All that would be needed is for the common ancestor of Old World monkeys and apes to have two chance mutations: one in alpha-synuclein and one in glucocerebrosidase. These chance mutations would be passed on to all daughter species. Although the Z-value analysis can compensate for more common patterns of chance correlations corresponding to deep divisions in the evolutionary tree, the patterns corresponding to more recent divisions will be less common, so additional criteria for assessment are needed.

To be confident that a correlation is genuine, the ideal pair of mutations should pass the phylogenetic bias test, which requires that the pair of mutated positions arise independently at least twice: that is, in at least two unrelated ancestral species. The top correlated pair for alpha-synuclein and glucocerebrosidase failed the test, because the pair of mutations apparently occurred just once, in the ancestor of Old World monkeys and apes. What would be required is for a correlated pair of mutations at the same amino acid sites to have occurred in some unrelated branch of the vertebrate evolutionary tree, as in birds, for instance.

As an aside, the alpha-synuclein mutation at position 53 is also seen in the squirrel and capuchin monkeys, both New World species, and most likely it arose independently of the one that occurred in the ancestor of Old World monkeys and apes. This is because all other New World monkeys tested so far lack the mutation. However, because squirrel and capuchin monkeys do not have the corresponding mutation in glucocerebrosidase, this does not prove or disprove whether the top correlated alpha-synuclein glucocerebrosidase pair is genuine. The only other vertebrates to have a different amino acid in position 53 of alpha-synuclein are sharks, for which four species have had the genomes sequenced. The full glucocerebrosidase sequence is known only for three of the four species, and they do not have a corresponding mutation at position 115 of glucocerebrosidase. Another surprise is that mandrills, which are Old World apes, appear to have reverted back to the ancestral amino acid at position 115, even though baboons and mangabeys, their close relatives, have not.

An additional issue with the top correlated pair of mutations is that the alpha-synuclein amino acid mutation at position 53 was far from the region I had seen to interact with glucocerebrosidase by NMR spectroscopy, which included amino acids in positions 118—137. In the presence of lipid vesicles, both proteins bind to the vesicles, and using fluorescence spectroscopy, the region near alpha-synuclein amino acid 53 was observed to interact with glucocerebrosidase, but that did not help my case [5]. In the Jim hypothesis, I claimed that the interaction with glucocerebrosidase seen by NMR was specific to alpha-synuclein amino acids 118—137. What I needed to support the hypothesis was a correlated pair that had the alpha-synuclein amino acid in this region.

Finally, taking off my bioinformatics hat and putting on my molecular modeler hat, there was something fishy about the top correlated pair. The amino acid in glucocerebrosidase that changed, in position 115, is located on the side of the protein opposite the side that binds to lipid vesicles. The first 100 amino acids of alpha-synuclein bind to lipid vesicles, so if amino acid 53 of alpha-synuclein is at the lipid vesicle surface, and glucocerebrosidase amino acid 115 lies far from the vesicle surface, how could the two amino acids be in contact? The evidence against the top correlated pair being a real correlation (i.e., a false positive) was piling up.

I checked the other top correlated pairs in the Z-value analysis. Curiously, another two top correlated pairs involved other Parkinson's disease—causing mutations. In fact, this pattern has been seen in mutation analyses of other disease-related proteins, in which the amino acid change in evolution mirrors the human disease mutation [19]. Four more pairs in the top 10 failed the phylogenetic bias test, but the number two ranked correlated pair passed the test with flying colors. The same pair of mutations had arisen independently in bats, amphibians, and an elephant shrew. Even better, the alpha-synuclein mutation was at position 124, within the region where I had seen specific interaction with glucocerebrosidase by NMR spectroscopy. But wait, there's more! The corresponding glucocerebrosidase amino acid was in a position close to a mutation known to interfere with binding to alpha-synuclein. My ulterior motive had paid off; the two proteins appeared to have coevolved, and the interaction involved the same region I had observed by NMR spectroscopy.

Additional evidence supporting coevolution between alpha-synuclein and glucocerebrosidase is detailed in my paper "Did α-Synuclein and Glucocerebrosidase Coevolve? Implications for Parkinson's Disease," published in the online journal *PLoS One*. The editor for the paper was Philipp J. Kahle, already mentioned in Chapter 4, one of first researchers of synuclein overexpression in transgenic mice. He suggested I add analyses of beta-synuclein with alpha-synuclein and glucocerebrosidase. I did, and the results were intriguing. Beta-synuclein showed somewhat weaker correlation with glucocerebrosidase, but its correlation with alpha-synuclein was strong, and the strongest correlations involved alpha-synuclein amino acid 53, the same amino acid as in the top alpha-synuclein glucocerebrosidase pair.

Alpha- and beta-synuclein interact, at least in vitro. Beta-synuclein is able to block alpha-synuclein fibril formation in test tubes, and when both synucleins are overexpressed in transgenic mice, they do not develop the pathology seen when overexpressing alpha-synuclein alone [20]. The beta-synuclein mutations of the top correlated pairs arose in the ancestor of all monkeys and apes. This suggests the following chain of events: the mutation of alpha-synuclein at amino acid 53 in Old World monkeys and apes arose in response to the mutations in beta-synuclein, and the mutation in glucocerebrosidase arose in response to the mutation in alpha-synuclein. The same alpha-synuclein mutation in squirrel and capuchin monkeys might also have arisen in response to the mutations in beta-synuclein.

Although this sequence of mutation events could just be chance occurrences, the top alpha beta-synuclein correlated pairs might actually be genuine, because the same alpha-synuclein mutation appears to have arisen independently twice, first in Old World monkeys and then in squirrel and capuchin monkeys. Also, two amino acids need not necessarily be in direct contact with each other to be interacting. Two amino acids might influence each other through intervening amino acids, or their interaction might be mediated by a third interacting protein partner. The bottom line is that the implied interaction between glucocerebrosidase and alpha-synuclein amino acid 53 might be real, but until unrelated species are found with mutations of the same pair of amino acids, correlated mutation analysis cannot prove it.

Things were looking good for my hypothesis that there was a beneficial, specific interaction between alpha-synuclein and glucocerebrosidase.

The correlated mutation analysis suggested that alpha-synuclein and glucocerebrosidase had coevolved. The top correlated pair involved a mutation that mirrored a familial Parkinson's disease mutation, although admittedly, it could have arisen by chance. On the other hand, the second top correlated pair was far less likely to be due to chance, and it included an amino acid in the region of alpha-synuclein seen to interact with glucocerebrosidase by NMR spectroscopy, bolstering the evidence that the interaction was specific.

My satisfaction was cut short when my collaborator Ellen Sidransky informed me of another amino acid change that forced me to put on my pesky molecular modeler hat once again. The amino acid was in position 326 of glucocerebrosidase, changing a negative amino acid, glutamate, to a positive one, lysine. This change was considered a polymorphism because it was not associated with Gaucher disease. However, several studies, including the genome-wide association studies mentioned in Chapter 2, showed that the polymorphism is associated with increased risk for developing Parkinson's disease.

Why was I so bothered by this polymorphism? The amino acid happened to be right in the middle of where I predicted the C-terminal region of alpha-synuclein would interact with glucocerebrosidase. Recall that this region of alpha-synuclein has many negatively charged amino acids, so the replacement of a negative with a positive amino acid on glucocerebrosidase should increase the strength of electrostatic attraction between the two proteins. If the interaction were beneficial, as my hypothesis proposed, how could strengthening it have the negative effect of increasing Parkinson's disease risk? Maybe my hypothesis was wrong.

It dawned on me that there was a flaw in the logic of the Jim hypothesis. Because of the connection between mutations in glucocerebrosidase and increased risk for Parkinson's disease, I had assumed that the specific interaction between them, which apparently has been maintained throughout vertebrate evolution, was beneficial because it had some role in preventing the pathology associated with Parkinson's disease. The problem is that only humans develop Parkinson's disease. Therefore, alpha-synuclein clearance and aggregation probably have not been a significant issue in the evolution of vertebrate species in general, and thus one should not expect evolutionary selection favoring the interaction if its only role were to promote the clearance of alpha-synuclein from neurons.

Still, alpha-synuclein and glucocerebrosidase appeared to have a specific interaction. The correlated mutation analysis supported the notion that they had coevolved, which implied that their interaction must have benefited vertebrates in some way. Yap and Velayati showed that alpha-synuclein inhibits glucocerebrosidase activity [5]. Glucosylceramide is the main building block for the synthesis of more complex glycolipids called gangliosides [21]. In synuclein triple knockout mice, ganglioside levels are lower than in controls [22], although as far as I know, no study has looked for this in mice with only alpha-synuclein knocked out. Could this inhibition of glucosylceramide degradation somehow be the evolutionarily beneficial interaction? The outer leaflet of the cell membranes of neurons are particularly enriched in gangliosides. The abundance and differing profiles of particular gangliosides are likely part of the suite of biochemical markers that identifies the components of neurons, such as presynaptic boutons, as such to interacting neurons and glial cells.

Lysosomal enzymatic activity, including that of glucocerebrosidase, decreases with age, especially for those with Parkinson's disease [6], which is bad news because the lysosome is one of the major sites for degradation of alpha-synuclein. For mutant *GBA1* carriers, the situation is even worse; activity of glucocerebrosidase is as much as 50% lower than in age-matched noncarriers. Despite this, neuronal levels of glucosylceramide in carriers lie in the normal range [8,9]. Where does the expected excess of glucosylceramide go? The blood of type I Gaucher patients has elevated levels of glucosylceramide and a particular ganglioside, GM3; even more intriguingly, the levels of these lipids also appear to be elevated in the blood of Parkinson's patients in general [23,24]. This suggests that not only does some of the excess glucosylceramide become converted into gangliosides, excess levels of both might somehow be transferred from neurons to the bloodstream in *GBA1* mutation carriers and in noncarrier Parkinson's patients alike.

Glucocerebrosidase breaks down glucosylceramide into glucose and ceramide, and many studies have shown that ceramide levels are reduced in brain tissue of Parkinson's disease patients, including those without *GBA1* mutations [25,26]. Ceramide is an important messenger molecule in a wide array of cell types, and increased levels are involved in inflammation and apoptosis [27]. More important for Parkinson's disease, ceramide contributes to the maturation of lysosomal proteases, called cathepsins, that degrade synuclein, so it makes sense that reduced ceremide levels inside lysosomes could lead to a reduction in synuclein degradation [28,29].

To recap, this is the vicious cycle of the Arash hypothesis: reduced glucocerebrosidase activity results in reduced ceramide, which leads to lower lysosomal protease activity, reducing alpha-synuclein degradation, resulting in a higher level of alpha-synuclein, which then further inhibits glucocerebrosidase activity.

A study from the laboratories of Virginia Lee and John Trojanowski at the University of Pennsylvania School of Medicine presented evidence that the Arash hypothesis might need to be modified. They found that only after being exposed to aggregated alpha-synuclein does inhibition of glucocerebrosidase cause the level of alpha-synuclein to rise in neurons [30]. Their results show that aggregated alpha-synuclein inhibits glucocerebrosidase even more than the monomer, so perhaps this enhanced level of inhibition is what triggers the vicious cycle and propagation of Lewy pathology. In other words, their work suggests that reduced glucocerebrosidase activity does not initiate Lewy pathology, but enhances its spread.

And what of the Jim hypothesis? I have not completely abandoned it; I still think the proteins coevolved, perhaps to help achieve the optimum balance of gangliosides at presynaptic boutons, but I no longer believe their interaction is beneficial with regard to Parkinson's disease.

There is also a variant of the Wendy hypothesis that I should mention; that is, there is another possible explanation for why mutant *GBA1* carriers have increased risk for Parkinson's disease in which no direct interaction between alpha-synuclein and glucocerebrosidase is required. In 1995, in the laboratory of Anthony H. Futerman at the Weizmann Institute of Science, Israel, researchers discovered that cultured rat neurons treated with an inhibitor of glucocerebrosidase developed longer, more branched axons [31]. Perhaps something similar happens in the neurons of *GBA1* mutation carriers, with lower glucocerebrosidase activity leading to longer, more branched axons with the potential to form more synapses. With more synapses per neuron, the total amount of alpha-synuclein per neuron would have to go up to maintain the same concentrations within the presynaptic boutons. As part of the normal aging process, more and more synapses are lost with time. If the neuron fails to slow alpha-synuclein production, as evidence suggests [32], alpha-synuclein levels in the remaining synapses and other parts of the neuron would rise, increasing its propensity to aggregate.

At this time, it is not known whether reduced glucocerebrosidase activity leads to higher synapse density in actual brainsof humans or any other animal, so the variant Wendy hypothesis should be considered highly speculative. We shall muse on these and other neuronal architecture issues in subsequent chapters. However, it is known that synuclein levels rise in neurons with aging, likely owing, at least in part, to reduced lysosomal activity. It could be that the Arash and Wendy hypotheses even work synergistically in promoting Parkinson's disease.

Let us return to the putative beneficial interaction between alpha-synuclein and glucocerebrosidase. Remember how alpha-synuclein knockout mice developed movement and learning deficits as they aged [33]? I suspect that gangliosides might be involved. In the developing mouse brain, sufficient levels of gangliosides are apparently present in both knockout and control mice, because no gross differences are seen in the brain, and no differences are seen in movement and learning for young mice. I wonder whether once the mouse brain has developed, the primary pathways of ganglioside synthesis switch to being more reliant on membrane recycling via endocytosis and lysosomal processing. Without alpha-synuclein to reduce the amount of glucosylceramide that becomes degraded, levels of gangliosides in presynaptic boutons could gradually fall. By adulthood, the levels might have fallen so much in the knockouts that synapses start to lose full functionality or lose the ability to remodel themselves. That is, the ability to break old synapses or form new ones might erode, and thus the mice have difficulty learning.

I propose that the reduced activity of glucocerebrosidase mutants, combined with the inhibition of glucocerebrosidase by alpha-synuclein, cause a vicious cycle, the Arash hypothesis, which leads to higher alpha-synuclein levels, and which increases the chance of developing Parkinson's disease. It is also possible that reduced glucocerebrosidase activity leads to higher levels of gangliosides that promote higher synapse density, the Wendy hypothesis, also leading to higher levels of alpha-synuclein as the brain ages and synapses are lost. Finally, I propose that in vertebrates, the interaction between alpha-synuclein and glucocerebrosidase evolved to help maintain healthy levels of neuronal gangliosides in adults, the Jim hypothesis. All of these hypotheses are highly speculative, and I invite readers to regard them with due skepticism. Obviously, more experimental studies are needed, especially with regard to ceramide and ganglioside levels and their impact on lysosomal activity and synapse density.

Before I end this chapter, I have an update on correlated mutation analysis. While I was finishing my analysis of alpha-synuclein–glucocerebrosidase coevolution, two papers with new correlated mutation analyses came out: one from the laboratory of David Baker, a leading researcher in protein structure prediction, and another from the laboratory of Debora S. Marks, one of the creators of the earlier direct coupling analysis that I had used for comparison in my study. In the months that followed, both laboratories created websites in which one could submit protein sequences for analysis. I was eager to check how well they performed with alpha-synuclein and glucocerebrosidase, but both new methods had strict requirements with regard to the number of sequences across species, and my proteins had too few sequences available.

Still, I wanted to compare the new methods with my mutual information Z-value analysis, so I chose a different set of proteins to compare the methods. The proteins I chose were those of the respiratory chain complexes, which are similar in bacteria and eukaryotes (organisms whose cells have nuclei). Because thousands of bacterial genomes have been sequenced, I was able to get more than enough sequences of the respiratory chain proteins to satisfy the new methods. Better yet, the websites generated the sequence alignments automatically and provided them as well, so I could use the same alignments with my own program.

Using my program, an average of 30% of the top 10 ranked correlated amino acid pairs were actually in contact in respiratory chain protein complexes. This was better than my results with hemoglobin, 20%, but it was unsurprising because the respiratory chain sequence alignments typically consisted of several thousand sequences, compared with the roughly 300 I had for hemoglobin. The two new methods performed much better, however, both getting 60%–70% true contacts in their top 10 correlated pairs. What was so different about the new methods? They both were still based on direct coupling analysis, though with improvements, but they had replaced mutual information with a simpler covariance analysis.

In designing the best correlated mutation analysis, earlier researchers, including myself, had made a faulty assumption: to wit, that for the purpose of the analysis, interacting amino acids could be considered to mutate at the same rate as noninteracting ones. Recall that one of the supposed strengths of using mutual information was that it gives stronger correlation to pairs with multiple correlated mutations, and because multiple correlated mutations were less likely to occur by chance, pairs with higher mutual information values should have a greater chance of having a real interaction. Unfortunately, we had neglected something that studies had shown ever since the first coevolution studies: that amino acids at interaction interfaces mutate at slower rates than noninterface amino acids. It turns out that this slower mutation rate more than negates any benefit of using the mutual information approach. You win some, you lose some.

References

[1] Shindyalov IN, Kolchanov NA, Sander C. Can three-dimensional contacts in protein structures be predicted by analysis of correlated mutations? Protein Engineering March 1994;7(3):349–58.

[2] Göbel U, et al. Correlated mutations and residue contacts in proteins. Proteins April 1994;18(4):309–17.

[3] Yap TL, et al. Alpha-synuclein interacts with Glucocerebrosidase providing a molecular link between Parkinson and Gaucher diseases. Journal of Biological Chemistry August 12, 2011;286(32):28080–8.

[4] Sidransky E, et al. Multicenter analysis of glucocerebrosidase mutations in Parkinson's disease. New England Journal of Medicine October 22, 2009;361(17):1651–61.

[5] Yap TL, et al. Membrane-bound α-synuclein interacts with glucocerebrosidase and inhibits enzyme activity. Molecular Genetics and Metabolism January 2013;108(1):56–64.

[6] Dehay B, et al. Lysosomal impairment in Parkinson's disease. Movement Disorders June 2013;28(6):725–32.

[7] Mazzulli JR, et al. Gaucher disease glucocerebrosidase and α-synuclein form a bidirectional pathogenic loop in synucleinopathies. Cell July 8, 2011;146(1):37–52.

[8] Gegg ME, et al. No evidence for substrate accumulation in Parkinson brains with GBA mutations. Movement Disorders July 14, 2015;30(8):1085–9.

[9] Boutin M, et al. Tandem mass spectrometry multiplex analysis of glucosylceramide and galactosylceramide isoforms in brain tissues at different stages of Parkinson disease. Analytical Chemistry February 2, 2016;88(3):1856–63.

[10] Zunke F, et al. Reversible conformational conversion of α-synuclein into toxic assemblies by glucosylceramide. Neuron January 3, 2018;97(1):92–107.

[11] Nilsson O, Svennerholm L. Accumulation of glucosylceramide and glucosylsphingosine (psychosine) in cerebrum and cerebellum in infantile and juvenile Gaucher disease. Journal of Neurochemistry September 1982;39(3):709–18.

[12] Giulian D, et al. The role of mononuclear phagocytes in wound healing after traumatic injury to adult mammalian brain. Journal of Neuroscience December 1989;9(12):4416–29.

[13] Rosenbloom B, et al. The incidence of Parkinsonism in patients with type 1 Gaucher disease: data from the ICGG Gaucher registry. Blood Cells, Molecules and Diseases January 15, 2011;46(1):95–102.

[14] Thaler A, et al. Parkinson's disease phenotype is influenced by the severity of the mutations in the GBA gene. Parkinsonism and Related Disorders October 2018;55:45–9.

[15] Lee BC, Park K, Kim D. Analysis of the residue–residue coevolution network and the functionally important residues in proteins. Proteins: Structure, Function, and Bioinformatics August 15, 2008;72(3):863–72.

[16] Tufts DM, et al. Epistasis constrains mutational pathways of hemoglobin adaptation in high-altitude pikas. Molecular Biology and Evolution February 2015;32(2):287–98.

[17] Morcos F, et al. Direct-coupling analysis of residue coevolution captures native contacts across many protein families. Proceedings of the National Academy of Sciences of the United States of America December 6, 2011;108(49):E1293–301.

[18] Hamilton BA. alpha-Synuclein A53T substitution associated with Parkinson disease also marks the divergence of Old World and New World primates. Genomics April 2004;83(4):739–42.

[19] Kowarsch A, et al. Correlated mutations: a hallmark of phenotypic amino acid substitutions. PLoS Computational Biology September 16, 2010;6(9):e1000923.
[20] Hashimoto M, et al. beta-Synuclein inhibits alpha-synuclein aggregation: a possible role as an anti-parkinsonian factor. Neuron October 25, 2001;32(2):213–23.
[21] Yu RK, et al. Structures, biosynthesis, and functions of gangliosides–an overview. Journal of Oleo Science 2011;60(10):537–44.
[22] Guschina I, et al. Triple-knockout, synuclein-free mice display compromised lipid pattern. Research Square 2020;Vols. https://doi.org/10.21203/rs.3.rs-37981/v1.
[23] Mielke MM, et al. Plasma ceramide and glucosylceramide metabolism is altered in sporadic Parkinson's disease and associated with cognitive impairment: a pilot study. PLoS One September 2013;8(9):e73094.
[24] Chan RB, et al. Elevated GM3 plasma concentration in idiopathic Parkinson's disease: a lipidomic analysis. PLoS One February 2017;12(2):e0172348.
[25] Murphy KE, Halliday GM. Glucocerebrosidase deficits in sporadic Parkinson disease. Autophagy July 2014;10(7):1350–1.
[26] Wang P, et al. Blood plasma of patients with Parkinson's disease increases alpha-synuclein aggregation and neurotoxicity. Parkinson's Disease 2016:7596482.
[27] Hannun YA, Luberto C. Ceramide in the eukaryotic stress response. Trends in Cell Biology 2000;10(2):73–80.
[28] Heinrich M, et al. Cathepsin D targeted by acid sphingomyelinase-derived ceramide. EMBO Journal October 1, 1999;18(19):5252–63.
[29] Liu F, et al. Ceramide activates lysosomal cathepsin B and cathepsin D to attenuate autophagy and induces ER stress to suppress myeloid-derived suppressor cells. Oncotarget December 20, 2016;7(51):83907–25.
[30] Henderson MX, et al. Glucocerebrosidase activity modulates neuronal susceptibility to pathological α-synuclein insult. Neuron December 2019:004.
[31] Schwarz A, et al. A regulatory role for sphingolipids in neuronal growth. Inhibition of sphingolipid synthesis and degradation have opposite effects on axonal branching. Journal of Biological Chemistry May 5, 1995;270(18):10990–8.
[32] Chiba-Falek O, Lopez GJ, Nussbaum RL. Levels of alpha-synuclein mRNA in sporadic Parkinson disease patients. Movement Disorders October 2006;21(10):1703–8.
[33] Kokhan VS, Afanasyeva MA, Van'kin GI. α-Synuclein knockout mice have cognitive impairments. Behavioural Brain Research May 16, 2012;231(1):226–30.

CHAPTER 10

Monkey Brains – Evolution of the human brain and the vulnerability of the substantia nigra

Why do only humans get Parkinson's disease? It could be just because we live so long, longer than any other primates, but there are plenty of non-primate vertebrates that live as long or longer (elephants, whales, parrots, tortoises, sharks [and probably coelacanths!], to name a few), and nothing like Parkinson's disease has been observed among them. Nonhuman primates can show motor impairments as they reach old age, correlated with a decline in function in dopaminergic neurons of the substantia nigra, but this is not accompanied by the Lewy body and neurite pathology that defines Parkinson's disease [1]. In my Web searches researching for this

Synuclein and the Coelacanth
ISBN 978-0-323-85707-9
https://doi.org/10.1016/B978-0-323-85707-9.00016-2

book, I came across one anecdotal report of a rhesus monkey named Sophie at a Boston University facility that apparently spontaneously developed parkinsonian symptoms, but apparently her brain was not autopsied, or perhaps the autopsy was performed, but nothing particularly noteworthy was found.

Chimpanzees and bonobos are our closest living relatives, and they can have life spans that can extend 50 years or more, with chimps living a bit longer on average. Many elderly chimp brains have been examined, and although diffuse amyloid plaques have been observed reminiscent of early stages of Alzheimer's disease [2], no traces of Lewy pathology have been seen. In humans, Lewy pathology can start years or even decades before the movement-related symptoms of Parkinson's disease arise, so in theory, 50 years should be long enough for the pathology to develop in chimpanzees. On the other hand, Parkinson's disease is less common than Alzheimer's, so many more elderly chimp brains will have to be examined before we can be certain that Lewy pathology never occurs in chimps.

If it turns out to be true that chimps never, or very rarely, develop Lewy pathology, what differences between chimp and human brains might explain this? The most obvious difference is brain size: human brains are 3.5 times larger on average. As described in *The Human Advantage: How Our Brains Became Remarkable* by Suzana Herculano-Houzel of Vanderbilt University, Tennessee, the human brain is basically a scaled-up version of the primate brain. A few areas of the human cortex are slightly larger than one might expect, thought to be involved in speech, planning, and perhaps manual dexterity. Because they are 3.5 times larger in size, one might also expect human brains to have 3.5 times more neurons than chimp brains, but actually humans have only 3.1 times as many. Part of the reason is that the axons of many neurons, so-called projection neurons, have to reach farther in the enlarged human brain to form connections between the various brain regions, and in doing so, their axons take up more space. These long axons, often surrounded by the white, lipid-containing sheaths of myelin, make up the white matter of the brain.

Another reason for the mismatch between brain size and neuron number ratios is that humans might have more synapses per neuron than chimps. The gray matter of the neocortex, or cerebrum, is roughly 50% thicker in humans than in chimps, and this region contains a large fraction of the synapses in the brain. This increase mirrors the increase in brain size, but because the number of neurons is less than that predicted solely based on the increased brain size, this makes the neuron density a bit lower, too. Thus, it is plausible that the number of synapses per neuron increases a little, assuming synapse density, the number of synapses per unit volume, does not change. Pinning down values for the number of synapses per neuron must be experimentally challenging, however, because I cannot find any values for chimpanzees in the literature. Also, any increase in the synapses per neuron might not be uniform throughout the brain, so even if neocortical neurons do have more synapses, it is not necessarily true for other regions, such as the substantia nigra. Still, if human brain neurons really do have more synapses, this could have important implications regarding the Lewy pathology of Parkinson's disease.

Human and chimp brains also differ in their neurochemistry. The neurons in chimp brains express significantly less messenger RNA for proteins involved with energy metabolism, which suggests that they use energy at a slower rate [3]. In contrast to this, studies of neuron metabolism by Herculano-Houzel in other mammals, including primates, suggest a relatively constant metabolic rate per neuron across species. In any case, the much greater number of neurons makes human brains the gas-guzzlers of the neural world, using roughly 25% of the total calories we eat, even though the brain makes up only 2% of total body weight. The brains of chimps use only around 7%—10% of their total calorie intake, and likewise for the other great apes, gorillas and orangutans [4].

Curiously, the brains of the smallest primates, squirrel monkeys and mouse lemurs, for instance, appear to use percentages of their caloric intake comparable to humans [5]. One might say the gas-guzzler brains of humans are typical of primates, and the brains of other great apes are the exceptions. The switch to a diet rich in meat, along with the discovery of cooking with fire, have been hypothesized to have allowed our human ancestors to evolve, or perhaps reevolve, our large, energy-hungry brains, starting about 2 million years ago [6].

A more intriguing difference with regard to Parkinson's disease is that human brains appear to have more dopamine in part of the striatum. In studies of monkeys, increased dopamine is linked to more social interactions and cooperative behavior. The substantia nigra pars compacta supplies the striatum with dopamine via its axonal synapses, primarily to the dorsal, top part of the striatum. However, the region supposedly involved in social behavior is the ventral, bottom part of the striatum. Loss of dopamine-releasing synapses in the dorsal striatum leads to the movement-related symptoms of Parkinson's disease. Loss is also often seen in the ventral portion of the striatum as well, although to a lesser degree and later in the disease [7]. Although asocial behavior is not a prominent symptom of Parkinson's disease, some studies suggest a reduction in social awareness, in particular theory of mind, the ability to put yourself in someone else's shoes. Perhaps the loss of facial expression as well, which is so important for social interactions, and the apathy typical of the disease, might be related to pathology in the ventral portions of the striatum, although this is speculation on my part.

The striatal dopamine study was published in 2018 by a team headed by Mary Ann Raghanti of Kent State University, Ohio. The part of the striatum where they measured higher dopamine in humans was in the medial caudate, which lies between the ventral and dorsal parts of the striatum (for some reason, they did not look at the ventral part of the striatum). They also looked at two parts of the dorsal striatum, the dorsal caudate and the putamen. In these regions, humans had dopamine levels comparable to and slightly lower on average than chimps, but higher than any other primates. Looking over their experimental details, I noticed that the human brain samples had been screened for senile plaques and neurofibrillary tangles, but not for Lewy pathology. I cannot help but wonder whether some of their samples were already showing loss of dopamine synapses, because two of the seven human brains came from men in their fifties. Hopefully, future studies on dopamine levels in humans and chimps will help reinforce their results.

Based on what I have presented so far, there is no obvious difference between human and chimp neurochemistry that could explain why only humans develop Parkinson's disease. Both humans and chimps appear to have similar energy use per neuron and similar levels of dopamine in their dorsal striatum regions. However, there is one more piece of evidence, known for decades, that proves human substantia nigra neurons really are different from those of chimps. The difference is the pigment that makes the substantia nigra dark, neuromelanin. Human substantia nigra neurons are packed with more of this pigment than any other animal. Chimps have less, monkeys have even less than chimps, and in nonprimate mammals, little or no neuromelanin is observed [8].

Scientists have long puzzled over neuromelanin. Its chemical structure is similar to the pigment melanin, which is responsible for hair, skin, and eye color, but what function could it have in the brain? It is enriched in metals, and some have posited that it might have a protective role, sopping up toxic metal ions [9]. However, it could also be that neuromelanin is simply a waste by-product of neurotransmitter metabolism, which the neuron packs into granules to keep out of mischief. In fact, neuromelanin in the substantia nigra is made from dopamine and its metabolites, although in other brain regions, such as the locus coeruleus, the pigment arises from the neurotransmitter noradrenaline. Noradrenaline and dopamine are chemically related, both belonging to a group of neurotransmitters called the catecholamines.

Recall that dopamine is especially vulnerable to oxidation by reactive oxygen species, such as those arising as a by-product of mitochondrial energy production, so it might be beneficial to sequester the reactive dopamine products into granules [10]. Thus, it could be that neurons have evolved cellular processes to facilitate this. Why is more neuromelanin made in human versus chimp substantia nigra neurons? Because their dopamine levels appear comparable in the dorsal striatum, the implication is that the human neurons might have more reactive oxygen species, which in turn implies greater mitochondrial activity.

As mentioned, the energy used per neuron, at least on average, appears to be relatively constant across species. However, there is good reason to believe this does not apply to all classes of neurons. Nature has evolved two kinds of projection neurons in vertebrates, myelinated and nonmyelinated. Myelinated neurons have myelin sheaths surrounding the length of the axon, although some neurons are only partially myelinated. Jawless vertebrates, lampreys and hagfish, lack myelin sheaths, but it is unclear whether they branched off the vertebrate tree before myelin sheaths evolved or subsequently lost them [11]. Also, some invertebrates have myelin sheaths. Neurons with myelin sheaths evolved as part of the fight-or-flight survival response. The speed with which the electrical signal of the firing neuron moves down a myelinated axon is much faster than for a nonmyelinated one. During the course of evolution, when an organism was in immediate danger, the faster it could react, the more likely it survived to pass on its genes. Neurons involved in these reactions (vision, fear response, and muscle movement, to name a few) evolved thick myelin sheaths.

The reason why axons with myelin sheaths are faster is that the electrical signal, a voltage pulse, causes electric fields around the axon. Ions, atoms and small molecules that are electrically charged, in the fluid surrounding the axon get accelerated by the electrical fields, colliding with their neighbors and generating heat. This interaction with the surrounding ions not only slows the voltage pulse, it reduces its strength. Myelin consists primarily of sphingolipids arranged in bilayers that repeatedly wrap around the axon, forming a thick barrier between the axon and outside. The myelin sheath keeps the outside ions farther from the electrical fields, which allows the voltage pulse to travel faster with less attenuation.

In nonmyelinated axons, the electrical signal needs to be boosted to counteract attenuation. Voltage-gated ion channel proteins sense the electrical signal and allow more charged ions, sodium and calcium, for instance, to flow into the axon and increase the signal strength. After the voltage pulse passes, other ion channel proteins must pump the ions back out, and this requires energy, so mitochondria have to produce more of the energy molecule adenosine triphosphate to power the ion pumps. Because of this, the axons of nonmyelinated neurons require more energy, and the amount of energy needed increases with the length of the axon.

The dopaminergic neurons of the substantia nigra are mostly nonmyelinated. Their main function is to provide the striatum with a steady supply of dopamine; hence, fast signal conduction is not required. The substantia nigra dopaminergic neurons innervating the striatum have highly branched axonal arbors, ending in roughly a million synapses [12,13]. This is orders of magnitude more than other human central nervous system neurons, which typically have around 7000 synapses, and I know of no other neuron types that have as many. The electrical signal also becomes attenuated at branch points, so with more branches, even more energy is needed to maintain the axon in a firing-ready state [14].

The linear dimensions of the human brain are about 50% greater than those of the chimpanzee brain.[1] This increase implies a similar increase in the length of substantia nigra axons and a comparable increase in axonal energy use. This alone could explain the greater density of neuromelanin granules in human substantia nigra neurons. If it also happens to be true that humans have more synapses per neuron than chimps, this would further increase the energy cost of axonal signal transduction.

A confounding factor is that neurons of species with larger brains tend to be larger themselves, with larger cell bodies and wider dendrites and axons. Attenuation of the electric signal in axons decreases with increased axon diameter. I have not come across literature suggesting any gross difference between chimp and human neuron size, however. A couple of studies of axon diameters comparing humans and chimps have looked at a different part of the brain, the corpus callosum, which connects the left and right lobes of the neocortex. One study saw no significant difference [15], but in another study, the largest axon diameters were 50% wider than those of the chimps in the single human male studied, although the diameters in the single human female were comparable to chimps. However, in both studies, a wide range of diameters was seen, and much smaller axons were predominant [16]. It seems evolution is stingy, perhaps widening only those axons most critical to survival. The nonmyelinated axons of the substantia nigra are of the narrowest sort, so I suspect their width is similar in chimps

[1] Fifty percent, or a factor of 1.5 times longer, comes from taking the cube root of the human–chimp volume ratio of 3.5.

and humans. If this is true, then the increased neuromelanin density seen in humans could simply be due to greater mitochondrial activity necessitated by longer axonal length.

Up to this point, we have learned that human substantia nigra neurons have longer axons and appear to have greater mitochondrial activity than chimps, but where does alpha-synuclein enter the picture? In previous chapters, I have made the case that increased levels of alpha-synuclein are what ultimately allow Parkinson's disease to develop. Unfortunately, I could find no information in the scientific literature regarding alpha-synuclein concentrations in chimp neurons, so it is not known whether the level in humans is greater or the same. Synuclein concentrations have been quantified in humans, macaques, and rats, however. In tissue from the striatum, alpha-synuclein constituted 0.7% of the total protein present in human patients without neurodegenerative disease, 0.3% in macaques, and 0.2% in rats [17—19]. Recall that the axons of the dopaminergic neurons of the substantia nigra innervate the striatum, so this is where the bulk of their alpha-synuclein lies. Although these numbers suggest that humans have more alpha-synuclein in their dopaminergic neurons, they come from just three studies using nonidentical protocols; more studies are needed to confirm the results.

Nevertheless, I think there is a good reason to suspect humans might have higher alpha-synuclein levels in their dopaminergic neurons than chimps: namely, our gas-guzzler brains. Remember how alpha-synuclein knockout mice had lower mitochondrial activity? [20] If ramping up alpha-synuclein revs up mitochondrial activity, and this is a big if, maybe humans evolved higher alpha-synuclein levels to help mitochondria in nonmyelinated projection neurons provide the extra energy needed to maintain their longer axons. Of course, boosting the number of mitochondria could also achieve the same goal as would boosting levels of other mitochondrial activity-promoting proteins, but higher alpha-synuclein levels could have a role. The basic tenet of evolution is that nature will select for the optimum balance of all factors for maximum survival, and the level of alpha-synuclein could be one of these factors.

I suspect increased activity of mitochondria in human substantia nigra neurons is what makes them especially vulnerable to alpha-synuclein pathology. Once the pathology spreads to the substantia nigra, the oligomer-induced dysfunction could result in the revved-up mitochondria producing even more reactive oxygen species and inflammatory signals than in other neuron types, triggering speedier synaptic degeneration and ultimately, neuronal destruction. Dopamine might also have a role, because it has been shown to alter how alpha-synuclein aggregates, perhaps leading to more toxic oligomers, and alpha-synuclein has been shown to be present in neuromelanin granules [21].

Nevertheless, even if my suspicions are true, this still does not provide a complete explanation for why chimps do not develop Lewy pathology. In humans, the pathology typically shows up first in other areas of the brain, such as the brain stem or olfactory bulb, and then spreads to the substantia nigra. Perhaps there is something different about chimp brains that prevents the spread of the pathology, although the protein sequence of alpha-synuclein for chimps and humans is exactly the same. However, there is evidence to dispute this, because studies have shown that Lewy pathology *can* spread in three other nonhuman primates: baboons, macaques, and marmosets.

In the macaque and baboon studies, Lewy body extracts purified from human brains were injected into the striatum [22,23]. In another study using macaques, fibrils were made in the laboratory with human alpha-synuclein purified from engineered bacteria and injected into the striatum [24]. A similar experiment was carried out on marmosets, using laboratory-purified fibrils, except that the mouse sequence of alpha-synuclein was used [25]. In all four studies, the injections were done on one side of the striatum so that researchers could compare pathology in the injected and non-injected hemispheres of the brain. In all studies, Lewy pathology spreading from the site of injection was observed, including the formation of Lewy bodies and the loss of dopaminergic neurons in the substantia nigra of the affected hemisphere. The mouse and human alpha-synuclein sequences differ at seven amino acids, which suggests that any alpha-synuclein fibrils, regardless of their particular sequence, can trigger spreading Lewy pathology.

So, we seem to be back at square one with regard to why humans develop Parkinson's disease and chimps do not. However, there is one more piece of information that further incriminates our gas-guzzler brains in triggering the disease. The gray mouse lemur of Madagascar is one of the smallest primates, weighing in at about 2 oz, and it also has a gas-guzzler brain. The brains of aged mouse lemurs, with life spans of up to 15 years in captivity, show signs of alpha-synuclein aggregation, with amyloid-beta and tau aggregation, too [26]. Although mouse lemurs lack Lewy bodies and neurites, the smaller-scale alpha-synuclein aggregation occurs in many of the same parts of the brain affected in humans, in particular the vagus nerve, which spreads down from the brain stem into organs such as the heart and intestines. To reiterate from earlier chapters, pathology in this nerve is likely related to constipation and other gastrointestinal symptoms in those who go on to develop Parkinson's disease. Similar to the nerves of the substantia nigra, the axons of the vagus neurons are nonmyelinated. They are a part of the body's parasympathetic nervous system, the so-called rest and digest circuitry of the nervous system, because they calm the body's functions in the absence of threats and stimulate slow processes such as digestion.

At the time of writing, except for humans and mouse lemurs, no naturally occurring alpha-synuclein aggregation has been observed in any primates, including those with brains that use large percentages of their caloric intake. What can account for the aggregation seen in humans and aged mouse lemurs? The easiest explanation would be that they have the highest levels of alpha-synuclein, but the concentration in mouse lemurs has not yet been measured. Another possibility is that humans and mouse lemurs have less beta-synuclein, allowing alpha-synuclein to aggregate more easily. Looking at their amino acid sequences, there does not appear to be anything special about human or mouse lemur alpha-synuclein. In fact, their sequences differ significantly, at nine positions, which is even more than between humans and mice.

In the brains of aged squirrel monkeys, small primates that also have gas-guzzler brains, posttranslational modifications of alpha-synuclein are seen that are typical of Parkinson's disease, phosphorylation, for example, despite the lack of any detectable aggregation [27]. In addition, the levels of alpha-synuclein in the substantia nigra of squirrel monkeys rise with age, something that has been seen in other monkeys as well [1]. In normal, nonaged substantia nigra neurons, most alpha-synuclein is located at the ends of their axons in the presynaptic boutons in the striatum, whereas little is seen in their cell bodies. As the monkeys aged, the amount of alpha-synuclein seen in the cell bodies increased. Could it be that the alpha-synuclein level rises in the cell body because there are fewer remaining presynaptic boutons to go to? In other words, the alpha-synuclein pathology of Parkinson's disease might not be the root cause of the neurodegeneration in the substantia nigra; rather, it speeds the rate of neurodegeneration that naturally occurs in primate brains.

In the chapter after next, we will turn our attention to the vagus nerve and the olfactory bulb, the two places where Lewy body pathology usually arises first in idiopathic Parkinson's disease. However, before we move on, I would like to touch upon another aspect of the human brain that might be related to Parkinson's disease: intelligence.

References

[1] Emborg ME, et al. Age-related declines in nigral neuronal function correlate with motor impairments in rhesus monkeys. Journal of Comparative Neurology November 16, 1998;401(2):253–65.
[2] Edler MK, Sherwood CC, Meindl RS, Hopkins WD, et al. Aged chimpanzees exhibit pathologic hallmarks of Alzheimer's disease. Neurobiology of Aging November 2017;59:107–20.
[3] Bauernfeind AL, et al. Evolutionary divergence of gene and protein expression in the brains of humans and chimpanzees. Genome Biology and Evolution July 10, 2015;7(8):2276–88.
[4] Pontzer H, Brown MH, Raichlen DA, Dunsworth H, Hare B, Walker K, Luke A, Dugas LR, et al. Metabolic acceleration and the evolution of human brain size and life history. Nature May 19, 2016;533(7603):390–2.
[5] Boyer DM, Harrington AR. Scaling of bony canals for encephalic vessels in euarchontans: implications for the role of the vertebral artery and brain metabolism. Journal of Human Evolution January 2018;114:85–101.
[6] Aiello LC, Wheeler P. The expensive-tissue hypothesis: the brain and the digestive system in human and primate evolution. Current Anthropology April 1995;36(2): 199–221.
[7] Morrish PK, Sawle GV, Brooks DJ. An [18F]dopa-PET and clinical study of the rate of progression in Parkinson's disease. Brain April 1996;119(2):585–91.
[8] Marsden CD. Pigmentation in the nucleus substantiae nigrae of mammals. Journal of Anatomy 1961;95:1080–9.
[9] Zecca L, et al. A proposed dual role of neuromelanin in the pathogenesis of Parkinson's disease. Neurology October 10, 2006;67(7 Suppl. 2):S8–11.
[10] Bisaglia M, Mammi S, Bubacco L. Kinetic and structural analysis of the early oxidation products of dopamine: analysis of the interactions with alpha-synuclein. Journal of Biological Chemistry May 25, 2007;282(21):15597–605.
[11] Werner HB. Do we have to reconsider the evolutionary emergence of myelin? Frontiers in Cellular Neuroscience November 15, 2013;7:217.
[12] Matsuda W, et al. Single nigrostriatal dopaminergic neurons form widely spread and highly dense axonal arborizations in the neostriatum. Journal of Neuroscience January 14, 2009;29(2):444–53.
[13] Bolam JP, Pissadaki EK. Living on the edge with too many mouths to feed: why dopamine neurons die. Movement Disorders October 2012;27(12):1478–83.
[14] Pissadaki EK, Bolam JP. The energy cost of action potential propagation in dopamine neurons: clues to susceptibility in Parkinson's disease. Frontiers in Computational Neuroscience March 18, 2013;7:13.
[15] Caminiti R, et al. Evolution amplified processing with temporally dispersed slow neuronal connectivity in primates. Proceedings of the National Academy of Science United States of America November 17, 2009;106(46):19551–6.
[16] Phillips KA, Stimpson CD, Smaers JB, Raghanti MA, et al. He corpus callosum in primates: processing speed of axons and the evolution of hemispheric asymmetry. Proceedings of the Royal Society B: Biological Sciences November 7, 2015;282:1818. 20151535.

[17] Tong J, et al. Brain alpha-synuclein accumulation in multiple system atrophy, Parkinson's disease and progressive supranuclear palsy: a comparative investigation. Brain January 2010;133(1):172–88.
[18] Koprich JB, et al. Towards a non-human primate model of alpha-synucleinopathy for development of therapeutics for Parkinson's disease: optimization of AAV1/2 delivery parameters to drive sustained expression of alpha synuclein and dopaminergic degeneration in macaque. PLoS One November 30, 2016;11(11):e0167235.
[19] Landeck N, et al. A novel multiplex assay for simultaneous quantification of total and S129 phosphorylated human alpha-synuclein. Molecular Neurodegeneration August 2016;11(1):61.
[20] Ellis CE, et al. Mitochondrial lipid abnormality and electron transport chain impairment in mice lacking alpha-synuclein. Molecular and Cellular Biology November 2005;25(22):10190–201.
[21] Halliday GM, et al. Alpha-synuclein redistributes to neuromelanin lipid in the substantia nigra early in Parkinson's disease. Brain November 2005;128(11):2654–64.
[22] Recasens A, et al. Lewy body extracts from Parkinson disease brains trigger α-synuclein pathology and neurodegeneration in mice and monkeys. Annals of Neurology March 2014;75(3):351–62.
[23] Bourdenx M, et al. Identification of distinct pathological signatures induced by patient-derived α-synuclein structures in non-human primates. Science Advances 2020;20(6):eaaz9165. (preprint).
[24] Chu Y, et al. Intrastriatal alpha-synuclein fibrils in monkeys: spreading, imaging and neuropathological changes. Brain October 3, 2019. https://doi.org/10.1093/brain/awz296.
[25] Shimozawa A, et al. Propagation of pathological α-synuclein in marmoset brain. Acta Neuropathologica Communications 2017;5(1):12.
[26] Canron MH, et al. Age-dependent α-synuclein aggregation in the *Microcebus murinus* lemur primate. Scientific Reports 2012;2:910.
[27] McCormack AL, Mak SK, Di Monte DA. Increased α-synuclein phosphorylation and nitration in the aging primate substantia nigra. Cell Death and Disease May 2012;3:e315.

CHAPTER 11

A Troublesome Subject – Exploring the possible connection between intelligence and Parkinson's and Alzheimer's diseases

Back when I was first working on the interactions between alpha-synuclein and glucocerebrosidase, I happened to be reading a book, *Before the Dawn—Recovering the Lost History of Our Ancestors*, by Nicholas Wade, a science writer for *The New York Times*. It was a fascinating read, using genetics and anthropology to piece together the forces that shaped human society in the millennia before recorded history. Toward the end, I was startled to read a section involving the protein I was studying, glucocerebrosidase. Even more startling was what Wade was saying the protein was implicated in; but first, some background.

Synuclein and the Coelacanth
ISBN 978-0-323-85707-9
https://doi.org/10.1016/B978-0-323-85707-9.00007-1

Among people of Ashkenazi Jewish ancestry, a number of genetic diseases involving recessive mutations occur at higher frequency than in the general population. Four of the diseases arise from mutations in proteins concerning lysosomal processing of sphingolipids, causing lysosomal storage diseases, which involve the accumulation of molecules such as sphingolipids in lysosomes [1]. The diseases are Gaucher disease, Tay-Sachs disease, Niemann-Pick disease type A, and mucolipidosis type IV, with carrier frequencies in the Ashkenazi population of about 1 in 14, 1 in 29, 1 in 90, and 1 in 110, respectively. Recall that mutations in glucocerebrosidase gene *GBA1* cause Gaucher disease and are also risk factors for developing Parkinson's disease. These carrier frequencies are much higher than the corresponding frequencies in non-Jewish Europeans living in the same regions; the carrier frequency for *GBA1* mutations in non-Jewish British people is around 1 in 85 [2] and about 1 in 200 for mutations in the Tay-Sachs gene (*HEXA*) in Americans of Irish descent [3].

Ashkenazi populations have been genetically isolated from their European neighbors by centuries of endogamy, proscribed from marrying outside their faith. For instance, one study estimated that after the founding population, gene flow from Western and Eastern Europe ultimately contributed roughly 20% of their genomes [4]. This translates to an average gene flow of less than 1% each generation in the roughly 1100 years since the founding population. Genetic diseases can easily be established in isolated populations by the founder effect, but a cluster of genetic diseases involving genes in a specific biological pathway can indicate some sort of genetic selection that has occurred.

Dozens of proteins are involved in lysosomal degradation of lipids, proteins, and their attached carbohydrates. These include transporter proteins, degradation enzymes, and their cofactors. Mutations in roughly 40 of them have been associated with lysosomal storage diseases [5]. However, there are approximately 20,000 genes in the human genome. In addition, for both Gaucher and Tay-Sachs diseases, multiple common mutations are found at much higher frequencies in Ashkenazi populations: four for Gaucher and two for Tay-Sachs [6,7]. In a letter to the journal *Nature* in 1994, Jared M. Diamond, author of *Guns, Germs, and Steel*, likened the situation to a house being struck by lightning eight times.[1]

[1] I assume that for his eight "lightning strikes," Diamond is including all of the common mutations that lead to the four diseases.

I cranked through the math, and the situation might not be as unlikely as Diamond paints it. First, we need some starting numbers. In a general population, it has been estimated that individuals carry on average one recessive lethal gene mutation in their genomes [8]. Among people with Ashkenazi ancestry, a milder, nonlethal form of Gaucher disease is common, so to adjust roughly for this, let us increase the average number of recessive mutations to two, to include both lethal and nonlethal but still serious disease mutations. I myself am a carrier for two recessive diseases, cystic fibrosis and hemochromatosis, and who knows what other, rarer recessive mutations I might harbor, so setting the number to two seems reasonable (to me, at least). There are about 6000 described genetic conditions in humans [9]; some of these involve mutations in more than one gene, many are not lethal, and many are dominant, not recessive, so to adjust roughly for this, let us reduce the number to 4000 for genes with the potential for recessive lethal and serious disease mutations.

The most recent estimates by geneticists put the Ashkenazi founding population at about 350 individuals living around AD 900, based on the distribution of lengths of identical DNA stretches in present-day Ashkenazi genomes [10]. The actual founding Ashkenazi population might have been larger, containing extended family groups, but the number of founders predicted by geneticists is meant to reflect how many non—closely related individuals would be needed to yield the genetic diversity seen in Ashkenazi populations today [11].

How many lysosomal sphingolipid processing proteins would we expect to have mutations just by chance? With an average number of two recessive mutations per individual, this would give an estimated 700 mutations distributed among the 350 individuals in the founding population. Some of these mutations would be in the same genes in different people, so the total number of genes with mutations would be a bit less, closer to 600. Therefore, about 15% (600/4000) of the recessive gene mutation candidates would be present in the population. Likewise, 15% of the 40, or six lysosomal proteins involved in sphingolipid processing, would be expected to have one or possibly two mutations harbored by various individuals in the population. That is, six "lightning strikes" are predicted just by chance.

Of course, my calculation is meant to give only a rough estimate, and perhaps I was too liberal in my adjustment of the average recessive gene number per individual and in the number of genes with the potential for recessive lethal or serious disease mutations. If I use more conservative numbers, an average recessive mutation number per individual of one, and up the gene number to 6000, only two mutations in the 40 lysosomal proteins would be predicted. Given this uncertainty, it is wise also to explore the possibility that some of the mutations might have arisen after the founding of the Ashkenazi population, and then rose to relatively high frequencies owing to some advantage conferred on the carriers.

How plausible is it that a mutation arising after the time of the founding population could reach the frequencies seen for the lysosomal protein mutations? As an example, let us imagine the Ashkenazi population has grown by a factor of 100, to 35,000, some 200 years or so after its founding. The most common lysosomal recessive mutations have about a 1 in 30 frequency in Ashkenazi populations. If a lysosomal protein mutation spontaneously occurred in an Ashkenazi individual 200 years after the founding population, how much of an evolutionary advantage would be needed for that mutation to reach a frequency of 1 in 30 in the present-day Ashkenazi population? Carriers would need to have 53% more offspring than average, half of whom would also be carriers that survived to pass on the mutation. With this advantage, in the roughly 30 generations between then and now, the frequency of the mutation would grow from 1 in 35,000 in AD 1100 to 1 in 30 in AD 2000.

Admittedly, 53% more offspring is a pretty steep advantage. Using an analysis called linkage disequilibrium, geneticists usually can tell whether a particular mutation confers such a large advantage. When the advantage is large, other genes near the mutated gene in carriers will tend to be inherited along with the mutated gene. In fact, linkage disequilibrium analysis shares similarities with correlated mutation analysis, and covariance has a central role. Linkage disequilibrium studies of the lysosomal storage disease mutations have suggested that if they confer any advantage on carriers, the advantage is probably much smaller than 53% [12]. Although it is conceivable that one or more of the lysosomal protein mutations could have arisen soon after the founding population, it is more likely that the mutations, especially the most common ones, were already present. Thus, if the mutation were present in the founding population, how much of an advantage would be needed to boost the frequency to the levels seen today? With a 14% boost in offspring, in the roughly 37 generations between AD 900 and AD 2000, the frequency of the mutation would grow from 1 in 350 to 1 in 30. A small boost of 14% is more consistent with disequilibrium analyses [13].

Is it possible that, presuming they confer no advantages, the frequency of the mutations could be boosted by pure chance alone (i.e., random genetic drift excluding founder effects) to the frequencies seen today? Almost certainly not. In each generation, mutation frequencies fluctuate by small amounts even without selective advantages or disadvantages. The fluctuations are related to something in statistics called standard error, which for a low-frequency mutation is roughly equal to the frequency divided by the square root of the population. Once the population became sufficiently large, greater than 10,000, the mutation frequency would vary by less than

1/100th of its value each generation, and this variation would continue to shrink as the population grew to hundreds of thousands to the millions of today. In any case, with lethal recessive mutations, such as those giving rise to Tay-Sachs disease, and assuming no carrier advantage, negative selection should progressively lower the frequencies, not raise them.

There are a couple of loopholes, however, one of which was alluded to earlier; if the founding population included extended families that harbored the mutations, the corresponding frequencies would be higher in the founding population. There are even records of a Jewish family with the surname Kalonymos, that immigrated to the Rhineland, the proposed Ashkenazi homeland, at approximately the same time as was predicted for the founding population, so the idea that extended families might have been part of the founding population is not far-fetched [14]. The other loophole is that the founding population, or some significant fraction of it, could have come from another genetically isolated population that already had elevated frequencies of certain mutations owing to an earlier founder effect. Some studies have explored the possibility that the earlier founder effect might have arisen during or soon after the Jewish diaspora of the first century AD [15].

To get a better idea how much an extended family might boost a mutation frequency, consider a scenario in which one great-grandparent had the mutation. Let us further imagine that the extended family was fairly prolific, with an average of five children per couple each generation who went on to start their own families. About one-eighth of the great-grandchildren, roughly 16 out of the 125 of them, would be expected to carry the mutation. If this extended family of 125 great-grandchildren were part of the founding population, the frequency of their mutation in the population would be boosted to 1 in 30.[2]

I have glossed over other details relevant to the founder effect debate. For example, the second common mutation for Tay-Sachs disease arose long after the time of the Ashkenazi founding population, when part of the population split off and migrated to Eastern Europe [6]. Some have speculated that a second bottleneck occurred in the Ashkenazi population, around the time of this migration, owing to the Black Death in AD 1348–50, and that the high frequency of the second mutation might be the

[2] 1 in 30 is what you get when you take 16 and divide it by 350 plus 125; or better, it should be 350 plus 117, because the Ashkenazi founding population estimate should already reflect the genetic diversity of the eight great-grandparents of the extended family.

result of a founder effect in the surviving Eastern Europe population. The bottom line is that although it is likely that the lysosomal mutation cluster in Ashkenazi populations arose by chance founder effects, it is also plausible that evolutionary selection has increased the carrier frequencies of lysosomal protein mutations since the founding event.

Two types of evolutionary selection can increase gene frequencies: natural and sexual selection. Natural selection favors mutations that lead to a higher chance of survival, whereas sexual selection favors mutations that make one more attractive to potential mates. There can be overlap between the two types; larger body size can increase survival odds and be more sexually attractive, for instance. In general, natural selection should weed out recessive disease-causing genes. However, when recessive mutations persist in a population, it can mean that the carriers have, or had in the past, some survival or sexual advantage, compensating for the chance that their offspring might get the disease. This situation is called balancing selection. The sickle cell disease mutations and their protective effect in carriers against malaria are the best-known examples. The persistence and frequency of lysosomal protein mutations in Ashkenazi populations suggest that some sort of balancing selection might be responsible.

A hint that evolutionary selection might be involved in lysosomal protein mutations in people of Ashkenazi ancestry came from a questionnaire in 1972 given to the parents of children who had Tay-Sachs disease about the cause of death of the child's grandparents [16]. Of 254 grandparents who had died of natural causes, only one had died of tuberculosis. In the early 1900s, up to 20% of deaths in cities in Eastern Europe, from which many of the grandparents had come, were due to tuberculosis. Because Ashkenazi Jews were often forced to live in densely populated ghettoes, they were more exposed to tuberculosis, and thus were under more selective pressure to evolve resistance to the disease, or so the theory goes.

Debate about whether the tuberculosis—Tay-Sachs connection was real or just a statistical fluke has gone on for decades; later studies were unable to reproduce the initial results [17]. Further fueling the debate was that the second mutation in the gene for Tay-Sachs disease occurred in Eastern Europe centuries after the founding of the Ashkenazi population, which I mentioned earlier. This new mutation also achieved a relatively high frequency, which is consistent with the notion that the mutations had some beneficial effect in carriers, but it could also have been caused by a chance founder effect. Although some scientists still favor the Tay-Sachs tuberculosis connection, definitive proof is lacking.

Researchers have also sought to discover whether Gaucher disease mutations offer protection against diseases common in Europe, but to my knowledge, no convincing candidates have emerged. Instead, some have proposed an altogether different selective advantage, the startling implication I alluded to earlier, that the mutations are connected to higher intelligence in people of Ashkenazi ancestry. The hypothesis Wade described has been championed by Gregory Cochran and Henry Harpending of the University of Utah, as described in detail in their book *The 10,000 Year Explosion: How Civilization Accelerated Human Evolution.*

On IQ tests, those with Ashkenazi ancestry score 8–15 points higher on average than their non-Ashkenazi European neighbors [18]. Higher intelligence is also implied by the relatively large percentage of scientists and mathematicians of Ashkenazi ancestry, and even more by the disproportionately high impact their work has had. Of the Nobel Prizes won by Americans, 27% have been won by people of Ashkenazi ancestry, even though they make up less than 2% of the population. Socioeconomic status is strongly correlated with the results of IQ tests, but even when this is corrected for, the difference in IQ persists. Intelligence is also highly heritable; twin studies and other analyses suggest that by adulthood, genes account for 60%–80% of one's intelligence [19].

The average increase in IQ for those of Ashkenazi ancestry is not uniform across every type of cognitive ability tested, however. Visuospatial ability, which includes the ability to rotate objects in one's mind to predict how they appear from different angles, is typically slightly poorer on average in those of Ashkenazi ancestry compared with their non-Ashkenazi European neighbors [18]. Curiously, in Parkinson's patients with *GBA1* mutations, including the E326K polymorphism, there is also evidence of cognitive impairments in working memory, executive function and visuospatial ability [20]. Coincidence?

Ideas regarding how Ashkenazi populations evolved a higher average intelligence have been debated for some time. One theory holds that the centuries-long reverence for religious scholarship and the high status of rabbis drew the brightest students to rabbinical schools, and drew the best daughters from wealthier families to wed them. This resulted in relatively large, wealthy and highly regarded families, with offspring who were more likely to go on to found their own successful families, thus enlarging the genetic footprint of the best and brightest. Another theory is that during the repeated, deadly persecutions of Jews in the Middle Ages, the cleverer ones were more likely to escape [21].

Cochran and Harpending proposed a different theory, that Ashkenazi Jews evolved their higher intelligence owing to centuries of their primary occupation as moneylenders, expanding in later centuries to other areas of finance and number-crunching professions, such as the management of wealthy estates. An intriguing hypothesis, but what really caught my attention was the apparent connection between intelligence and Gaucher disease. The researchers observed among 255 adult, nonretired Gaucher patients at the Shaare Zedek Medical Center in Jerusalem that engineers and scientists were heavily overrepresented, with roughly 11 times the number one would predict based on the total number of engineers and scientists in the general population in Israel. Although not a direct measure of IQ, these are occupations in which the average IQ tends to fall far above the norm. In these patients, both copies of the *GBA1* gene had mutations, but what about carriers with only one mutated *GBA1* gene? There is no information regarding the mental abilities of carriers, except for the unfortunate few who went on to develop Parkinson's disease [20], but it is conceivable that their (pre-Parkinson's) IQs might be higher, too, although perhaps to a lesser degree.

Excited about what I had read, I asked my National Institutes of Health collaborator in the alpha-synuclein glucocerebrosidase studies, Ellen Sidransky, about the possible connection to IQ. An MD, pediatrician, and clinical geneticist, she assured me she had seen no evidence of increased IQ in patients or their immediate families. In fact, what she often saw was the opposite: patients with significant mental deficits. I was chagrined; I should have known better, because I already knew Gaucher disease often causes neurological symptoms, including mental deficits, in particular in processing speed, memory, and visuospatial ability (hmm ...) [22].

Gaucher patients are classified into three categories: types 1, 2, and 3. Type 1 Gaucher disease patients have the mildest form of the disease, typically with no overt neurological symptoms, and can live full life spans, especially with the advent of enzyme replacement therapy. Type 1 Gaucher disease is common among those with Ashkenazi ancestry. Types 2 and 3 are termed neuronopathic, because in addition to more severe symptoms, those who have it also have neurological problems. Tragically, for type 2 and 3 patients, intravenously delivered glucocerebrosidase cannot cross the blood–brain barrier. Type 2 patients have the most severe form of the disease and often do not make it past age 2. Type 3 patients have a slightly less severe form and can survive into young adulthood. Type 3 is perhaps the most common form of the disease in non-Ashkenazi populations, although I wonder whether type 1, with its milder symptoms, might often go undiagnosed in populations where Gaucher disease is rare.

The more I thought about my excitement about a possible Gaucher disease connection to intelligence, the more foolish I felt. When it comes to polygenic traits, intelligence is the granddaddy of them all. Although decades of study have shown time and again that intelligence is highly heritable, studies to find single gene mutations responsible for high IQ have failed (with the possible exception of torsion dystonia, a rare condition found in people with Eastern European Ashkenazi ancestry [23], but even this connection is disputed [21]). This is in stark contrast to mutations causing mental retardation, which can be caused by single mutations in hundreds of candidate genes. I should not have expected that a single mutation in the *GBA1* gene could alone be responsible for an increase in IQ.

While doing my preliminary review of Parkinson's disease literature in preparation for this book, several years after my Gaucher IQ gaffe, I came across a moderately cited study from the Mayo Clinic in Rochester, Minnesota, "Education and Occupations Preceding Parkinson Disease: a Population-Based Case-Control Study." [24] The study looked at data from the Rochester Epidemiology Project, established in 1966, which consists of all medical records of about 500,000 volunteers from Olmstead County, Minnesota. They identified all subjects who developed Parkinson's disease in 1976–95. Their results suggested that subjects with 9 or more years of education were at double the risk for Parkinson's disease. For those with over 16 years of education, the risk was even higher, 2.5. How had I missed this paper before? (To be fair to myself, entering "Parkinson's disease" into an online search engine gives around 1.8 million hits.)

The Mayo Clinic study saw the greatest increased risk in physicians and the lowest risk in occupations such as construction and factory workers. Mathematicians and scientists also had greater risk, although their numbers were too low to yield a statistically robust result. At first, i suspected they did not correct for smoking, because smoking correlates with a lower Parkinson's risk, and physicians have one of the lowest percentages of smokers, around 3% [25]. However, the study had corrected for smoking, and the risk, although somewhat reduced, still held true. On the other hand, their data on the smoking habits of the subjects were pretty drastic; they only considered whether the subjects had never smoked, or not. What about heavy smokers versus light, or those who had quit versus those who never did?

I continued to dig for more studies, to see whether the Mayo Clinic study was supported or disputed. Several studies have come from Sweden combining census data with the National Patient Register, identifying all those with Parkinson's disease in the entire population. One study saw a modest but statistically robust correlation between Parkinson's disease and socioeconomic status, with those of lower status having reduced risk for the disease, but the study did not correct for smoking [26].

The other Swedish study looked for correlations between job complexity and Parkinson's disease risk [27]. Jobs that were more complex in terms of data handling, such as scientists, and dealing with people, such as lawyers, were modestly correlated with higher Parkinson's disease risk, whereas those more complex in terms of dealing with things, such as master craftsmen, showed no correlation. This study corrected for smoking, using current, instead of never versus ever, smoking status. A study of two regions in Western Scotland also saw a correlation between Parkinson's disease risk and whether one lived in a more or less affluent region; those in the more affluent region had a higher incidence [28]. The more affluent region had a higher education level, but also lower smoking rates. Despite this, the authors thought the difference in disease incidence was larger than could be explained by smoking alone.

Although none of the studies seemed to be in conflict with the Mayo Clinic study, none were overwhelmingly convincing, in my opinion at least, with regard to a potential connection between intelligence and Parkinson's disease. IQ shows a clear correlation with socioeconomic status, but there are too many other potentially confounding variables that are also associated with socioeconomic status, such as physical activity levels, time spent outside, air quality, and prenatal care and nutrition, including breastfeeding, just to name a few. I find the modest correlations with education level and job complexity rather more compelling, but I wonder whether accounting for job income level, with its obvious connection to socioeconomic status, might weaken the connection.

I might have dropped the idea of a possible connection between Parkinson's disease and intelligence if it were not for another neurodegenerative disease with a much clearer connection to intelligence: Alzheimer's. In 1986, the Nun Study of Aging and Alzheimer's Disease was initiated by David Snowdon and collaborators at the University of Minnesota [29]. The study observed over 678 Catholic nuns over the following decades, assessing their cognitive abilities and with the nuns allowing their brains to be autopsied after their deaths. The researchers were also given access to the convent archives.

When the nuns entered the convent around age 22, they were asked to write autobiographical essays. The researchers assessed these essays for their level of idea density, a measure of linguistic ability [30]. For those with the lowest idea density, about 80% went on to develop Alzheimer's disease in old age, whereas for those with the highest scores, only 10% developed the disease. An additional, surprising find was that for many of those with the highest scores, the hallmark pathological signs of Alzheimer's, neurofibrillary tangles and senile plaques, were present in their autopsied brains even though they did not have cognitive impairment before they died.

In the late 1990s, a massive survey was rediscovered that included data on every student born in Scotland in 1921 and in 1936, the Lothian birth-cohorts, consisting of over 150,000 individuals combined. Eleven years later, the students were tested using a standard IQ test as part of the Scottish Mental Surveys of 1932 and 1947. The story of these cohorts and what has been learned observing them into late adulthood is described in the book *A Lifetime of Intelligence*, by Ian J. Deary, Lawrence J. Whalley, and John M. Starr.

Following up on the 1921 Scottish cohort, studies were able to correlate IQ scores at age 11 directly with the development of dementia as the students aged into late adulthood [31]. The researchers evaluated the data in a number of ways, including county-level variation in socioeconomic status as one of the variables. Women in the lowest scoring group, with IQ scores of less than 85, were 50% more likely to develop dementia as they aged than those in the highest scoring group, with scores greater than 114. The effect was more pronounced in women than in men, possibly because men have increased rates of dementia from causes other than Alzheimer's disease.

The connection between Alzheimer's disease and intelligence supports the notion that other neurodegenerative diseases might also have a connection, although it could be that for Parkinson's disease the connection is more subtle. Reconsidering the Israeli Gaucher disease patients whose occupations skewed strongly toward the brainier side, they were working adults, and so they almost certainly had the milder type 1 form of the disease, and probably many, if not most, had Ashkenazi ancestry (remember that type 1 Gaucher disease is more common in those of Ashkenazi ancestry). On the other hand, the patients seen by my colleague Sidransky included many pediatric cases with the more severe, neuronopathic form of the disease. It is possible that only the milder form of Gaucher disease, perhaps combined with the genetic background of Ashkenazi populations, increases intelligence.

Intelligence is a polygenic trait; thus, if milder *GBA1* mutations promote intelligence, it would make sense that they do so combined with other mutations and polymorphisms that might also have evolved to higher frequencies in Ashkenazi populations. However, what would really be convincing is if there were biological evidence that *GBA1* mutations affected neurophysiology in a way that might lead to higher intelligence. It turns out that there is such evidence. As mentioned two chapters ago, in 1995, Anthony H. Futerman tested what happened when cultured neurons were treated with a compound that inhibits glucocerebrosidase. After a few days, the neurons developed longer, more branched axonal arbors. They also tested what happened when glucosylceramide synthesis was blocked, and they saw the opposite effect [32].

Recall that glucosylceramide is the starting point in the synthesis of gangliosides, a diverse collection of lipid molecules consisting of complex carbohydrates attached to the sphingolipid glucosylceramide. Gangliosides has roles in axon and dendrite growth and maintenance. By inhibiting the degradation of glucosylceramide, more glucosylceramide was made available for ganglioside synthesis in the cultured neurons, although perhaps glucosylceramide itself stimulated the additional axonal growth. On the other hand, as also mentioned a couple of chapters back, in neurons with a single *GBA1* mutation, glucosylceramide levels do not appear to be especially elevated, perhaps because any excess becomes converted to gangliosides or transported away in the blood.

It could be that something similar to the cultured neurons happens in the brains of people with *GBA1* mutations, and their neurons end up with larger, more branched axonal arbors. With larger arbors, the axons could form more synapses with more dendrites, increasing the complexity of the brain and thus somehow boosting intelligence. However, no obvious proliferation of axonal arbors has been seen on autopsy in *GBA1* carriers or those with type 1 Gaucher disease, so any effect must be relatively subtle. In fact, in those with type 2 and 3 Gaucher disease, axonal dystrophy and degradation are observed, pretty much the opposite of axonal proliferation [33].

Intriguingly, the mutations that cause Tay-Sachs disease and Niemann-Pick type A are correlated with increased levels of a particular ganglioside (GM2) that is associated with the excessive formation and growth of dendrites [34]. Perhaps carriers of these diseases develop additional or more branched dendrites, increasing the number of synapses in the brain and boosting intelligence. However, like Gaucher carriers, any effect must be subtle, because no gross difference in the brain anatomy of carriers has been noted in the literature, although I suppose it is possible no one has bothered to look carefully at the brains of carriers or patients of Gaucher, Tay-Sachs, or Niemann-Pick type A diseases who happened to die of non-neurodegenerative causes.

Does having longer, more branching axons or dendrites actually lead to higher intelligence in humans? Surprisingly, a neuroimaging study suggested the opposite: that looser, less-branched arbors are associated with higher IQ [35]. The study exploited the fact that areas of less dense neural connections allow extracellular fluid to move more freely, which can be measured using magnetic resonance imaging. How can a brain with less neural connections (i.e., less complexity) perform more intelligently? The answer seems to have something to do with the synapse pruning that occurs as the brain matures.

The gray matter of neocortices of children is thicker than those of adults, thinning with the synaptic pruning that accompanies maturation. Perhaps, if one starts with more extensive axonal and/or dendritic arbors, as the maturing juvenile brain responds to life's lessons, there are more options to choose from to yield the optimum set of synapses to retain, pruning the rest. It could be that the optimum set of synapses, given a more extensive set of synapses to start with, actually ends up being smaller in number than for those who start with smaller arbors. But here, I admit, I am speculating far out of my depth. In any case, I should not assume that the pruning that occurs in the neocortex of juveniles is mirrored in other parts of the brain, such as the striatum. In fact, the evidence in human and other primates is that the synapses in the striatum coming from substantia nigra neurons are lost steadily throughout adult life, with no mention in the literature of accelerated loss during juvenile development [36].

Is there really a connection between intelligence and Parkinson's disease? The evidence is consistent with such a connection, from brainily skewed Gaucher patients in Israel, from the population studies in Minnesota, Sweden, and Scotland, and from longer, more branched axons in cultured neurons; however, none of these correlations is definitive. In Chapter 6, it was noted how Parkinson's patients with *GBA1* mutations show a pattern of Lewy body pathology that does not match well with the typical Braak stages seen for idiopathic Parkinson's disease. It could be that for most cases, especially for idiopathic Parkinson's disease with no connection to any known mutations, the disease has nothing to do with intelligence; instead, it may being due to environmental factors or to biological factors unrelated to cognitive ability.

Finally, I have to wonder why, given the Scottish Mental Surveys, no study has emerged exploring the potential connection between IQ and developing Parkinson's disease later in life. Could it be that such a study using the Scottish cohort data has already been attempted, but that the statistics were not conclusive? Personally, I suspect that if there is any increased risk of Parkinson's disease owing to higher intelligence, the risk is relatively small.

References

[1] Ferreira CR, Gahl WA. Lysosomal storage diseases. Translational Science of Rare Diseases 2017;2(1−2):1−71.
[2] Neumann J, et al. Glucocerebrosidase mutations in clinical and pathologically proven Parkinson's disease. Brain July 2009;132(7):1783−94.
[3] van Bael M, et al. Eterozygosity for Tay-Sachs disease in non-Jewish Americans with ancestry from Ireland or Great Britain. Journal of Medical Genetics October 1996;33(10):829−32.
[4] Xue J, et al. The time and place of European admixture in Ashkenazi Jewish history. PLoS Genetics April 2017;13(4):e1006644.
[5] Futerman AH, van Meer G. The cell biology of lysosomal storage disorders. Nature Reviews Molecular Cell Biology July 2004;5(7):554−65.
[6] Risch N, et al. Geographic distribution of disease mutations in the Ashkenazi Jewish population supports genetic drift over selection. The American Journal of Human Genetics April 2003;72(4):812−22.
[7] Velayati A, Yu WH, Sidransky E. The role of glucocerebrosidase mutations in Parkinson disease and Lewy body disorders. Current Neurology and Neuroscience Reports May 2010;10(3):190−8.
[8] Gao Z, et al. An estimate of the average number of recessive lethal mutations carried by humans. Genetics April 2015;199(4):1243−54.
[9] Chial H. Rare genetic disorders: learning about genetic disease through gene mapping, SNPs, and microarray data. Nature Education 2008;1(1):192.
[10] Carmi S, et al. Sequencing an Ashkenazi reference panel supports population-targeted personal genomics and illuminates Jewish and European origins. Nature Communications September 9, 2014;5:4835.
[11] Palamara PF, et al. Length distributions of identity by descent reveal fine-scale demographic history. The American Journal of Human Genetics November 2, 2012;91(5):809−22.
[12] Diaz GA, et al. Gaucher disease: the origins of the Ashkenazi Jewish N370S and 84GG acid beta-glucosidase mutations. The American Journal of Human Genetics June 2000;66(6):1821−32.
[13] Boas FE. Linkage to Gaucher mutations in the Ashkenazi population: effect of drift on decay of linkage disequilibrium and evidence for heterozygote selection. Blood Cells, Molecules and Diseases August 2000;26(4):348−59.
[14] Ben-Sasson H. A history of the Jewish people. Cambridge: Havard University Press; 1976.

[15] Behar DM, et al. MtDNA evidence for a genetic bottleneck in the early history of the Ashkenazi Jewish population. European Journal of Human Genetics May 2004;12(5):355–64.
[16] Myrianthopoulos NC, Aronson SM. Population dynamics of Tay-Sachs disease. II. What confers the selective advantage upon the Jewish heterozygote?. In: Proceedings of the 4th international symposium on Sphingolipidoses; 1972. p. 561–70.
[17] Spyropoulos B, et al. Heterozygote advantage in Tay-Sachs carriers? The American Journal of Human Genetics May 1981;33(3):375–80.
[18] Lynn R. The intelligence of American Jews. Personality and Individual Differences 2004;36:201–6.
[19] Plomin R, Deary J. Genetics and intelligence differences: five special findings. Molecular Psychiatry 2015;20:98–108.
[20] Davis MY, et al. Association of GBA mutations and the E326K polymorphism with motor and cognitive progression in Parkinson disease. JAMA Neurology 2016;73(10):1217–24.
[21] Ferguson RB. How Jews became smart: anti-"Natural history of Ashkenazi intelligence". 2008. http://pdfs.semanticscholar.org/1eed/b19bcf7c059a4b10a9ed8c58027d9ed22bae.pdf.
[22] Lopez G, et al. Clinical course and prognosis in patients with Gaucher disease and parkinsonism. Neurology Genetics March 4, 2016;2(2):e57.
[23] Riklan M, Cullinan T, Cooper IS. Psychological studies in dystonia musculorum deformans. Advances in Neurology December 31, 1975;14:189–200.
[24] Frigerio R, et al. Education and occupations preceding Parkinson disease: a population-based case-control study. Neurology November 22, 2005;65(10):1575–83.
[25] Nelson DE, et al. Trends in cigarette smoking among US physicians and nurses. Journal of the American Medical Association April 27, 1994;271(16):1273–5.
[26] Yang, et al. Socioeconomic status in relation to Parkinson's disease risk and mortality. A population-based prospective study. Medicine July 2016;95(30):e4337.
[27] Valdés EG, et al. Occupational complexity and risk of Parkinson's disease. PLoS One September 8, 2014;9(9):e106676.
[28] Newman EJ, Grosset KA, Grosset DG. Geographical difference in Parkinson's disease prevalence within West Scotland. Movement Disorders February 15, 2009;24(3):401–6.
[29] Snowdon DA. Healthy aging and dementia: findings from the Nun Study. Annals of Internal Medicine September 2, 2003;139(5 Pt 2):450–4.
[30] Riley KP, et al. Early life linguistic ability, late life cognitive function, and neuropathology: findings from the Nun Study. Neurobiology of Aging March 2005;26(3):341–7.
[31] Russ TC, et al. Childhood cognitive ability and incident dementia: the 1932 Scottish Mental Survey cohort into their tenth decade. Epidemiology May 2017;28(3):361–4.
[32] Schwarz A, et al. A regulatory role for sphingolipids in neuronal growth. Inhibition of sphingolipid synthesis and degradation have opposite effects on axonal branching. Journal of Biological Chemistry May 5, 1995;270(18):10990–8.
[33] Zigdon H, et al. Altered lysosome distribution is an early neuropathological event in neurological forms of Gaucher disease. FEBS Letters March 2017;591(5):774–83.
[34] Walkley SU. Neurobiology and cellular pathogenesis of glycolipid storage diseases. Philosophical transactions of the Royal Society of London. Series B, Biological Sciences May 2003;358(1433):893–904.
[35] Genç E, et al. Diffusion markers of dendritic density and arborization in gray matter predict differences in intelligence. Nature Communications May 15, 2018;9(1):1905.
[36] Emborg ME, et al. Age-related declines in nigral neuronal function correlate with motor impairments in rhesus monkeys. Journal of Comparitive Neurology November 16, 1998;401(2):253–65.

CHAPTER 12

Appendix — Alpha-synuclein in the gut, Crohn's disease, and the possible protective role of the synucleins in innate immune response

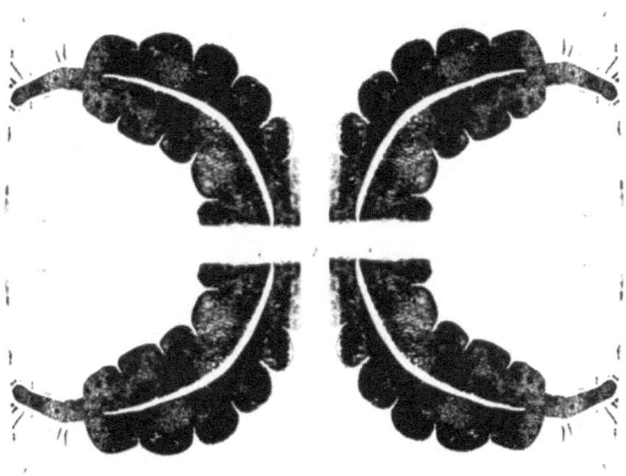

When I was first planning the chapters for this book, I thought I might include one near the end with leftover scientific tidbits pertinent to Parkinson's disease, but ones that did not fit neatly into the other chapters, an appendix of sorts. Plans change, and it turns out this chapter is not an appendix, but commences with an intriguing announcement regarding the appendix, the vestigial organ that protrudes from the beginning of the large colon. In 2019, an abstract for a talk at a conference on gastrointestinal surgery presented a startling and confusing observation. The study reviewed medical records of over 62 million Americans, and found that those who had undergone an appendectomy, a total of 448,190 cases, were over three times more likely to develop Parkinson's disease [1].

The startling part of this study was that such a significant risk for Parkinson's disease had not been noticed before. About 10% of the

Synuclein and the Coelacanth
ISBN 978-0-323-85707-9
https://doi.org/10.1016/B978-0-323-85707-9.00004-6

population has had or will get their appendix removed at some point in their lives. More confusing is that a number of previous studies have looked at patients with and without appendectomies, and nothing even close to such a large increase in Parkinson's disease risk had been observed. In fact, the largest previous study, which followed the entire population of Sweden (just under 10 million, for over 50 years, with even more appendectomy cases, 551,647), found that those who had the appendix removed had a 20% *reduced* risk for developing Parkinson's disease [2]! Looking at even earlier studies further muddies the picture; some researchers found that appendectomies were associated with a slightly increased risk for Parkinson's disease, although nowhere near a threefold increase [3].

In the Swedish population study, performed by Viviane Labrie and the Van Andel Research Institute in Grand Rapids, Michigan, two controls were chosen at random from the nonappendectomy group for each individual who had had an appendectomy, but matched for birth year, gender, and geographical location. In the American population study, done by Mohammed Z. Sheriff and colleagues at Case Western Reserve University in Cleveland, Ohio, a commercial database was used, created by mining the Internet cloud for deidentified clinical data, containing records from 26 US health care systems, although the time period was not specified. However, rather than finding age-, gender-, and location-matched controls, Sheriff and colleagues simply used all individuals who had not undergone appendectomies as the control group.

Using everyone who had never undergone an appendectomy as the control is problematic, however. The cumulative chance of having an appendectomy increases as you age, and it follows that for the general population, including infants through centenarians, those who have had an appendectomy will be older on average than those who have not. Recall that age is the strongest risk factor for developing Parkinson's disease. Sheriff and colleagues addressed this concern in their abstract, however, stating that when including data only for those aged 65 years or older, the actual risk was only two times that of the age-matched, nonappendectomy group. They even provided a graph that supported the factor of two increase in risk. Personally, I think they should only have announced the twofold, age-corrected increase in risk, but that is just my opinion.

Even after correcting for age bias, the discrepancy between the twofold increase in risk of developing Parkinson's disease compared with earlier studies begs explanation. I wish I could provide one. When I first read the news blurbs on the 2019 announcement (yes, it was all over the news, not just in the United States, but worldwide), I was certain I would find some flaw in the study. However, after carefully rereading their conference abstract, which is a short summary of results to be presented, I could find no obvious flaws other than the small down-adjustment correcting for age bias that they already mentioned.

I decided to read up on appendicitis epidemiology, looking for possible clues, and learned some interesting facts. For reasons not well-understood, appendicitis rates appeared to have dropped in the United States during the period from 1970 to 1984, becoming stable at the lower rates since then [4]. Elective procedures to remove healthy appendices, so called incidental appendectomies, because they are typically piggybacked onto abdominal surgeries for unrelated causes, have become less popular, too. Does this mean we should expect a decrease in new Parkinson's disease cases in the next few decades as people born in 1970–84 reach old age? Another interesting fact is that the regions of the United States with the lowest rates of appendicitis are the East North Central, Middle Atlantic, and South Atlantic states. These regions include the Rust Belt states, where the rate of Parkinson's was found to be about twofold higher than the rest of the United States in the Medicare study described in Chapter 2 [5]. This is the opposite of what one would predict if having your appendix out increased Parkinson's risk. Hmm ...

Sheriff and colleagues have not yet published a peer-reviewed report on their study. It is hoped that when they do, they will employ the same birth year, gender, and geographical location matching for controls that was employed in the Swedish population study. If such a study is published before this book goes to press, I will write an update here ... Months go by ... Well, no update, except there seems to be a growing consensus that the perceived correlations between appendicitis and Parkinson's disease could be statistical anomalies [6].

Why all the interest in appendectomies? Those studies were inspired by the 2013 discovery in the laboratory of John M. Woulfe at the Ottawa Hospital Research Institute of large amounts of alpha-synuclein in the appendices of neurologically healthy patients, including alpha-synuclein in an aggregated, nonfibrillar form [7]. In contrast, Lewy bodies and neurites, which contain alpha-synuclein in its fibrillar form, had already been observed in the nerves of the intestines, but only in patients who also had Lewy pathology in the central nervous system, in the dorsal motor nucleus of the vagus nerve [8].

The presence of aggregated alpha-synuclein in the appendix suggested it might be a starting point for alpha-synuclein pathology spreading up the vagus nerve and into the brain. The Van Andel Research Institute study cited earlier also compared the aggregated alpha-synuclein from appendices with that from the substantia nigra, both from neurologically healthy controls and Parkinson's patients. In contrast to the substantia nigra, they found that alpha-synuclein in the appendix was highly enriched in truncated forms of the protein, with forms missing the last 14—15 amino acids at the C-terminus being predominant. Loss of these amino acids greatly speeds the aggregation process of alpha-synuclein in the test tube [9].

Why should alpha-synuclein be present at high levels in a truncated form prone to aggregation in what is supposedly a vestigial organ? The appendix is thought to be the evolutionary remnant of the cecum, a chamber that might have helped our early primate ancestors digest leaves. New theories challenge the idea that the appendix no longer has an important function, however. One theory proposes that the appendix serves as a reservoir for the healthy gut biome [10]. The idea is that when a dangerous microbe or toxin is ingested, the body responds with diarrhea, flushing the microbe or toxin out of the system. Because the appendix forms a small, dead-end offshoot of the intestines, it would retain its biota, which could then spread to repopulate the intestine after the diarrheal episode concludes. The appendix is also an important part of the immune system of the intestines, involved in fighting bacterial and viral infections. If the function of the appendix is to serve as a reservoir of healthy gut microbiota, it makes sense that the immune system should be especially vigilant there.

Could it be that alpha-synuclein has some role in immune function in the appendix? Using purified forms of alpha-synuclein, in 2016, a Korean research team led by Mi-Kyeong Jang and Jung Ro Lee tested the effect of the protein on bacteria and fungi [11]. Alpha-synuclein had antibacterial activity against four common bacterial strains, including *Escherichia coli* and *Staphylococcus aureus*, and against several yeast and mold strains. I regret to say that this exciting result should be viewed with caution, because another laboratory tried to replicate the results of the Korean team and failed [12].

Another reason for caution is that the purified alpha-synuclein used in scientific studies is almost always produced in special laboratory-grown strains of *E. coli* bacteria using artificially introduced DNA, so it is puzzling how the bacteria can survive with plentiful alpha-synuclein on the inside, but not when added to the medium outside the bacteria, as was done in the Korean study. On the other hand, extracellular alpha-synuclein is already known to be toxic to cultured cells, compromising membrane impermeability [13]. In fact, a study led by Kun Huang of Huazhong University, Wuhan, China saw that a truncated form of alpha-synuclein missing the last 19 amino acids, similar to that seen in the Grand Rapids study, rapidly formed nonfibrillar aggregates and was even more toxic to cultured neural cells than the full-length aggregated protein [14]. Perhaps the alpha-synuclein used in the Korean study was already aggregated, at least in part, but nowhere is this mentioned in their paper.

For alpha-synuclein to be an effective microbicide, it would somehow need to exit the cell to reach microbes outside. In fact, there is abundant evidence that alpha-synuclein can be secreted from cells [13]. The dominant mode of secretion appears to be exocytosis, where, upon fusion of a late endosome with the plasma membrane, alpha-synuclein within is released directly into the extracellular medium. "Late" means that they are not the immediate products of endocytosis (i.e., "early" endosomes), or the unmodified products of other vesicle-producing mechanisms within the cell. Late endosomes have been processed by the cell, with their contents modified, as well as the proteins and lipids that make up their membranes. One of these processes, chaperone-mediated autophagy, transports alpha-synuclein inside the endosomes, the same process that transports alpha-synuclein into lysosomes.

Cells can also release vesicles into the extracellular medium, either by vesicles directly budding from the plasma membrane or by exocytosis of multivesicular bodies, which are special late endosomes that contain numerous small vesicles within. A small amount of alpha-synuclein has been detected inside these extracellular vesicles [15]. Extracellular vesicles can enter other cells by fusing with the plasma membrane or via endocytosis, providing a pathway for aggregated alpha-synuclein to enter new cells.

Neighboring neurons can also connect to each other via long thin hollow tubes called nanotube tunnels, through which proteins such as alpha-synuclein, and even small organelles, can be transferred [16]. On the other hand, for microbes outside cells, only the alpha-synuclein released directly into the extracellular medium can interact with the outside of microbial plasma membranes, as in the antimicrobial experiments performed by the Korean research team.

There is one more obstacle to overcome before alpha-synuclein could conceivably attack microbes in the gut: the neurons of the vagus nerve are not in direct contact with the contents of the intestines. They are separated by the intestinal epithelium, a layer of cells that form the intestinal surface. One type of cell in this layer, called an enteroendocrine cell, has neuron-like qualities and has been found to produce alpha-synuclein and to form synapse-like connections with vagus nerve neurons [17]. If extracellular alpha-synuclein has a role in immune defense in the intestines, perhaps these enteroendocrine cells are the main source of the secreted protein.

The putative antimicrobial activity of alpha-synuclein would make it part of the innate immune system, the body's all-purpose, first line of defense against microbial invaders. Interestingly, DJ-1, encoded by the *PARK7* gene and associated with recessive familial Parkinson's disease, also appears to have an important role in innate immunity of the gut [18]. For years, researchers have suspected that secreted alpha-synuclein, especially in aggregated forms, might have some role in stimulating innate immune responses in the brain, causing microglia to attack nearby neurons, especially neurons already afflicted with alpha-synuclein pathology, thereby accelerating Parkinson's disease [19]. This stimulatory aspect of alpha-synuclein, if real, could conceivably complement its intrinsic antimicrobial activity; thus, the protein could have a dual role in the innate immune system, attacking microbes directly while stimulating nearby innate immune cells to attack.

There is additional, compelling evidence that alpha-synuclein really has an important role in immune response, ironically from alpha-synuclein transgenic and knockout mice. The irony is that it took so long for scientists to discover it. There are tens of thousands of papers on alpha-synuclein transgenic mice, and although a fraction of them have hinted at alterations in innate immune response, it was only recently discovered that pathology is much worse in alpha-synuclein transgenic mice exposed to normally occurring gut microbiota. How did such an important observation evade notice for so long? Transgenic and knockout mice are expensive to make and maintain, and they are typically raised in sterile facilities to minimize the chance that infectious agents might compromise the various ongoing studies on the transgenic and knockout mice housed there. Unless the gene being studied is known or suspected to be important for an immune response to infectious agents, tests involving dangerous pathogens are avoided.

The role of the gut microbiome in promoting alpha-synuclein pathology was explored in 2016 by a team led by Sarkis K. Mazmanian at the California Institute of Technology [20]. When transgenic mice that overexpressed alpha-synuclein were raised with normal gut microbiota, they had more motor dysfunction and more aggregated alpha-synuclein in their brains, even after treatment with antibiotics, compared with mice of the same transgenic strain raised in a germ-free environment. More intriguingly, when the germ-free transgenic mice were inoculated with human fecal samples (by gavage, using a tube inserted down the esophagus), the mice receiving samples from Parkinson's disease patients developed more severe motor impairment compared with those who received samples from healthy human controls.

The authors of the transgenic mouse study noted alterations in fatty acid chain length in the more debilitated animals, and they proposed this could account for the motor deficits. Shorter chain length lipids of the brain can promote activation of microglia in the brain, which can attack vulnerable neurons, such as those of the substantia nigra. However, in humans, alpha-synuclein—related pathology in the intestines typically occurs years, even decades, before movement-related symptoms occur. Chronic constipation likely has major impacts on the microbiome of the gut. In effect, the authors suggest that alterations in the microbiome change the profile of fatty acids extracted from food, and that the altered fatty acid profiles enhance microglia activation, thus promoting Parkinson's disease pathology in the brain.

I wonder, with regard to the altered fatty acid hypothesis, whether there might be an alternate explanation. Perhaps the intestinal innate immune response of patients has already been impaired by alpha-synuclein pathology, allowing more virulent microbiota to persist in the gut. More virulent microbiota could account for the more dramatic response observed in transgenic mice gavaged with samples from Parkinson's disease patients. In other words, the Parkinson's disease fecal samples could have induced a stronger innate immune response, resulting in more aggregated alpha-synuclein in the gut, which subsequently spread to the brain.

There is another part of our anatomy where the central nervous system is just one cell away from the dangerous microbes and toxins of the outside world: our noses. At the roof of the nasal cavity is the olfactory epithelium, a mucous membrane embedded with millions of scent-detecting neurons. The axons from these neurons bundle together and cross through small holes in the bone separating the nasal cavity from the brain, forming synapses with the central nervous system neurons of the olfactory bulb. Like the appendix, the olfactory epithelium and bulb are loaded with alpha-synuclein [21] and with beta- and gamma-synuclein as well.

In a 2017 report by a team led by Julianna J. Tomlinson and Michael G. Schlossmacher of the University of Ottawa, Canada, they used a microscope to look at thin slices from the heads of mice [12]. The nerve connections between the olfactory epithelium and bulb cross bone, which normally prevents obtainment of the razor-thin slices needed for the microscope slides, so they devised a new procedure. After treating the mouse heads with formalin to preserve them, they soaked them in 15% formic acid to dissolve the calcium in the bones, followed by more treatments with ethanol and other chemicals to remove water. Finally, the heads were soaked in paraffin and carefully cut into 5-μm-wide slices.

Although alpha-synuclein is usually observed to be most concentrated in presynaptic regions, they saw the protein throughout olfactory sensory neurons of the epithelium, even in the dendrite-like cilia extending to the mucous membrane surface of the nasal cavity. As with the appendix studies, the Ottawa team detected aggregated alpha-synuclein in the olfactory bulb and in the axons of the sensory neurons, but not in the cilia. They did not assess whether alpha-synuclein was present in a truncated form, however. If, compared with the appendix, olfactory alpha-synuclein is in a nontruncated form or modified in a different manner, this raises an intriguing possibility. Perhaps the alternate form of alpha-synuclein could result in the hypothetical differing strain of alpha-synuclein amyloid responsible for dementia with Lewy bodies (DLB) mentioned in Chapter 6.

The Ottawa team also looked at alpha-synuclein knockout mice. They wanted to see whether alpha-synuclein influenced immune defenses, so they performed two experiments, one exposing the mice to a bacterium and the other exposing them to a virus. The bacterium was *Salmonella typhimurium*, which in mice induces lethal sepsis and is closely related to the bacterium that causes typhoid fever in humans. The number of bacteria in the spleens of knockout mice exceeded that seen in controls, which implied that the bacteria multiplied more easily in the knockouts. The virus they used was reovirus-T3D, which can cause fatal encephalitis. The knockout mice developed encephalitis more often and had decreased survival compared with wild-type controls.

The Ottawa team was not the first to notice a protective effect from alpha-synuclein against viral infections. A group of researchers led by J. David Beckham at the University of Colorado School of Medicine, Aurora, Colorado, saw a similar effect in alpha-synuclein knockout mice infected with West Nile virus [22]. The virus was far more lethal to the knockout mice, and the amount of virus they saw in their brains was orders of magnitude higher than that seen in their littermates with one or two wild-type copies of the alpha-synuclein gene. To make sure this was not a fluke result specific to West Nile virus, they also infected knockout mice with Venezuelan equine encephalitis virus and saw similar results. The amount of virus outside the central nervous system was the same in knockout mice and controls; whatever the effect of alpha-synuclein, it seemed to be specific to the neurons of the central nervous system.

The Colorado team made another important observation. In brains from human victims of West Nile virus encephalitis, they saw higher levels of alpha-synuclein compared with controls. In fact, a report describes a patient who went on to develop DLB after contracting West Nile virus [23]. Similarly, in cultured brain neurons from mice, those exposed to the virus had double the amount of alpha-synuclein compared with mock-infected neurons (cultured neurons subjected to the same inoculation protocol, but using a solution lacking West Nile virus). The increased alpha-synuclein they saw was inside the neurons, especially near the endoplasmic reticulum and Golgi apparatus, which are involved in expressing and modifying proteins destined for the various lipid membranes of the cell.

West Nile virus and Venezuelan equine encephalitis virus are membrane-enveloped RNA viruses, and, as you might guess from their name, their genetic information is encoded in strands of RNA. The Colorado team hypothesized that the antiviral role of alpha-synuclein might be to interfere with the expression and processing of viral membrane proteins. In particular, they pointed to previous research showing that alpha-synuclein can interfere with a protein, Rab1, involved in the transport of newly expressed membrane proteins between the endoplasmic reticulum and Golgi apparatus. In contrast, reovirus-T3D used in the Ottawa study, although an RNA virus, is not enveloped by a membrane. Thus, interfering with membrane protein expression would not appear to be a plausible mechanism for how alpha-synuclein protects against reovirus-T3D infection.

Despite the lack of a membrane for reovirus-T3D, it has something in common with West Nile and Venezuelan equine encephalitis viruses: it enters the cell via the same pathway. Viruses enter cells by one of three pathways. Many membrane-enveloped viruses fuse directly with the plasma membrane, such as HIV and influenza viruses, spilling their contents directly into the cellular interior. Bacteriophages lack a plasma membrane, but they can inject their genetic contents through the plasma membrane using a molecular apparatus resembling a cross between an oil derrick and a hypodermic needle (as I said before, do an Internet search on them!).

The third way viruses can enter the cell is via endocytosis. Binding to specific receptors at the cell surface, some bound viruses will be in membrane regions budding inward into the cell, eventually pinching off from the plasma membrane, releasing endosomes into the cell cytoplasm. For viruses with a membrane envelope, they can then release their contents into the cell by fusing with the endosomal membrane. Such is the case for West Nile and Venezuelan equine encephalitis viruses. For viruses with no membrane envelope, such as reovirus-T3D, the exact mechanism of endosomal exit is less clear, perhaps through membrane rupture or via specific transport of genetic contents across the endosomal membrane. It could be that this is where alpha-synuclein might combat viral infection, interfering somewhere in the endocytosis viral entry pathway.

Regardless of the exact mechanism alpha-synuclein uses to combat infectious agents, I find it remarkable that the protein seems to do so both inside the cell, against viruses, and outside, against bacteria. It is so remarkable, in fact, that the skeptic in me questions whether it is too good to be true. Certainly, more research is needed, especially with respect to the putative antimicrobial activity of alpha-synuclein outside the cell. However, there is another surprising connection between another Parkinson's disease—associated protein, LRRK2, and immune response in the gut, in particular with respect to Crohn's disease.

Crohn's disease is an inflammatory bowel disease involving abdominal pain, diarrhea, weight loss, and fever. The symptoms can be severe and can lead to life-threatening complications, requiring surgery to remove the most inflamed intestinal sections. Like Parkinson's disease, it is thought to result from a combination of genetic and environmental factors, perhaps involving bacteria in the gut. Unlike Parkinson's disease, smoking increases the risk for developing the disease. Coincidently, like Parkinson's disease, it occurs more frequently in Ashkenazi Jewish populations.

Genome-wide association studies (GWAS) on Crohn's disease patients have identified genetic risk factors in several genetic loci, including the *LRRK2* gene. To pinpoint the actual mutations conferring risk for and protection from the disease, a study in 2018 including over 50 researchers, led by Judy H. Cho at the Yale University School of Medicine and Inga Peter at the Icahn School of Medicine at Mount Sinai, performed complete DNA sequencing of the protein coding regions of the genomes of 50 Ashkenazi Jewish individuals, identifying over 4000 new potential genetic variants [24].

The team then had a new gene chip created, including the coding region polymorphisms they discovered plus those previously studied, to look for these variants in a larger cohort of nearly 25,000 test subjects, with over 8000 Ashkenazi Jewish participants. Of the participants, roughly a quarter had Crohn's disease, another quarter had Parkinson's disease, and about half had neither disease. The large number of participants provided the statistical power needed to identify specific coding region variants responsible for conferring increased risk for developing Crohn's disease and other variants protecting against it.

Comparing the gene chip results of participants with Crohn's disease versus controls, they identified one variant that was strongly correlated with increased risk, and two others that together conferred a protective effect. Repeating the analysis for those with Parkinson's disease, the researchers were able to identify previously known variants known to affect risk, including the most common polymorphism, G2019S, in which the amino acid glycine in position 2019 of the LRRK2 protein sequence is replaced with serine. More exciting, they found that most of the risk modifiers for Crohn's disease were correlated with those for Parkinson's. Variants that decreased risk for developing Crohn's disease also reduced the risk for developing Parkinson's disease, although not always to the same degree, and likewise for variants that increased risk.

LRRK2 is a large, complex protein of over 2500 amino acids, with two enzymatic domains and several other regions likely involved in protein interactions. Most genetic variants associated with Crohn's and Parkinson's diseases occur in the enzymatic domains. The scientists sorted the disease-correlated mutants into two categories: loss-of-function and gain-of-function. For mutations in the protein-coding regions of genes, loss-of-function can result from the mutation reducing enzymatic activity or impeding important interactions with proteins or other biomolecules. It can also be the result of the mutation destabilizing the structure, interfering with correct folding of the protein, or reducing its lifetime in the cell, or it can even result in incomplete translation of the protein. Conversely, gain-of-function means the protein activity, stability, or interactions, are enhanced or altered in a damaging way. The mutations associated with increased risk for Crohn's and Parkinson's diseases are the gain-of-function type.

Inflammatory factors released by the immune system, including the protein interferon-gamma, strongly increase translation and expression of the LRRK2. One of the enzymatic domains of LRRK2 is a kinase, and the other domains of LRRK2 presumably help transport and target the protein to its various kinase substrates: that is, the proteins that become phosphorylated by the kinase. The other enzymatic domain of LRRK2, with the awkward name Ras Of Complex proteins (ROC), binds a molecule called guanosine triphosphate (GTP) and regulates the activity of the kinase domain. With GTP bound, LRRK2 is in its on state, and kinase activity increases in the cell. Typically for GTP-binding proteins, cleavage of phosphate from GTP leaves guanosine diphosphate (GDP) bound to the enzyme, putting the protein in its off state. However, the situation seems to be more complicated in the case of LRRK2.

A report by a Belgian and US team of researchers led by Jean-Marc Taymans, including my National Institutes of Health colleague Mark R. Cookson, found that purified LRRK2 kinase activity in a test tube was unaffected by either GTP or GDP binding [25]. However, when cell extract was added, activation by GTP was restored, indicating that an additional factor or factors in the cell, perhaps other proteins, also seem to be required for ROC function. Thus, the gain-of-functions mutations in the LRRK2 GTP-binding domain might not directly alter the enzymatic properties of the domain, but instead could affect interactions with additional cofactors necessary for its activity.

The connection between LRRK2 kinase activity and alpha-synuclein pathology is also more complicated than many might have guessed. The most common Parkinson's disease—related LRRK2 mutation, G2019S, causes greater kinase activity, as does the mutation most strongly correlated with Crohn's disease risk, N2081D, which suggests that phosphorylation by LRRK2 has a role in the pathology of both diseases [24]. That a significant amount of the alpha-synuclein in Lewy bodies is phosphorylated initially led many to hypothesize that LRRK2 might be the responsible kinase; however, subsequent studies showed that alpha-synuclein becomes phosphorylated by other kinases [26]. Instead, the most interesting target of the LRRK2 kinase domain is another GTP-binding protein, RAB7L1, and its gene happens to lie in a locus of increased risk for Parkinson's disease in GWAS.

In a 2018 study led by Cookson and Andrew B. West of the University of Alabama at Birmingham, the researchers discovered that not only was RAB7L1 a target for the LRRK2 kinase domain, it was much more readily phosphorylated by the mutant G2019S form [27]. In addition, interaction with RAB7L1 facilitated phosphorylation of the LRRK2 ROC domain by its own kinase domain. In an earlier study from the laboratory of Steven Finkbeiner at the University of California, San Francisco, which included the ubiquitous Cookson as a coauthor, the researchers presented evidence that autophosphorylated LRRK2 had a longer lifetime in cultured neurons, leading to higher levels of the protein [28]. The increased LRRK2 levels were toxic in cultured neurons and also promoted alpha-synuclein pathology. The fact that increased LRRK2 levels were toxic to cultured neurons, even without aggregated alpha-synuclein, might help explain why some Parkinson's disease patients with LRRK2 mutations have less Lewy pathology upon autopsy than others without mutations.

So, what are the connections among immune response, LRRK2, and Parkinson's disease? Both LRRK2 and RAB7L1 are part of a large fleet of proteins involved in vesicle trafficking inside cells. In particular, they affect autophagy, the process by which the cell engulfs parts of itself for recycling inside lysosomes. During an infection, an important part of the cellular immune response is to engulf the invaders using endocytosis, especially in macrophage cells. The resulting endosomes become transported to lysosomes to be digested, and many of their partially digested components get transported back to the cell surface, where they are displayed by cell surface proteins that help stimulate the adaptive immune response in white blood cells. Thus, a role for LRRK2 could be to inhibit autophagy so that more of the cell's lysosomal machinery can be devoted to digestion and processing of the engulfed invaders [29].

The activity levels of LRRK2 in Parkinson's patients are higher than in controls, even in patients with no LRRK2 mutations. In fact, a correlation between LRRK2 and alpha-synuclein protein levels has been observed in numerous in vivo and in vitro studies [30]. Increased LRRK2 levels caused increased expression of the *SNCA* gene in neuronal cell cultures [31]. Also, by impeding autophagy, less alpha-synuclein is degraded, which also boosts its levels in cells. The apparent protective role in innate immunity displayed by alpha-synuclein suggests that the coordinated increase in LRRK2 and alpha-synuclein levels could be part of the normal response by the innate immune system to infection.

LRRK2 appears to affect glial cells, too. A type of glial cell called an astrocyte supports neurons and their synapses. Normally, astrocytes can take up extracellular alpha-synuclein and degrade it, but in cell cultures with astrocytes containing the common G2091S mutation of LRRK2, which causes increased levels of the enzyme, the situation is reversed, with astrocytes accumulating alpha-synuclein [32]. Perhaps, astrocytes also have a similar role in innate immune response. Such a response could exacerbate Parkinson's disease progression, because neural cells in the same culture plate as the mutant astrocytes were observed to take up the alpha-synuclein released by the astrocytes.

We have already touched on a putative extracellular microbicidal role for alpha-synuclein, perhaps in an aggregated form, but what role might elevated intracellular levels of alpha-synuclein have in protecting against infection? My first guess was that because alpha-synuclein has a role in synaptic vesicle maintenance, higher levels might somehow redirect virus-containing endosomes toward the synaptic vesicle pathway, where they would sit inert until released back into the synapse, buying time for the adaptive immune system to mount a response. However, viral particles tend to be larger than synaptic vesicles, which makes this scenario questionable. A more plausible scenario is that excess alpha-synuclein could have a protective role by slowing or otherwise interfering with endosomal processing and trafficking. Many studies have noted disruption of various endosomal pathways by alpha-synuclein, such as vesicle trafficking between the endoplasmic reticulum and Golgi apparatus, as mentioned earlier in the West Nile virus study.

The olfactory and vagus neurons vulnerable to Lewy pathology have something in common: they are dopaminergic. Alpha-synuclein is expressed in many types of neurons throughout the brain, not just in dopaminergic neurons, but still, it makes me wonder whether dopamine might also have a coordinated role in the putative innate immune function of alpha-synuclein. As mentioned in earlier chapters, dopamine can easily be altered into toxic forms, by reactive oxygen species arising from mitochondria, for instance. Furthermore, increased alpha-synuclein levels can promote the production of these dangerous molecules by mitochondria. Perhaps in response to infection, dopaminergic neurons make and release toxic forms of dopamine to attack the microbial invaders, but this is speculation on my part.

The alpha-synuclein Lewy pathology leading to most Parkinson's disease cases almost certainly originates in the vagus nerve innervating the intestines, in the neurons of the olfactory epithelium and bulb, or in both locations at the same time, basically Braak stage 1. But again, we are left with the conundrum of why only humans, of all vertebrates, develop the disease. The vagus nerve axons that come from the vulnerable region of the brain stem, called the dorsal motor nucleus, are nonmyelinated, like the axons that go from the substantia nigra to the striatum. In contrast, olfactory sensory neuron axons are surrounded by a special type of glial cells, called olfactory ensheathing glia, but like the substantia nigra and dorsal motor nucleus axons, they have small diameters, which suggests that more energy is also needed to maintain them in a functioning state.

Perhaps the same modifications in human evolution that gave rise to the greater energy use in substantia nigra neurons also resulted in greater energy use in similar neurons in other parts of the nervous system. Although the work of Herculano-Houzel and others suggests that human brain neurons do not consume more energy on average than those of other primates, it could be that some classes of neurons have evolved to consume more whereas others, in particular the largest, most myelinated neurons, consume less, balancing average energy use. It is also possible that humans have higher levels of alpha-synuclein in their gut and olfactory neurons, but as far as I know, there have been no studies comparing alpha-synuclein levels with other species in these two regions.

The apparent role of alpha-synuclein in innate immunity also opens the possibility that pathogens specific to ancestral humans, following the split from chimpanzees, could have been the evolutionary factor that led to the alterations responsible for Lewy pathology. What that pathogen might be or have been, I have no idea. The skeptic in me wonders why, in the decades of molecular biology and genetics research, no evidence of such a human-specific culprit has surfaced. On the other hand, the protective role of alpha-synuclein in innate immunity was discovered only recently despite decades of molecular biology and genetics research, so who knows what amazing discoveries the future might hold?

Alpha-synuclein is not alone in its apparent role in innate immunity; amyloid-beta and prion protein (PrP) also appear to have antimicrobial roles [33,34]. Amyloid-beta has even been caught in the act, surrounding a brain-invading bacterium with amyloid plaque-like deposits [35]. The bacterium in that case was *Porphyromonas gingivalis*, one of the bacteria responsible for gum disease. Keep those gums healthy! For its part, PrP has been shown to protect against an impressive array of viruses in cell cultures and laboratory animals [36]. Curiously, there is one exception for PrP: it seems to promote infection by herpes simplex virus [37].

The spread of amyloid-beta pathology in the brain mirrors the routes by which alpha-synuclein pathology spreads from the olfactory bulb [38], which suggests that microbes might invade the brain along the same routes in Alzheimer's, Parkinson's, and DLB diseases. If the amyloid pathology along these routes is actually the remnant of the innate immune response of the brain, this could supplant the prion hypothesis, at least partially. On the other hand, numerous experiments have already established that injected alpha-synuclein fibrils can trigger the spread of Lewy pathology on their own. However, this does not rule out that invading microbes, traveling up the vagus nerve or invading the olfactory bulb, might invade further into the brain, triggering pathology as they go. Indeed, Braak and colleagues proposed just such a mechanism to explain their staging hypothesis [8].

Although the antimicrobial activity of amyloid-beta appears to involve the peptide in its amyloid form [39], for PrP the situation is less clear [40]. PrP is expressed in several isoforms [41], which are alternate translations of its gene, some of which are intracellular and some of which are found at the cell surface [42]. It is also not known which PrP isoforms are most important for its antimicrobial effects. Like alpha-synuclein, our knowledge of the antimicrobial mechanisms of PrP is still preliminary. We do not yet know the structure of their protective forms (monomeric, oligomeric, or amyloid), or whether their respective, protective forms differ inside and outside the cell.

Three different amyloid-forming proteins, alpha-synuclein, amyloid-beta, and PrP, all seem to be involved in innate immunity. Obviously, this hints that their protective roles might somehow be related to their amyloid, or perhaps oligomeric, forms, but I caution against over-generalizing. In the arms race between vertebrates and the microbes that infect them, evolution will favor any advantage irrespective of preconceived notions. Although the amyloid forms of amyloid-beta and PrP can occur naturally in nonhumans, no naturally occurring amyloid form of alpha-synuclein has been observed in nonhumans (naturally occurring aggregated alpha-synuclein has been observed in mouse lemurs, but it has not yet been determined whether it is in fibrillar form [43]). Despite my penchant for unbridled theorizing, in this case I prefer to wait for results from future experimental studies, especially with regard to alpha-synuclein.

References

[1] Sheriff MZ, Mansoor E, Cooper GS. Parkinson's disease is more prevalant in patients with appendectomies: a national population-based study. In: [Online] The Society for Surgery of the Alimentary Tract: 60th Annual Meeting; May 18–21, 2019. http://meetings.ssat.com/abstracts/2019/739.cgi.

[2] Killinger BA, et al. The vermiform appendix impacts the risk of developing Parkinson's disease. Science Translational Medicine October 31, 2018;10(465):eaar5280.

[3] Svensson E, et al. Appendectomy and risk of Parkinson's disease: a nationwide cohort study with more than 10 years of follow-up. Movement Disorders December 2016;31(12):1918–22.

[4] Addiss DG, et al. The epidemiology of appendicitis and appendectomy in the United States. American Journal of Epidemiology November 1990;132(5):910–25.

[5] Willis AW, et al. Geographic and ethnic variation in Parkinson disease: a population-based study of US medicare beneficiaries. Neuroepidemiology April 2010;34:143–51.

[6] Lu HT, et al. Lack of association between appendectomy and Parkinson's disease: a systematic review and meta-analysis. September 19, 2019. https://doi.org/10.1007/s40520-019-01354-9.

[7] Gray MT, et al. Alpha-synuclein in the appendiceal mucosa of neurologically intact subjects. Movement Disorders July 2014;29(8):991–8.

[8] Braak H, et al. Staging of brain pathology related to sporadic Parkinson's disease. Neurobiology of Aging March-April;24(2):197–211.
[9] Li W, et al. Aggregation promoting C-terminal truncation of alpha-synuclein is a normal cellular process and is enhanced by the familial Parkinson's disease-linked mutations. Proceedings of the National Academy of Sciences United States of America February 8, 2005;102(6):2162–7.
[10] Girard-Madoux MJH, et al. The immunological functions of the Appendix: an example of redundancy? Seminars in Immunology April 2018;36:31–44.
[11] Park SC, et al. Functional characterization of alpha-synuclein protein with antimicrobial activity. Biochemical and Biophysical Research Communications September 16, 2016;478(2):924–8.
[12] Tomlinson JJ, Shutinoski B, Dong L, Meng F, Elleithy D, et al. Holocranohistochemistry enables the visualization of α-synuclein expression in the murine olfactory system and discovery of its systemic anti-microbial effects. Journal of Neural Transmission 2017;124(6):721–38.
[13] Emmanouilidou E, et al. Cell-produced alpha-synuclein is secreted in a calcium-dependent manner by exosomes and impacts neuronal survival. Journal of Neuroscience May 19, 2010;30(20):6838–51.
[14] Ma L, et al. C-terminal truncation exacerbates the aggregation and cytotoxicity of α-Synuclein: a vicious cycle in Parkinson's disease. Biochimica et Biophysica Acta - Molecular Basis of Disease December 2018;1864(12):3714–25.
[15] Gustafsson G, et al. Secretion and uptake of α-synuclein via extracellular vesicles in cultured Cells. Cellular and Molecular Neurobiology 2018;38(8):1539–50.
[16] Abounit S, et al. Tunneling nanotubes spread fibrillar α-synuclein by intercellular trafficking of lysosomes. The EMBO Journal October 4, 2016;35(19):2120–38.
[17] Chandra R, et al. α-Synuclein in gut endocrine cells and its implications for Parkinson's disease. JCI Insight June 15, 2017;2(12):e92295.
[18] Singh Y, et al. DJ-1 (Park7) affects the gut microbiome, metabolites and development of Innate Lymphoid cells (ILCs). BioRxiv September 19, 2019. https://doi.org/10.1101/776005v1.
[19] Ferreira SA, Romero-Ramos M. Microglia response during Parkinson's disease: alpha-synuclein intervention. Frontiers in Cellular Neuorscience 2018;12:247.
[20] Sampson TR, et al. Gut microbiota regulate motor deficits and neuroinflammation in a model of Parkinson's disease. Cell December 1, 2016;167(6):1469–80.
[21] Duda JE, et al. The expression of alpha-, beta-, and gamma-synucleins in olfactory mucosa from patients with and without neurodegenerative diseases. Experimental Neurology December 1999;160(2):515–22.
[22] Beatman EL, et al. Alpha-synuclein expression restricts RNA viral infections in the brain. Journal of Virology March 15, 2016;90(6):2767–82.
[23] Segers K, et al. West Nile virus neuroinvasive disease accelerating probable dementia with lewy bodies. Alzheimer Disease and Associated Disorders August 17, 2020. https://doi.org/10.1097/WAD.0000000000000405.
[24] Hui KY, et al. Functional variants in LRRK2 confer pleiotropic effects on risk for Crohn's disease and Parkinson's disease. Science Translational Medicine January 10, 2018;10(423):eaai7795.
[25] Taymans JM, et al. LRRK2 kinase activity is dependent on LRRK2 GTP binding capacity but independent of LRRK2 GTP binding. PLoS One August 12, 2011;6(8):e23207.
[26] Mbefo MK, et al. Phosphorylation of synucleins by members of the Polo-like kinase family. Journal of Biological Chemistry January 22, 2010;285(4):2807–22.
[27] Liu Z, et al. LRRK2 phosphorylates membrane-bound Rabs and is activated by GTP-bound Rab7L1 to promote recruitment to the trans-Golgi network. Human Molecular Genetics January 15, 2018;27(2):385–95.

[28] Skibinski G, et al. Mutant LRRK2 toxicity in neurons depends on LRRK2 levels and synuclein but not kinase activity or inclusion bodies. Journal of Neuroscience January 8, 2014;34(2):418–33.
[29] Takagawa T, et al. An increase in LRRK2 suppresses autophagy and enhances Dectin-1-induced immunity in a mouse model of colitis. Science Translational Medicine June 6, 2018;10(444):eaan8162.
[30] Guerreiro PS, et al. LRRK2 interactions with α-synuclein in Parkinson's disease brains and in cell models. Journal of Molecular Medicine April 2013;91(4):513–22.
[31] Carballo-Carbajal I, et al. Leucine-rich repeat kinase 2 induces alpha-synuclein expression via the extracellular signal-regulated kinase pathway. Cellular Signalling May 2010;22(5):821–7.
[32] di Domenico A, et al. Patient-specific iPSC-derived astrocytes contribute to non-cell-autonomous neurodegeneration in Parkinson's disease. Stem Cell Reports February 12, 2019;12(2):213–29.
[33] Alais S, et al. Functional mechanisms of the cellular prion protein (PrP(C)) associated anti-HIV-1 properties. Cellular and Molecular Life Sciences April 2012;69(8):1331–52.
[34] Kumar DK, et al. Amyloid-β peptide protects against microbial infection in mouse and worm models of Alzheimer's disease. Science Translational Medicine May 25, 2016;8(340):340ra72.
[35] Dominy SS, et al. Porphyromonas gingivalis in Alzheimer's disease brains: evidence for disease causation and treatment with small-molecule inhibitors. Science Advances January 2019;5(1):eaau3333.
[36] Chida J, et al. Rion protein protects mice from lethal infection with influenza A viruses. PLoS Pathogens May 2018;14(5):e1007049.
[37] Thackray AM, Bujdoso R. PrP(c) expression influences the establishment of herpes simplex virus type 1 latency. Journal of Virology March 2002;76(5):2498–509.
[38] Rey NL, Wesson DW, Brundin P. The olfactory bulb as the entry site for prion-like propagation in neurodegenerative diseases. Neurobiology of Disease January 2018;109(Pt B):226–48.
[39] Gosztyla ML, Brothers HM, Robinson SR. Alzheimer's amyloid-β is an antimicrobial peptide: a review of the evidence. Alzheimer's Disease 2018;62(4):1495–506.
[40] Lathe R, Darlix JL. Prion protein PRNP: a new player in innate immunity? The aβ connection. Journal of Alzheimer's Disease Reports 2017;1(1):263–75.
[41] Makrinou E, Collinge J, Antoniou M. Genomic characterization of the human prion protein (PrP) gene locus. Mammalian Genome December 2002;13(12):696–703.
[42] Kikuchi Y, et al. Hypoxia induces expression of a GPI-anchorless splice variant of the prion protein. FEBS Journal June 2008;275(11):2965–76.
[43] Canron MH, et al. Age-dependent α-synuclein aggregation in the *Microcebus murinus* lemur primate. Scientific Reports 2012;2:910.

CHAPTER 13

Circuitry—The discovery of neural pathways involved in Parkinson's disease and deep brain stimulation

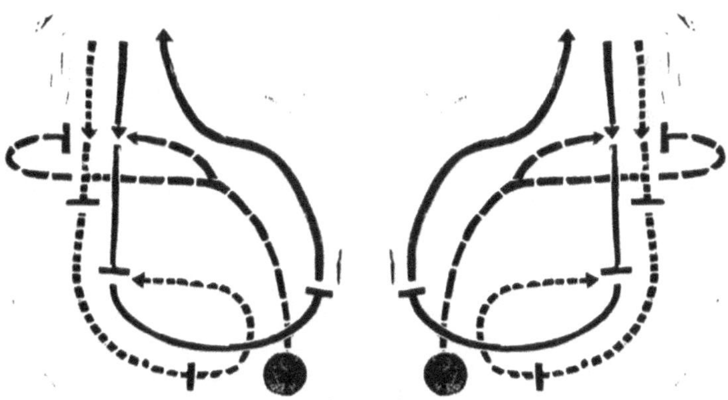

It was the golden age of neurological exploration. After Luigi Galvani's discovery in the late 1770s that electricity applied to the sciatic nerve of a dead frog caused the target leg muscles to contract, others started using electricity to probe animal brains, leading to the discovery of the motor cortex. By the late 19th century, neurosurgeons treating epilepsy were using electrical stimulation to pinpoint the motor regions of the human brain responsible for seizures in order to resect them.

Perhaps you have seen the cartoon diagram of the cortical mapping, the motor cortex homunculus. The diagram starts at the rearmost portion of the frontal lobe, inside the medial longitudinal fissure at the top of the brain that divides the left and right lobes, then going out and across the top of the cortex and down the side. Inside the fissure, the homunculus starts with the genitals, followed by the toes, feet, leg, and hip. Then, emerging from the fissure, the mapping continues with the trunk and shoulder, then down the arm to the elbow, wrist, hand, and fingers, starting with the pinky and

Synuclein and the Coelacanth
ISBN 978-0-323-85707-9
https://doi.org/10.1016/B978-0-323-85707-9.00003-4

ending with the thumb. Further continuing down the side of the frontal lobe is the motor cortex for the brow, eyes, face, lips, jaw and tongue, and swallowing control. The motor cortex of the right lobe of the brain controls the left side of the body, and vice versa for the left lobe.

Pioneering neurosurgeons at turn of the 19th into the 20th century started to perform brain surgeries under local anesthesia, which allowed patients to describe sensations associated with electrical stimulation of the cortex. Just to the rear of the motor cortex was the sensory cortex in the parietal lobe, with a similar homunculus diagram. One of these pioneering neurosurgeons was Victor Horsley of the University College Hospital in London. A staunch supporter of medical reform, women's suffrage, and the temperance movement, he was also one of the first surgeons to use electrical stimulation maps of the brain to improve epilepsy surgery. He died of heatstroke in Iraq while serving as a field surgeon for the British armed forces during World War I.

In 1908, Horsley, along with Robert H. Clarke, a physiologist and mathematician, invented the stereotactic frame, a metal frame attached to the head that allowed one to guide electrodes into the brain with unprecedented accuracy [1]. Before entering the military to serve his country, Horsley had used the frame to map the brain regions of nonhuman primates and cats. The first actual use of the frame in humans dates to 1933. German neurosurgeon Martin Kirschner used it to treat trigeminal neuralgia by inserting an electrode into the trigeminal nerve, which enervates the face, and ablating it, destroying it with the heat caused by the current in the electrode tip [2].

James Parkinson in his 1817 seminal work *"An Essay on the Shaking Palsy"* noted that the resting tremor in one patient disappeared after a stroke that left the patient half-paralyzed, which suggested that the part of the brain responsible for the tremor had been part of brain tissue damaged by the stroke. The first attempts to treat Parkinson's disease using surgery tried the same approach as used for epilepsy: finding the motor region of the cortex corresponding to the hand with the worst tremor and removing that region. Although this reduced the tremor, it left patients with a partially paralyzed hand. The surgeons also tried to reduce tremor by cutting the nerve roots from the spinal cord leading to the hand, but not only did this cause partial paralysis, it also resulted in partial loss of sensation in the hand [3]. Furthermore, neither approach improved the other movement-related symptoms, rigidity and bradykinesia.

In the late 1930s, neurosurgeon Russell Meyers at Kings County Hospital in New York suspected that the best target for reducing Parkinson's symptoms was somewhere in the basal ganglia, brain regions beneath the neocortex. The problem was that these regions were difficult to reach, requiring the outer brain regions to be cut through to get access to them. His fellow neurosurgeons warned him against such surgery, especially because one region he wanted to target, the caudate nucleus of the striatum, was thought at the time to be involved in consciousness. Then, in 1939, he chanced to examine an emergency room patient who had walked into an airplane propeller, which had cut through her brain, including her striatum [4]. She was conscious, able to answer questions. This convinced Meyers that his colleagues were probably wrong, and that surgery to remove part of the caudate nucleus might not affect consciousness after all.

Still, the surgery that Meyers proposed was dangerous, something to be undertaken only under desperate circumstances. Later that year, he found a candidate. The patient, a woman, had contracted encephalitis lethargica, leaving her with severe tremor in one arm. Her life was effectively ruined, her husband had left her, and she was unemployable. She considered suicide and had requested the arm be amputated. Meyers performed the neurosurgery, cutting through part of the cortex well away from the motor region, and removed the front two-thirds of the caudate nucleus on the side opposite the arm with the tremor. Once removed, the tremor immediately stopped, although some tremor returned in the following weeks.

In the next few years, Meyers operated on more Parkinson's patients, systematically targeting other basal ganglia regions in addition to the caudate nucleus. Eventually, he achieved the best reduction in symptoms by targeting other regions. One region was the thalamus, an important relay center for sensory and motor signaling in the brain. Along with the thalamus, symptoms could be reduced by cutting another region, the ansa lenticularis, part of the nerve tract between the thalamus and another nearby region, the globus pallidus. Not only was tremor reduced by targeting these regions, rigidity and bradykinesia were often reduced as well.

Meyers had demonstrated the importance of the basal ganglia in Parkinson's disease, but the operations he and others performed had too many complications and a 12% death rate, so neurosurgeons needed a safer alternative. Ironically, the safer alternative, stereotactic surgery, had been known for decades. In 1947, Ernest A. Spiegel and Henry T. Wycis at Temple Medical School in Philadelphia published the first paper on deep brain stereotactic (they preferred the alternate term "stereotaxic") surgery in humans [5].

The stereotactic technique required only a small hole drilled through the skull, through which a small hollow tube could be targeted to areas deep in the brain. Inside the tube would be a thin electrode, used both to stimulate potential targets and to ablate them once the desired target region had been located. Using x-ray imaging, surgeons could plan out the best angles to insert the electrodes, avoiding blood vessels and taking advantage of the fluid-filled ventricles. Early innovators used curved tubes, devising arced trajectories that allowed even less disruption of intervening neural tissue. X-ray images could be taken during the surgery, to make sure the surgeons were on target.

Stereotactic neurosurgery took off in the 1950s, with surgeons around the world using the technique to treat psychiatric disorders, pain, tumors, and movement disorders such as Parkinson's disease [6]. Eventually, the globus pallidus won out as the most popular target for treating tremor in Parkinson's disease and essential tremor. This location had been discovered by a fortunate accident. In 1952, neurosurgeon Irving S. Cooper inadvertently severed a brain artery supplying blood to the globus pallidus in a Parkinson's patient, causing the brain tissue to die [7]. The patient awoke with the tremor gone but with no serious side effects. Finally, Parkinson's patients had a proven, effective treatment. Then, in just a few short years, the situation changed completely. The golden age of neurosurgery was over.

The situation for Parkinson's patients did not change in a bad way. Quite the opposite; it was something of a medical miracle, at least at first. Levodopa arrived on the scene in the 1960s. Instead of potentially life-threatening brain surgery à la Frankenstein (the stereotactic frame was literally bolted to the skull for the operation), you could just pop a pill. Levodopa, which was converted to dopamine inside the brain, would quell Parkinson's disease symptoms for hours at a time. Physicians and patients alike embraced the new drug therapy[a]. There was a paradigm shift in how neurological diseases were viewed. Instead of faulty brain wiring, neurological diseases came to be seen as chemical imbalances, jokingly referred to as the "brain soup" picture of neurology. Neurosurgeons were abandoning the field. This exodus was not just owing to patients opting out of surgery because of the new miracle drug, however; public opinion had turned against neurosurgery in general.

The reason for the shift in public opinion had little to do with stereotactic surgery; it mostly had to do with lobotomies. The first frontal

[a] Another drug, apomorphine, was already being used to treat Parkinson's disease starting in the 1950s. It was just as effective as levodopa, but for a shorter duration, and it had to be injected, while levodopa can be taken orally. There has been renewed interest in the drug as a rescue treatment for acute "off" period symptoms, and an inhalable formulation has received FDA approval.

lobotomies, or leukotomies, as they were initially known, were done in the 1930s [8]. The procedure, which cut the white matter connecting the frontal lobes to the thalamus, could turn a violent psychotic into a docile patient, albeit with a somewhat cowlike personality. Until this time, there had been no effective treatment for most violently insane patients, and institutions and caretakers alike were desperate for any reduction in the burden that such patients placed on families and society in general. So great was the initial admiration for the procedure that its first practitioner, Antonio Egas Moniz of Lisbon, Portugal, received the Nobel Prize in Physiology or Medicine in 1949.

Unfortunately, lobotomies were touted as more than just a last-ditch therapy for incurably insane patients who were a threat to themselves and others. The procedure was promoted for patients with any distressing mental condition, those with depression, insomniacs, homosexuals, and even tantrum-prone children, promising to exchange their anguished mental state with placid serenity. The foremost promoter, and the one who coined the term "lobotomy," was an American, Walter J. Freeman [9].

Freeman modified the procedure, transforming it from a delicate hospital operation into a much simpler procedure that could be performed in a doctor's office. First, knocking out the patient with an electric shock to the brain (electroshock therapy was already well-established), he would maneuver around the eyeball to the rear of the eye socket with an ice pick, and then with a mallet, he would drive the ice pick through the bone into the brain to a predetermined depth and pivot the ice pick in specific directions and angles to cut the frontal lobe connections. Then, he would repeat the procedure with the other eye socket. He had determined the depth, directions, and angles earlier by experimenting on cadavers. Ultimately, he was to perform his new procedure on about 2500 patients (he charged just $25 per procedure). He called his procedure the transorbital lobotomy, although it was also commonly known as the ice pick lobotomy.

The biggest problem with lobotomies was not that you might be left with a cowlike personality, it was that you might not be left with any personality at all. Roughly a third of patients were turned into vegetables or died owing to complications from the procedure. The sister of John F. Kennedy, Rosemary, was left permanently incapacitated from a lobotomy performed by Freeman and his colleague James W. Watts. She was institutionalized for the rest of her life.

In the 1950s, the procedure lost its most ardent early supporters, medical personnel who cared for the insane. This was because of the discovery of chlorpromazine (Thorazine), the first antipsychotic medication. Lobotomies were no longer needed. Although the lobotomy had been hailed only a few years earlier as a medical breakthrough, the procedure came to be viewed by most in the medical field as unethical mutilation.

Eventually, despite proselytizing by Freeman and the like, the public became aware of the actual consequences of lobotomies. In 1967, Freeman's license to operate was revoked after the last patient on whom he performed surgery, her third lobotomy, died of a brain hemorrhage. By this time, many countries had banned the procedure altogether. However, despite all of the bad press, including an investigation during the Carter administration into how lobotomies might have been used preferentially on minorities, the procedure was never completely banned in the United States.

Thus, as people's awareness of the bad side of neurosurgery had grown in the 1960s, levodopa was embraced all the more enthusiastically in the Parkinson's community. Sure, they had noticed the growing side effects the longer one used the new drug, as well as the shortened efficacy, on—off episodes, and dyskinesia, but the dosing schedules were still new and had yet to be optimized, so doctors and patients alike were no doubt hopeful for future improvements. In fact, I imagine they might have expected even better drugs were just on the horizon.

By the 1970s, the honeymoon was over. Despite improvements in dosing, side effects from levodopa use inevitably surfaced, although how soon and how severe varied greatly from patient to patient [10]. New drugs did come on the scene. Carbidopa, which inhibits the breakdown of levodopa outside the brain, could be taken at the same time as levodopa, resulting in more levodopa reaching the brain [11]. Other drugs inhibiting dopamine breakdown, catechol-*O*-methyltransferase inhibitors, followed in the decades to come. Dopamine agonists such as bromocriptine, which bind and stimulate the same protein receptors that bind dopamine, could be used instead of, or in combination with, levodopa. The dopamine agonists used alone were not as effective as levodopa at ameliorating the movement-related symptoms but had less dyskinesia side effects. Yet even when the additional drugs were used in combination with levodopa, the side effects would eventually appear[b]. Also, the additional drugs also could have their own side effects, such as impulse control disorder.

[b] I was surprised to learn there is a drug that can help with dyskinesia symptoms, amantadine. The surprise to me was that I had only known it as an antiviral drug formerly used to treat influenza. How intriguing that two such different biological situations can be targeted with the same molecule!

Often, the side effects of levodopa use, especially dyskinesia, could become worse than the actual symptoms the drug treated. Patients started to return to neurosurgeons (the ones who were left). Ablation of the globus pallidus internus, the portion of the globus pallidus nearer the center of the brain, could reduce not only tremor, bradykinesia, and rigidity, but also dyskinesia and on—off side effects [12]. The surgery also had potential side effects in addition to the risks that any major surgery poses. The procedure, especially bilateral pallidotomy, where the globus pallidus internus regions on both sides of the brain were ablated, could sometimes cause other problems, such as drooling and vision issues, or worsen other symptoms, including hypophonia, low sound (i.e., soft speech), and micrographia (small writing).

Even before the levodopa revolution in Parkinson's therapy, innovations in neurosurgery were occurring. In 1963, Natalia Petrovna Bekthereva, a neuroscientist at the Institute of Experimental Medicine and the Academy of Medical Sciences in Leningrad, published the first paper on deep brain stimulation as a permanent treatment for hyperkinesia, a class of disorders involving repetitive, involuntary muscle contractions [13]. Instead of ablating the brain target region, wire electrodes were inserted into the target regions and oscillating voltage was applied, in effect short-circuiting the neurons in the target region (more on this later). Researchers had known since the early days of cortical mapping of epileptics that low-frequency oscillating voltage applied to the affected brain regions could trigger seizures. On the other hand, applying higher-frequency voltage could help movement orders such as hyperkinesia, reducing the abnormal movement.

Unfortunately for Bekthereva and for people with Parkinson's disease, her research was done behind the Iron Curtain, published in a Russian language journal. Her work languished unnoticed by the free world. Already in 1960, the first heart pacemaker with an implantable battery had been used successfully in a patient [14]. The electronic technology needed for deep brain stimulation was practically the same as for the heart pacemaker. But a decade passed until finally, in the 1970s, attempts at permanent deep brain stimulation started to be made in the Western world.

Some of the first attempts used deep brain stimulation to try to revive patients in a persistent vegetative state [15]. Deep brain stimulation was also tested on patients with chronic pain [16]. Epilepsy and cerebral palsy were also targeted [17,18]. Although the initial results of these early deep brain stimulation procedures were promising, careful follow-up studies often failed to confirm them, instead showing a strong placebo effect. Finally, in

1987, deep brain stimulation was targeted to the ventral intermediate nucleus of the thalamus in Parkinson's disease patients [19]. This study was done by Alim Louis Benabid, Pierre Pollak, and colleagues at the University of Grenoble, France. They were successful; permanent deep brain stimulation of the thalamus reduced symptoms in patients.

In 1993, working with nonhuman primates with MPTP-induced (1-methyl-4-phenyl-1,2,3,6-tetrahydropyridine) parkinsonism, Abdelhamid Benazzouz and colleagues at the University of Bordeaux, France, found that stimulating a small region just below the thalamus, the subthalamic nucleus (STN), was just as effective at reducing movement-related symptoms as stimulating the thalamus or globus pallidus [20]. By 1995, Benazzouz, working with Benabid and Pollak, showed that bilateral stimulation of the STN worked in human Parkinson's patients as well [21]. Because bilateral stimulation of the STN caused fewer side effects than bilateral stimulation of the globus pallidus internus, the STN became the favored target for deep brain stimulation in Parkinson's patients.

The popularity of deep brain stimulation for treating Parkinson's disease quickly grew. The surgery was reversible, unlike the previous ablation procedures. If the electrodes somehow become mistargeted, they could be repositioned. Double-blind studies could be done easily, with neither patients nor the doctors examining them knowing when the electrodes were turned on. These studies proved the worth of the procedure, and in the United States, Food and Drug Administration approval of the surgery came quickly, in 1997, for both Parkinson's and essential tremor patients [22].

In the 30 years since deep brain stimulation was first used to treat Parkinson's disease, over 100,000 patients have undergone the procedure worldwide [23]. Whereas the STN is still most common region targeted, in some cases the globus pallidus internus is unilaterally targeted instead. Neurologists decide which target is most appropriate based on the details of the patient's symptoms and response to levodopa. For instance, if the primary symptoms are tremor in one hand accompanied with dyskinesia, unilateral stimulation of the globus pallidus internus might be preferred.

Unlike pallidal stimulation, STN stimulation does not directly reduce dyskinesia. Instead, by relieving bradykinesia, rigidity, and tremor, STN stimulation allows the patient to reduce the levodopa dose, which results in reduced side effects. STN stimulation also has the practical benefit of requiring less power, because the target area is smaller, which means fewer battery changes. The battery is implanted under the skin, near the collarbone usually, and a battery change requires the nuisance of a minor surgery, on average every 3–5 years, although newer rechargeable batteries can last up to 9 years [24]. Most Parkinson's patients keep the stimulation system turned on around the clock, but some patients, as well as those with essential tremor, can turn them off when sleeping, to conserve power.

So how does deep brain stimulation actually work? A neuron fires; that is, it sends a voltage pulse down its axon to trigger neurotransmitter release, using voltage-gated ion channels in the cell membrane of the axon. As the voltage pulse travels, the channels open, allowing the charged ions to flow into the axon, and the resulting change in electric field maintains the voltage pulse as it propagates down the axon. In deep brain stimulation, the electrodes create their own oscillating electric fields, sending voltage waves into the surrounding neural tissue. These waves also cause the voltage-gated ion channels to open, hijacking the neural machinery and triggering neurotransmitter release at the frequency of the stimulator, although probably at lower levels than if the neuron fired in its normal manner.

Neurons have rhythms; that is, they are designed to fire in a concerted way with the neurons to which they are connected. You have probably heard the term "brain waves." These are the actual electrical signals coming from large groups of networked neurons firing in a regular pattern at a regular frequency. The different types of waves have associated frequency bands. Delta waves are the slowest, occurring during stage 3 sleep, with a frequency of about once per second, whereas gamma waves are the fastest, typically linked to cognitive processes, with frequencies up to 150 Hz, or 150 times per second. The basal ganglia neurons associated with movement typically fire in the beta (13–30 Hz) and low gamma (30–70 Hz) frequency bands [25].

One of the cool things about the electrodes used for deep brain stimulation is that you can also use them to measure brain waves in the tissue they target. Researchers have compared the signals from the STN of Parkinson's patients with those of volunteers without the disease [26]. In patients with bradykinesia and rigidity, the beta wave signals are stronger than in controls, whereas in patients with a strong tremor, signals in the low gamma band are stronger. When deep brain stimulation is turned on, the stimulation frequency is adjusted to achieve the best suppression of symptoms. Typically, this occurs at frequencies above 100 Hz.

The results of these deep brain stimulation experiments suggest that in Parkinson's disease, the STN neurons are excessively locked into networks firing in the beta and low gamma wave frequency bands, and that deep brain stimulation disrupts these networks by forcing the neurons to fire at higher frequency. Another reason why I think that deep brain stimulation must disrupt the network is that its effect is so similar to ablation of the same region.

What are these networks that deep brain stimulation disrupts? At a neurological disorders conference in 2015, I asked a dinner mate who was a neurologist how deep brain stimulation works. Even though he had patients who had undergone the procedure, he admitted he did not know. It turns out that neural pathways are complicated, and current theories regarding the function of basal ganglia are incomplete and speculative.

The best-understood basal ganglia networks are those involved in voluntary motion. Voluntary movement arises as a consequence of frontal lobe activity, where we think, plan, and imagine in our mind's eye. The thoughts and desires involving movement get processed and relayed through the premotor and supplementary motor regions of the neocortex, which then send excitatory signals to the primary motor cortex, but these signals alone are not sufficient to trigger movement. The same motor regions of the frontal lobe also send excitatory signals to the basal ganglia, primarily the dorsal striatum. In fact, mapping of the dorsal striatum, which in primates is divided into the putamen and caudate nucleus, has revealed it has a spatial organization, a homunculus, similar to the primary motor cortex [27].

Voluntary motion involves two overlapping neural networks, the direct and indirect pathways [28]. Two types of dopamine responsive neurons reside in the dorsal striatum: ones that dopamine helps to excite, which are part of the direct pathway, and ones that dopamine inhibits, which are part of the indirect pathway. Here is where things get really complicated. Both types of dorsal striatal neurons are inhibitory. That is, upon firing, they inhibit neurons with which they synapse. The key to this setup is that the axons of the two types of striatal neurons target different parts of the basal ganglia.

The direct pathway dorsal striatal neurons synapse with and inhibit firing of neurons in the globus pallidus internus. The globus pallidus internus neurons are also inhibitory and synapse with neurons in the thalamus. However, upon inhibition of the globus pallidus internus neurons by direct pathway striatal neurons, the thalamic neurons with which they synapse become disinhibited, so they can fire. These thalamic neurons loop back to synapse with the same motor-associated frontal cortex neurons that sent their signals to the dorsal striatum. The thalamic neurons combine their excitatory signal with those from frontal cortex, achieving the threshold necessary to excite the primary motor cortex neurons to fire, resulting in the execution of the desired movement.

The indirect pathway suppresses movement. The motor-associated neuron regions also send signals to the neurons in the dorsal striatum that dopamine inhibits. When the frontal lobe stimulation of these dorsal striatum neurons is strong enough to overcome the inhibition by the dopamine released by substantia nigra neurons, they fire, sending inhibitory signals to inhibitory neurons of the globus pallidus externus. Normally, these neurons suppress firing of neurons in the STN. However, when the globus pallidus externus neurons are inhibited, the subthalamic neurons can fire, exciting neurons in the globus pallidus internus, which in turn suppress neurons in the thalamus, preventing them from sending excitatory signals to the motor-associated neurons, squelching any motion associated with the particular motor-associated neurons that sent the signal to the dopamine-inhibiting striatal neurons.

Confused? I don't blame you. To help, think of the steps of the pathway as *dos* and *don'ts*. Whereas a single *don't* negates, a double *don't* cancels the negation, and a triple *don't* makes it negative again. The indirect pathway is like a triple *don't*, with a *do* thrown in between the second and third *don't*, just to confuse things more. The *do* does not affect the net outcome of the pathway; it is the number of *don'ts* that matters. An odd number of *don'ts* equals *don't*, and an even number equals *do*.

The direct—indirect pathway system has been theorized to operate in a center-surround model [29,30]. When a voluntary motion in a particular body region is desired, the direct and indirect pathway cortical neurons target the same region of the striatum, that is, the same part of the striatal homunculus. In this model, the indirect pathway targets a larger region of the striatum, suppressing both the desired movement and similar movements involving in the same target region of the body. The direct pathway targets a more focused region, stimulating just the desired motion, enough to overcome suppression by the indirect pathway, allowing the desired motion to occur. For example, moving your pointer finger would involve stimulating direct pathway neurons targeting the pointer finger, and indirect pathway neurons targeting not just the pointer finger, but perhaps also the middle finger and thumb.

In addition, part of the indirect pathway takes a shortcut directly through the globus pallidus, and there is yet another pathway, one that bypasses the striatum, the hyperdirect pathway. This pathway goes from the motor and motor-associated cortices to the substantia nigra pars reticulata (the nondopaminergic part), and from there to the globus pallidus, and thence the thalamus. Because this pathway bypasses the striatum with its associated homunculus, the hyperdirect pathway should affect all possible voluntary motions. This pathway has an inhibitory effect (maybe it should be called hyperindirect) that suppresses all motion right before the desired motion occurs, or at least that is the current theory.

In Parkinson's disease, the lack of dopamine in the striatum prevents the signals coming from the frontal lobe motor-associated neurons to proceed along the direct pathway, while it strengthens the indirect pathway signals. That explains bradykinesia, but it does not seem to explain rigidity and tremor, yet ablation and deep brain stimulation of the basal ganglia help these symptoms, too. Obviously, there must be more to the picture.

Rigidity involves isometric muscle contractions, in which the muscles contract to keep the target body region from moving. For example, holding your arms out still in front of you involves isometric contractions, not just in your arms but in your shoulders and even in your postural muscles, because your center of gravity is slightly changed with your arms out. Because the indirect pathway supposedly inhibits the firing of motor neurons, it should not cause muscle contractions, isometric or otherwise. However, I suspect this picture is incomplete and that the indirect pathway might also be involved in isometric muscle contractions.

The signals from the thalamus to the motor and motor-associated cortices are excitatory, but it is possible that some of the targeted cortical neurons are inhibitory. Indeed, inhibitory neurons in the motor cortex are known to have an important role in voluntary movement [31]. In particular, it could be that these inhibitory cortical neurons are responsible for inhibiting isometric contractions in target muscles. Thus, this putative aspect of the indirect pathway would involve four *don'ts* resulting in a *do*. Without sufficient dopamine in the striatum, this part of the indirect pathway would not be adequately suppressed, resulting in isometric contractions throughout the body, in other words, rigidity. However, I repeat, this explanation is just a suspicion on my part.

In addition, the direct–indirect pathway picture does not explain tremor, which also involves muscle contractions, involuntary ones that occur rhythmically. Still, even if tremor involves a distinct basal ganglia pathway, this pathway must also pass through the STN and the globus pallidus to the thalamus, given that deep brain stimulation and ablation of any of these regions can reduce tremor. For action tremor, the pathway might also involve the cerebellum and locus coeruleus, because pathology in these regions is often seen in essential tremor patients.

What is the actual purpose of the basal ganglia? Without dopamine, one of the symptoms they cause is bradykinesia; thus, it is natural to assume they have a major role in controlling voluntary movement. However, ablation of particular basal ganglia, the STN or the globus pallidus internus, for instance, restores voluntary movement, at least partially. The basal ganglia do not seem that critical for voluntary movement after all. Then, why do we have these brain regions? Well, there is more to the story.

Parkinson's patients have problems learning new motor skills [32]. In particular, they have problems with what is called implicit, or procedural, learning, the subconscious learning that occurs when repeating a motor task. One of the major roles of the basal ganglia, along with the cerebellum, seems to be that they help us learn new skills, not only movement-related, but also cognitive and behavior related skills, such as social interactions [33].

The key to the learning function of the basal ganglia is dopamine. The neurons of the substantia nigra pars compacta have two modes of dopamine release in the striatum: tonic and phasic firing. In tonic firing, steady waves of electric potential course down the axonal arbor, triggering a moderate release of dopamine. Tonic firing is largely independent of neurochemical signals received by dendrites of the firing neuron. Phasic firing occurs when the signals received by the dendrites pass a threshold, triggering a much stronger wave down the axonal arbor, resulting in much more dopamine being released.

The dopamine circuitry of the brain is always on the lookout for unexpected rewards. For example, say you have been trying to learn a particular step in dance class, and finally you get it right for the first time. The dopamine circuitry, including the substantia nigra pars compacta, responds to this reward by producing an increase in the frequency, a burst, of phasic firing [28]. The release of dopamine in other brain regions contributes to feelings of pleasure. The burst of phasic firing by the substantia nigra floods the striatum with extra dopamine. The extra dopamine could signal the particular sets of motor and motor-associated neurons that have just fired in the striatum to strengthen their synapses, helping to lock in desired motor skill, although this idea is speculative.

On the other hand, if one gets an unexpected, undesirable outcome from a particular action, the opposite occurs. Say you are playing a song on the piano that you have not played in a while, and you hit a wrong note.

The dopamine circuitry responds to this undesired event by suppressing phasic firing. Dopamine levels in the striatum drop, perhaps signaling the particular sets of motor and motor-associated neurons that have just fired there to weaken their synapses. One of the purposes of the direct and indirect basal ganglia pathways could be to relay the effect of altered striatal synapse strength back to corresponding cortical motor regions, so that the neurons there can also alter their synaptic connections accordingly.

The loss of the dopaminergic neurons of the substantia nigra also means a loss of the dopamine level response to unexpected favorable and unfavorable outcomes in Parkinson's patients. So, even though levodopa can supply the brain with dopamine, it cannot replace the additional dopamine level changes triggered by the reward circuitry of the brain. This can explain the difficulty with learning motor skills, and I wonder if it might also explain some of the side effects of levodopa, such as dyskinesia, although I am going to stop myself from speculating further here. There is too much about the basal ganglia and their pathways that is still not well-understood, especially their interactions with regions deeper in the brain.

The arrival of levodopa in the 1960s, combined with surgical techniques such as deep brain stimulation, has changed Parkinson's disease from a debilitating condition to a manageable one for most Parkinson's patients. The future will almost certainly bring additional improvements. For example, in mice in which the dopaminergic neurons of the STN have been killed by a neurotoxin, scientists have shown they can convert STN astrocytes into dopaminergic neurons, and they have evidence that the axons of the new neurons extend to and release dopamine in the dorsal striatum [34]. For many, however, even these improvements will not be enough. For patients with dementia with Lewy bodies and Parkinson's disease dementia, the prognosis is not good, with an average of only 5–6 years from onset of dementia to death. Levodopa and deep brain stimulation provide no protection from or reduction of dementia symptoms. Clearly, additional therapies are needed to treat dementia and stop the spread of synuclein-related pathology. In the next chapter, we turn our attention to the search for new drugs to combat Parkinson's disease.

References

[1] Horsley V, Clarke RH. The structure and functions of the cerebellum examined by a new method. Brain 1908;31:45–124.

[2] Kirschner M. Die Punktionstechnik und die Elektrokoagulation des Ganglion Gasseri. Archiv für Klinische Chirugie 1933;176:581–620.
[3] Hariz M. History of neurosurgery for Parkinson's disease. Revue Neurologique December 2017;173(10):607.
[4] Abel TJ, Walch T, Howard 3rd MA. Russell Meyers (1905-1999): pioneer of functional and ultrasonic neurosurgery. Journal of Neurosurgery December 2016;125(6):1589–95.
[5] Spiegel EA, et al. Stereotaxic apparatus for operations on the human brain. Science October 10, 1947;1106(2754):349–50.
[6] Gildenberg PL. Spiegel and Wycis - the early years. Stereotactic and Functional Neurosurgery 2001;77(1–4):11–6.
[7] Cooper IS. Ligation of the anterior choroidal artery for involuntary movements; parkinsonism. Psychiatric Quarterly April 1953;27(2):317–9.
[8] Tan SY, Yip A. António Egas Moniz (1874–1955): lobotomy pioneer and Nobel laureate. Singapore Medical Journal April 2014;55(4):175–6.
[9] Caruso JP, Sheehan JP. Psychosurgery, ethics, and media: a history of Walter Freeman and the lobotomy. Neurosurgical Focus September 2017;43(3):E6.
[10] Thanvi B, Lo N, Robinson T. Levodopa-induced dyskinesia in Parkinson's disease: clinical features, pathogenesis, prevention and treatment. Postgraduate Medical Journal June 2007;83(980):384–8.
[11] Tolosa E, et al. History of levodopa and dopamine agonists in Parkinson's disease treatment. Neurology June 1998;50(6 Suppl. 6):S2–10.
[12] Favre J, et al. Outcome of unilateral and bilateral pallidotomy for Parkinson's disease: patient assessment. Neurosurgery February 2000;46(2):344–53.
[13] Bekthereva NP, et al. Utilisation of multiple electrodes implanted in the subcortical structure of the human brain for the treatment of hyperkinesis. Zhurnal Nevrologii i Psikhiatrii Imeni SS Korsakova 1963;63:3–8.
[14] Aquilina O. A brief history of cardiac pacing. Images in Paediatric Cariology Apr-Jun;8:2.
[15] Sturm V, et al. Chronic electrical stimulation of the thalamic unspecific activating system in a patient with coma due to midbrain and upper brain stem infarction. Acta Neurochirurgica 1979;47(3–4):235–44.
[16] Levy RM, Lamb S, Adams JE. Treatment of chronic pain by deep brain stimulation: long term follow-up and review of the literature. Neurosurgery December 1987;21(6):885–93.
[17] Koy A, et al. Effects of deep brain stimulation in dyskinetic cerebral palsy: a meta-analysis. Movement Disorders May 2013;28(5):647–54.
[18] Loddenkemper T, et al. Deep brain stimulation in epilepsy. Journal of Clinical Neurophysiology November 2001;18(6):514–32.
[19] Benabid AL, et al. Combined (thalamotomy and stimulation) stereotactic surgery of the VIM thalamic nucleus for bilateral Parkinson disease. Applied Neurophysiology 1987;501(1–6):344–6.
[20] Benazzouz A, et al. Reversal of rigidity and improvement in motor performance by subthalamic high-frequency stimulation in MPTP-treated monkeys. European Journal of Neuroscience April 1, 1993;5(4):382–9.
[21] Limousin P, et al. Effect of parkinsonian signs and symptoms of bilateral subthalamic nucleus stimulation. Lancet January 14, 1995;345(8942):91–5.
[22] Gardner J. A history of deep brain stimulation: technological innovation and the role of clinical assessment tools. Social Studies of Science October 30, 2013;43(5):707–28.
[23] Sugiyama K, et al. The present indication and future of deep brain stimulation. Neurologia Medico-Chirurgica May 2015;55(5):416–21.

[24] Park K, et al. Battery life matters in deep brain stimulation. Sterotactic and Functional Neurosurgery 2018;96(1):65–6.
[25] Blenkinsop A, Anderson S, Gurney K. Frequency and function in the basal ganglia: the origins of beta and gamma band activity. The Journal of Physiology July 1, 2017;595(13):4525–48.
[26] Herrington TM, Cheng JJ, Eskandar EN. Mechanisms of deep brain stimulation. Journal of Neurophysiology January 1, 2016;115(1):19–38.
[27] Hunnicutt BJ, et al. A comprehensive excitatory input map of the striatum reveals novel functional organization. eLife November 28, 2016;5:e19103.
[28] Kandel ER, et al. Principles of neural science. New York: McGraw Hill; 2013. p. 985–98.
[29] Nambu A, Tokuno H, Takada M. Functional significance of the cortico-subthalamo-pallidal 'hyperdirect' pathway. Neuroscience Research June 2002;43(2):111–7.
[30] Schroll H, Hamker FH. Computational models of basal-ganglia pathway functions: focus on functional neuroanatomy. Frontiers in Systems Neuroscience 2013;7:122.
[31] Estebanez L, et al. Parvalbumin-expressing GABAergic neurons in primary motor cortex signal reaching. Cell Reports July 11, 2017;20(2):308–18.
[32] Pascual-Leone A, et al. Procedural learning in Parkinson's disease and cerebellar degeneration. Annals of Neurology October 1993;34(4):594–602.
[33] Roca M, et al. Impairments in social cognition in early medicated and unmedicated Parkinson disease. Cognitive and Behavioral Neurology September 2010;23(3):152–8.
[34] Qian H, Kang X, Hu J, et al. Reversing a model of Parkinson's disease with in situ converted nigral neurons. Nature 2020;582:550–6.

CHAPTER 14

Seeking a Magic Bullet—the search for new Parkinson's disease therapeutic agents

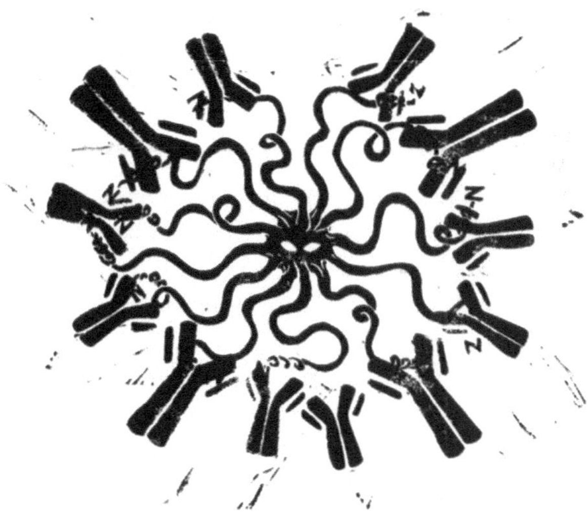

Two chapters ago, we saw how the synucleins seem to have a protective role in the innate immune system, the body's first line of defense against microbial invaders. The innate immune system is only part of the body's immune defenses, however. There is another branch, the adaptive immune system. This branch includes T cells. Each individual T cell can recognize a particular foreign bit of protein displayed on infected cell surfaces. Once a T cell detects its specific, foreign bit of amino acid sequence, it multiplies and its daughter cells go on to attack infected cells displaying the same foreign sequence.

The adaptive immune system includes B cells, each of which makes its own particular antibody. Antibodies are protein complexes that function primarily outside cells. Each antibody recognizes a particular bit of a foreign

Synuclein and the Coelacanth
ISBN 978-0-323-85707-9
https://doi.org/10.1016/B978-0-323-85707-9.00009-5

molecule, an epitope, on the outer surfaces of invading pathogens. Like T cells, once an antibody on a B cell discovers its foreign molecule epitope, the B cell starts to divide and multiply, turning into plasma cells, releasing more and more of the antibody. The antibodies coat the invading pathogens, blocking them from infecting cells, making it easier for macrophage cells to engulf and destroy them.

The potential number of different B cells, each with its own particular antibody specific for a particular epitope, is staggering, around a trillion possibilities, although only about a tenth that number of B cells is around in an individual at any particular time [1]. Moreover, there are still more kinds of cells in the adaptive immune system, not just seek and destroy cells, but also helper cells to modulate the immune response and memory cells that persist after an infection, ready to kick-start the adaptive immune response in case the pathogen invades again.

Currently, dozens of drug trials are testing therapies to combat Parkinson's disease, not just the symptoms, but the underlying causes. Many of these are antibody-based. Antibody-based therapies have had many successes, targeting protein epitopes specific to particular cancers as well as inflammatory proteins in immune-related conditions such as Crohn's disease and rheumatoid arthritis, to name a few [2]. Technically speaking, therapies that involve the injection of antibodies produced outside the body are termed passive immune therapies (we will get to active immune therapy in a bit).

These days, antibodies used in nearly all passive immune therapies are monoclonal; that is, their amino acid sequences are identical and therefore have the same epitope specificity, just as if they had come from the daughter cell clones of a single initial B cell. Most are produced by mammalian cells engineered to produce human antibodies, in many cases from a cell line initially derived in 1957 from an ovary taken from a Chinese hamster (CHO cells). CHO cells are able to grow and multiply in a liquid medium in machines called bioreactors. Bioreactors provide the optimized balance of nutrients, temperature, and solution conditions to maximize cell multiplication and protein production.

The antibody-based therapies being tested in Parkinson's patients are all targeted toward alpha-synuclein, in particular, its aggregated forms, oligomeric and fibrillar [3]. Theoretically, high enough levels of the injected antibodies could help prevent the spread of aggregated, pathogenic alpha-synuclein between neurons, thus halting or slowing disease progression. One of the issues with this theory is that antibodies delivered intravenously normally cannot pass through the blood—brain barrier. However, in trials for antibodies targeting amyloid-beta in Alzheimer's patients, intravenously delivered antibodies, when given at sufficiently high levels, were able to get by the barrier, although not always without unpleasant, dangerous side effects such as vasogenic edema, in which excess fluid builds up in the brain [4]. Nevertheless, the ability of antibodies to cross the blood—brain barrier encouraged researchers to initiate similar trials in Parkinson's patients.

Unfortunately, even the most promising of the antibody-based therapies for treating Alzheimer's disease ultimately failed in the final, phase III part of their clinical trials [5]. This has been a devastating blow, because after a long string of other failed trials, these passive immune therapies seemed to be the final hope for a treatment, at least for the near future. Does this mean that all antibody-based therapies for Alzheimer's or other neurodegenerative diseases are doomed to fail? Not necessarily. A reexamination of the latest trial suggests that at the highest doses, the antibody therapy might have slowed progression of the disease, so the trial has been revived using higher dosages. Moreover, in 2019, researchers dropped a bombshell.

For years, researchers had noted cases that were presumed to be Alzheimer's disease, but upon autopsy, they had far less amyloid-beta plaques in their brains than was typical for Alzheimer's patients. The researchers sought other possible causes of the Alzheimer-like symptoms and ultimately discovered that a different protein was to blame. This new disease, with symptoms nearly identical to Alzheimer's, is called limbic-predominant age-related TDP-43 encephalopathy (LATE) [6]. TDP-43 is a protein normally found in the nuclei of neurons and many other cells. Pathological deposits of this protein outside the nucleus were first identified in the motor neurons of patients with amyotrophic lateral sclerosis, or Lou Gehrig's disease [7]. TDP-43 pathology is also the culprit in another neurodegenerative disease, frontotemporal dementia.

The researchers estimate that roughly a third of patients with Alzheimer's symptoms actually have LATE, making it the second most common neurodegenerative disease, ahead of Parkinson's disease. With so many of the Alzheimer's trial participants misdiagnosed, it could be that small benefits for patients with bona fide Alzheimer's disease were obscured. Hope remains for antibody-based Alzheimer's treatments. Another avenue that researchers are pursuing that could boost the success of future Parkinson's disease trials is to use antibodies specially designed to cross the blood—brain barrier more easily.

Antibodies consist of two identical heavy chains and two identical light chains. They come together to form a *Y* structure in which each of the two branches at the top can bind the target epitope independently. It is possible, however, to construct an artificial antibody in which each branch has a different target. If you make one of those targets specific to a protein that transports molecules across the blood—brain barrier, much more of the antibody might enter the brain [8]. The other branch would be engineered to target alpha-synuclein in its aggregated oligomeric or fibrillar form. Because this dual-target antibody should enter the brain more easily, lower doses would be needed, perhaps avoiding the unpleasant and potentially harmful side effects resulting from the large doses in previous studies.

Most of the antibody-related therapies for Parkinson's disease currently in human trials employ passive immune therapies, but active immune therapy is being pursued as well. You have almost certainly experienced active immune therapy; it is just another name for vaccination. In active immune therapy, killed or weakened viruses or other pathogens, or their components, are injected so that your immune system can mount a response to the myriad pathogen epitopes presented by the vaccine. Of course, this approach faces the same challenges that the passive therapies do. Can the antibodies that are generated get past the blood—brain barrier, and if they do, will they be sufficient to prevent the spread of aggregated alpha-synuclein between neurons?

There is an additional challenge with the active immune approach: namely, how do you make sure the vaccine does not cause an immune response to normal, monomeric synuclein? Recall that there are two branches of the adaptive immune system: B cells with their antibodies and T cells that recognize bits of foreign protein sequences displayed on cell surfaces. The challenge is thus to find an injectable peptide or other biomolecule whose shape and surface properties mimic a part of the surface of aggregated alpha-synuclein, either oligomeric or fibrillar, but not the monomer, stimulating production of antibodies just against the aggregated form.

In addition, the vaccine also must not stimulate T cells to attack cell healthy cells with normal alpha-synuclein. Many cells, macrophages in particular, take up extracellular proteins, degrade them, and display bits of them on their cell surfaces. However, early in development, T cells that recognize self-sequences (that is sequences from one's own proteins) are removed to prevent autoimmune reactions. The process is not perfect, however, and it can happen that a foreign sequence similar to a self-sequence might trigger the T cells to attack healthy cells. Therefore, a double trick is required to design the peptide, or peptides, in the vaccine to mimic aggregated alpha-synuclein outside the cell, yet be sufficiently different from normal human protein sequences such that no autoimmune response is triggered [9].

Currently, the hope is that active immune therapy will combat the spread of aggregated alpha-synuclein, benefitting Parkinson's patients. As with passive therapies, the blood—brain barrier is problematic, because the antibodies stimulated by the vaccine will be generated in the blood. It could be that by the time alpha-synuclein pathology has spread to the brain, it is too late for the active immune approach to be effective. On the other hand, if alpha-synuclein pathology originates in the gut or nasal epithelium, as evidence presented in earlier chapters suggests, perhaps antibodies generated by active immune therapy could help block entry of aggregated alpha-synuclein from those locations into the central nervous system. Maybe someday, people will receive Parkinson's vaccines to help prevent the disease at its earliest stages.

Antibodies evolved to function outside cells, but the aggregation of alpha-synuclein and associated toxicity occurs inside cells, out of reach. For years, researchers have tried to come up with ways to get antibodies inside cells, where they could potentially block toxic interactions with intracellular targets such as aggregated alpha-synuclein [10]. The researchers have succeeded, at least in laboratory animals, using gene therapy. In gene therapy, a beneficial or corrected version of a gene is delivered to cells, often using engineered viruses. Nature has provided an uncannily well-suited virus for this gene therapy, the adenoassociated virus[11].

Why uncanny? Admittedly, I am being rather unscientific here, but I am not used to viruses being anything but harmful. However, infection with the adenoassociated virus typically produces only a mild immune response and does not seem to harm the cells it infects. In fact, it cannot even replicate on its own; it requires coinfection with adenovirus, using the proteins of the other virus to help assemble new adenoassociated virus particles; hence its name. The engineered versions of adenoassociated virus replace the viral DNA with the desired beneficial gene, which gets transferred to the nucleus as a small, circular DNA double helix that only rarely becomes incorporated into the DNA of the host's cells. The circular DNA does not become copied upon cell division, so in dividing cells, the original number of circular DNAs becomes diluted. This is a good thing; one need not worry that an accidently bad introduced gene will multiply out of control.

Of course, gene therapy using adenoassociated virus has limits. Getting the gene delivered to and expressed in the cells that need it can be technically challenging, and with the central nervous system, there is always that irksome blood—brain barrier to deal with. A more serious limitation is that the size of the gene is limited to small and medium-sized proteins. The protein chains that constitute typical antibodies are too large to encode with the limited amount of DNA that can fit inside the virus. But even here, nature has provided a work-around. In camelids, which include camels, llamas, and their kin, and also in sharks, smaller, simpler antibodies have been discovered [12]. When these are engineered to contain just the epitope binding section of the antibody, so called nanobodies, the DNA that encodes them is small enough to fit inside the virus. Smaller, simpler versions of human antibodies have also been engineered, called single-chain antibodies [13].

Antibodies that are engineered to remain inside the cells that express them are called intrabodies. The feasibility of intrabodies targeted to neurodegenerative proteins, delivered via adenoassociated virus, has been demonstrated in animal models, and human trials are on the horizon [14,15]. Previous gene therapy approaches used stereotactic brain surgery to deliver the gene therapy viruses to specific target tissues. However, because alpha-synuclein pathology affects numerous regions throughout the central nervous system, a more generalized delivery technique might be more effective for halting or slowing Parkinson's disease. Some strains of adenoassociated virus can actually pass through the blood—brain barrier; work is in progress to use these strains to deliver alpha-synuclein targeted intrabodies [16].

No financial gain, there will be pain—Ethics at NIH

The efforts to find new drugs to halt or slow Parkinson's disease have taken place in universities, hospitals, and research laboratories across the world, backed by government funding and nonprofit organizations such as the Michael J. Fox Foundation and the Cure Parkinson's Trust. Human trials are expensive, and they are often performed in partnership with pharmaceutical and biotechnology businesses that contribute both capital and expertise with the hope that the trial will be successful, leading to a profitable drug. My employer, the National Institutes of Health (NIH), encourages such partnerships, because transferring research from bench to bedside (that is, getting new therapies to patients as quickly as possible) is part of the primary mission of NIH.

On the other hand, when it comes to endorsing a particular drug over others, the legal team at NIH gets squirmy. In particular, there can be no financial connection between the NIH scientist and the company marketing the drug. The scientist can receive no consultation fees, no personal items or meals worth over $20 in value (just up from $15 last year; we're really rolling in the bucks now!), and no paid vacations disguised as special speaking engagements, and the scientist can own no stock in the drug company. In fact, the top scientists and physicians at NIH are forbidden from owning stock in *any* pharmaceutical, biotechnology, or health care company.

NIH did not always have such a strict policy. It was adopted after a scandal in the early 2000s, in which some senior NIH researchers involved in drug trials were receiving consultation fees from the companies developing the drug or from their competitors. They did not do it in secret; the fees were approved by the NIH administration. However, when the public learned that one of these researchers also participated in drafting national guidelines that advocated more aggressive use of one of the drugs (whose generic name is rosuvastatin, but you did not hear it from me), the resulting outcry convinced the NIH director to enforce new, strict conflict of interest rules.

I remember when the new rules were implemented. Because I was a junior researcher and did not own stock, I was more bemused than anything else, at least at first. At the same time, many of the senior researchers with whom I worked were fuming that government was dictating how they could invest their own private funds. Our branch chief, who was also my immediate supervisor, quit NIH because of it. That is when the seriousness of the situation hit home for me. To this day, I still hear rumors that the strict rules have hampered recruitment and retention of top scientists for senior posts at NIH. When politics are involved, there are no perfect solutions.

In the 1990s, the first clinical trials were undertaken for therapies attempting to treat the underlying cause of Parkinson's disease. The drugs tested were monoamine oxidase (MOA) B inhibitors such as deprenyl [17]. MOA B is an enzyme that breaks down dopamine in the brain. The drugs were already being used to treat the symptoms of Parkinson's disease, because inhibiting dopamine breakdown boosted its levels in the brain. One of the hypotheses driving these trials was the idea that by making dopamine last longer in the brain, this would reduce stress on the neurons making the neurotransmitter, the dopaminergic neurons of the substantia nigra, for instance, thus slowing their loss.

The challenge in these trials was that because the drugs already had a therapeutic effect in reducing symptoms, showing that they also protected dopaminergic neurons required a washout period. Let me explain. Say the trial involved a year of taking the drug or a placebo. Because the drug helped reduce symptoms, one would expect the drug group to have fewer symptoms regardless of whether the drug was actually neuroprotective. However, if after both groups stopped taking the drug or placebo for a time, the washout period, and the drug group continued to show reduced symptoms compared with the placebo group, this would provide evidence of a neuroprotective effect.

Although the early trials seemed to show a modest neuroprotective benefit, some researchers argued that the washout periods were too short. The most recent trial, for another MOA B inhibitor called rasagiline, used a different trial design that avoided the need for a washout period [18]. The study cohort of 1176 patients was divided into two groups: one that took the drug for the full 72 weeks of the trial and the other that started the trial with a placebo for the first 36 weeks and then took the drug for the remaining 36 weeks. In addition, both groups were divided into lower-dose and higher-dose groups.

At the end of the trial, both groups should be enjoying the same symptom reduction benefits from the drug, obviating the need for a washout period. However, if there truly were a neuroprotective effect, those who took the drug for the full 72 weeks should show less progression in the severity of symptoms compared with the 36-week group. For the lower-dose groups, the drug seemed to have a neuroprotective effect, but no significant difference was seen for the higher-dose groups, and because of this, the US Food and Drug Administration deemed the study inconclusive.

Another drug that has been proposed possibly to reduce stress on dopaminergic neurons is the blood pressure medication isradipine. The drug reduces the flow of calcium ions through voltage-gated calcium channel protein complexes at the neuron cell membrane. With less calcium entering the cell, less chemical energy, which is provided by mitochondria, is needed to pump them back out again. Physicians noticed that patients who used the drug seemed less likely to develop Parkinson's disease. In mouse models of alpha-synuclein pathology, the drug reduced the production of reactive oxygen species in neurons, further supporting a protective effect on mitochondria.

The initial trials of isradipine in Parkinson's patients were promising, demonstrating significant levels of the drug crossing the blood–brain barrier and possible neuroprotective effects. After a 3-year, phase III trial with over 300 participants divided into drug and placebo groups, the result of the trial was announced in May 2019 [19,20]. In fact, I delayed writing this chapter to wait for the trial result. Unfortunately, the trial flopped. Despite all the preliminary, promising results, there was no difference between the placebo and drug group in disease progression.

What went wrong? The usual explanation is that the initial, observed neuroprotective effects seen in the smaller, earlier trials might just have been statistical fluctuations. Or it could be that isradipine helps to prevent onset of the disease but does nothing to slow it once it is set in motion. A paper detailing the results of the phase III trial has not yet been published, but a year after the end of the trial, I found an editorial in a medical journal reviewing the trial results with various hypotheses about why the trial failed [21].

Not all of the drug trials have been discouraging. The ganglioside GM1, delivered intravenously or subcutaneously injected, not only reduces the severity of many Parkinson's disease symptoms, it appears to have a neuroprotective effect [22]. Recall that GM1 levels were lower in Parkinson's and DLB patients compared with controls not just in patients who were carriers of Gaucher disease, but also in those without *GBA1* mutations. The potential for GM1 treatment of Parkinson's disease has been demonstrated in mice treated with MPTP (1-methyl-4-phenyl-1,2,3,6-tetrahydropyridine) to induce parkinsonism [23]. The putative neuroprotective benefit has since been reproduced in several studies, although, as with the MOA B inhibitor trials, debate continues whether the washout periods have been sufficiently long. Taken orally, little GM1 reaches the bloodstream, and alternatives more readily absorbed through the digestive system are being explored, such as LIGA20, a version of GM1 that lacks one of its lipid chains [24].

One of the most intriguing candidates for Parkinson's disease therapy is sirolimus, perhaps better known by its original name, rapamycin. Rapamycin is a compound known as a macrolide, with a structure consisting of a complex arrangement of sugar-like carbohydrates connected in a large ring. It was discovered in 1972 as an antifungal agent isolated from a bacterium, *Streptomyces hygroscopicus*. The bacterium was found in a soil sample from Easter Island, or Rapa Nui in the Polynesian language of the island natives; hence the "rapa" part of the name. The "mycin" part is commonly given to the many macrolide compounds that have been isolated from other *Streptomyces* species.

Although originally studied as an antifungal agent, rapamycin was discovered to have potent immunosuppressive and anticancer activity [25]. However, what really put it on peoples' radars was a 2009 report that it extended the life span in mice [26]. Fantasies of a fountain of youth drug were soon dashed by its laundry list of adverse effects, including headache and abdominal pain, diarrhea and constipation (I assume not at the same time), increased cholesterol and triglycerides in the blood, increased blood pressure, increased susceptibility to infections ... the list goes on and on. Today, its medical uses are limited to prevention of the rejection of transplanted organs, treatment of a rare lung disease called lymphangioleiomyomatosis, and as a coating on coronary stents to prevent the stented blood vessels from renarrowing.

The name change from rapamycin to sirolimus had nagged my curiosity for some time. The names are so different, and what was wrong with "rapamycin" anyway? I finally looked into it. By the time that a drug candidate reaches a phase III clinical trial, it is supposed to be assigned a generic drug name. For new drugs in the United States, the US Adopted Names Council works with the World Health Organization to come up with an International Nonproprietary Name. The idea is to arrive at a name that indicates to what class of drugs the new drug belongs, preferably no longer than four syllables, and that is easy to pronounce and cannot easily be confused with other drug names.

The "imus" part of sirolimus identifies the drug as an immunosuppressant (IMmUnoSuppressant). I have no idea where "sirol" came from. A Web search of "sirol" turned up a road in India and a 24th-century Romulan commander in the TV show *Star Trek: The Next Generation.* Although I like the idea of naming a drug after a Romulan, my choice would have been to stick closer to the original name, calling it "rapanuimus." This would get the full name of the island into the drug name, although because of the four-syllable preference, some might pronounce the "nui" part as *nwee*, a pronunciation ambiguity that I imagine the council could find objectionable.

Back to Parkinson's disease. One of the ways in which rapamycin appears to help extend the life span of mice is to stimulate the autophagic lysosomal pathway in cells [27]. When nutrients are scarce, cells go into starvation mode, initiating a number of changes, one of which is to ramp up the autophagic lysosomal pathway. In this process, also called macroautophagy, a double-membraned organelle, the autophagosome, engulfs large portions of the cell's interior and fuses with a lysosome, which then degrades and recycles the engulfed contents to provide the nutrients the cell needs to survive. In neurons with alpha-synuclein pathology, macroautophagy helps to lower synuclein levels and sequester the toxic, aggregated forms of the protein.

The target of rapamycin in the cell is a protein called mechanistic (or mammalian, depending on who you ask) Target Of Rapamycin (mTOR), which rapamycin inhibits [28]. mTOR is a part of two protein complexes: mTORC1 and mTORC2. The important complex for macroautophagy is mTORC1. The complex is recruited to the lysosome surface by the protein Ras Homolog Enriched in Brain (RHEB). RHEB is similar to the Rab proteins mentioned in earlier chapters. When bound to the molecule guanosine triphosphate, they help recruit their target proteins to specific membrane locations on vesicles or organelles or at the cell membrane.

When nutrients are plentiful, such as the amino acids needed for building proteins, mTORC1 gets phosphorylated at the lysosome surface, which activates its own kinase, which then phosphorylates particular amino acids in its target protein Transcription Factor EB (TFEB). When phosphorylated by mTORC1, TFEB is sequestered in the cytosol. However, when nutrients become scarce, or when rapamycin binds mTOR, mTORC1 becomes inactive and the TFEB, no longer phosphorylated by mTORC1, is able to translocate to the cell nucleus and increase the expression of genes that stimulate macroautophagy.

The ideal drug for Parkinson's disease would target mTORC1 while avoiding other cellular targets leading to adverse effects. Currently, in addition to rapamycin, other drugs are being developed; some are aiming for greater specificity to mTORC1 [29] whereas others, such as the sugar trehalose, seek to stimulate macroautophagy independent of the mTORC1 pathway [30]. Another set of drug candidates seeks to stimulate lysosomal function via a different target protein, glucocerebrosidase [31]. We have already encountered this enzyme in previous chapters, in which inheriting two defective versions causes Gaucher disease and increases the risk for developing Parkinson's disease, not just for those with Gaucher disease, but for carriers as well.

To reiterate, reduced activity of glucocerebrosidase is correlated with impaired lysosomal function and increased alpha-synuclein levels. Furthermore, Parkinson's patients, even those without glucocerebrosidase mutations, have reduced activity of the enzyme compared with age-matched controls. Thus, it is natural to ask whether a drug that increases glucocerebrosidase activity might be beneficial to Parkinson's patients.

The glucocerebrosidase-targeted drug would need to cross the blood–brain barrier to reach the neurons in need of protection from alpha-synuclein pathology. In fact, such a drug is even more desperately needed by those with the neuronopathic form of Gaucher disease, who often do not make it past their first decade. For these patients, the standard treatment for Gaucher disease, enzyme replacement therapy, in which functioning glucocerebrosidase is delivered intravenously, does nothing for neuronopathic symptoms because the enzyme cannot cross the blood–brain barrier.

The first drug tested to increase glucocerebrosidase activity was isofagomine. The drug belongs to a class of molecules called iminosugars, which resemble normal sugars such as glucose, but with one carbon atom replaced by a nitrogen atom. The drug was initially designed to be an inhibitor of enzymes such as glucocerebrosidase; however, in 2002, in the laboratory of Jeffery W. Kelly at the Scripps Research institute, it was discovered that inhibitors of glucocerebrosidase could actually increase the activity of mutant forms of the enzymes inside cells [32]. This seeming paradox comes about because when they are bound inside the active site of the enzyme, the inhibitors help to stabilize its structure. Many glucocerebrosidase mutations destabilize its structure. The cell has protein quality control machinery, and when this machinery detects misfolded or unstable proteins, they get molecularly tagged for degradation. Because these compounds stabilize structure, the inhibitors are called chemical chaperones.

Initial tests of isofagomine in cell cultures and laboratory animal models of Gaucher disease showed promising increases in glucocerebrosidase activity, and the drug was well-tolerated. Initial tests also showed that the drug caused no serious side effects in humans, so a phase II clinical trial was initiated in 2009 with 19 Gaucher patients with the type I, non-neuronopathic, form of the disease. Of the 18 participants, only one showed significant improvement in symptoms. The trial was deemed a failure [33].

The disappointing results for isofagomine might have dampened interest in chemical chaperones for glucocerebrosidase, but since then, new compounds, ones that chaperone the mutant enzyme but inhibit it far less, have been developed. A team including the NIH laboratories of Ellen Sidransky at the National Human Genome Research Institute and Juan J. Marugan at the National Center for Advancing Translational Sciences (NCATS) used high-throughput screening to discover new compounds [34]. In an assay designed in part by Ehud Goldin (the double-hit hypothesis in Chapter 9), homogenized spleen tissue from a Gaucher disease patient was used to screen roughly 250,000 different compounds, identifying those that enhanced glucocerebrosidase activity.

I visited NCATS once and saw the robotic equipment used to perform the high-throughput screening. They had two large yellow robotic arms, like the kind you see in manufacturing factories, dwarfing the humans in the laboratory. They would smoothly pick out a tray from a large array of racks containing all of the 250,000 compounds and bring the tray to another machine that precisely injected tiny volumes of the compounds into another tray with 1536 individual wells, each containing a tiny volume of the spleen homogenate. The robots worked day and night, bringing new compounds to new trays. It was a lengthy process, because the contents of each tray were incubated for a half hour with the compounds before the enzyme activity was measured.

The process did not end with the high-throughput screen; in fact, it was just the beginning of years more of work. Medicinal chemists took the most promising compounds and modified them to optimize enzyme activity enhancement further [35]. In addition to activity enhancement, microscopic studies demonstrated that the new compounds indeed acted as chemical chaperones in cells, resulting in more glucocerebrosidase reaching the lysosomes. Using induced pluripotent stem cells from Gaucher disease patient donors, including some who also had Parkinson's disease, they delicately coaxed the cells, using different complex protocols and chemical cocktails, to become macrophages or dopaminergic neurons. Finally, they showed that not only did the new compounds increase glucocerebrosidase activity in the patient-derived cells, the alpha-synuclein levels were reduced.

As far as I know, no clinical trials are planned for the new NIH compounds, although there are several clinical trials for other glucocerebrosidase chemical chaperones. The most promising is ambroxol, a drug already approved as an expectorant for the treatment of coughs [36]. Ambroxol was identified in a glucocerebrosidase chemical chaperone screen using compounds that were already drugs approved to treat other conditions. In fact, screening using already-approved drugs is an excellent way to jump-start the search for drugs for other conditions, because the phase I trials, testing safety in humans, have already been done.

Another example of a previously approved drug possibly being repurposed to treat Parkinson's disease is sargramostim, which was originally intended to treat leukemia. It is currently being assessed in an NIH-funded trial in Parkinson's patients. The drug is actually a protein called granulocyte macrophage colony stimulating factor (GM-CSF). In a 2013 study performed in the laboratory of Howard E. Gendelman at the University of Nebraska Medical Center, when GM-CSF was given to rats before being given the neurotoxin MPTP, the dopaminergic neurons of the substantia nigra were protected [37].

Hearkening back to adenoassociated virus therapies, one company is testing a gene therapy approach, delivering the corrected *GBA1* gene to neurons as a potential therapy for neuronopathic Gaucher disease, and for Parkinson's disease as well [38]. Others have proposed an alternative to directly targeting glucocerebrosidase, aiming to inhibit the enzyme that makes glucosylceramide instead [39]. A drug, miglustat, that inhibits the enzyme that synthesizes glucosylceramide is already available as a treatment for Gaucher disease. However, in the more than a decade since miglustat has been approved, I know of no studies testing whether it can slow or halt Parkinson's disease pathology.

Another potential target for Parkinson's therapies is the protein LRRK2. Recall that increased levels of the protein are correlated with increased alpha-synuclein. Many of the mutations linked to Parkinson's disease increase the kinase activity of the protein, which suggests that drugs that reduce its kinase activity could counteract the effect of the mutations. The pharmaceutical industry has a lot of experience with kinase inhibitors, but attempts to find one for LRRK2 did not go well initially; the compounds inhibited not only LRRK2, but other kinases as well, and disappointing levels of the compounds reached the brain. However, these are problems that have been solved before, and by patiently tweaking the chemistry of lead compounds, a new generation of LRRK2 inhibitors, with greater specificity and blood–brain barrier penetration, are now entering clinical trials [40].

Yet another strategy to treat Parkinson's disease, as well as other neurodegenerative diseases such as Alzheimer's, is to target the brain's immune system. Interest in this approach got a big boost in 2011, when a report from a team led by Xiang Gao at Harvard University showed that long-term users of the nonsteroidal antiinflammatory drug (NSAID) ibuprofen, but not aspirin, acetaminophen, or other NSAIDs, had significantly lower risk for developing Parkinson's disease [41]. In all of these diseases, part of the problem is the strong immune response; glial cells in the brain seem to amplify inflammatory signals coming from the afflicted neurons, triggering more damage in neighboring, healthy neurons.

Perhaps the most ambitious strategy to treat Parkinson's and other neurodegenerative disease is designing drugs that block the aggregation of amyloid proteins. You are probably thinking, "Of course, blocking aggregation is an obvious strategy!" You are right, but the challenge is to achieve the specificity of the drug, especially its strength of binding to oligomers. Alpha-synuclein and many other amyloid-forming proteins have variable structures, and their oligomers are likewise inhomogeneous. Unlike enzymes such as glucocerebrosidase and LRRK2, alpha-synuclein has no well-defined active site pocket into which a drug can fit. Nevertheless, intrepid researchers employing various screening strategies have found compounds that block alpha-synuclein aggregation in a test tube [42]. Clinical studies are already under way testing how well the new compounds are tolerated in humans.

Will a cure for Parkinson's disease ever be found? I suppose a more realistic question is whether progression of the disease can be permanently halted before significant neurodegeneration has occurred. For this to happen, not only an effective therapy will need to be found, but also a reliable way to detect the disease in its earliest stages (unless the Parkinson's vaccine idea actually ends up working, but I would not bank on it). For Alzheimer's disease, significant progress in early detection has already been made, and this has inspired initiatives, especially at the Michael J. Fox

Foundation, to accelerate research into early detection of Parkinson's disease. Parkinson's disease may be closer to getting its magic bullet, but a magic detector will be needed, too.

References

[1] Alberts B, et al. The generation of antibody diversity & lymphocytes and the cellular basis of adaptive immunity. In: Molecular biology of the cell. New York: Garland Science; 2002.
[2] Chames P, et al. Therapeutic antibodies: successes, limitations and hopes for the future. British Journal of Pharmacology 2009:220−33.
[3] Sardi SP, Cedarbaum JM, Brundin P. Targeted therapies for Parkinson's disease: from genetics to the clinic. Movement Disorders May 2018;33(5):684−96.
[4] Pardridge WM. Alzheimer's disease drug development and the problem of the blood-brain barrier. Alzheimer's and Dementia 2009;5(5):427−32.
[5] Mehta D, Jackson R, Paul G, Shi J, Sabbagh M. Why do trials for Alzheimer's disease drugs keep failing? A discontinued drug perspective for 2010−2015. Expert Opinion on Investigational Drugs 2017;26(7):735−9.
[6] Nelson PT, et al. Limbic-predominant age-related TDP-43 encephalopathy (LATE): consensus working group report. Brain June 2019;142(6):1503−27.
[7] Neumann M, et al. Ubiquitinated TDP-43 in frontotemporal lobar degeneration and amyotrophic lateral sclerosis. Science October 6, 2006;314(5796):130−3.
[8] Pardridge WM. Re-engineering therapeutic antibodies for Alzheimer's disease as blood-brain barrier penetrating bi-specific antibodies. Expert Opinion on Biological Therapy December 2016;16(12):1455−68.
[9] Mandler M, et al. Next-generation active immunization approach for synucleinopathies: implications for Parkinson's disease clinical trials. Acta Neuropathologica June 2014;127(6):861−79.
[10] Lo AS, Zhu Q, Marasco WA. Intracellular antibodies (intrabodies) and their therapeutic potential. Handbook of Experimental Pharmacology 2008;181:343−73.
[11] Daya S, Berns KI. Gene therapy using adeno-associated virus vectors. Clinical Microbiology Reviews October 2008;21(4):583−93.
[12] Goldman ER, et al. Enhancing stability of camelid and shark single domain antibodies: an overview. Frontiers in Immunology 2017;8:865.
[13] Bannas P, Hambach J, Koch-Nolte F. Nanobodies and nanobody-based human heavy chain antibodies as antitumor therapeutics. Frontiers in Immunology November 22, 2017;8:1603.
[14] Southwell AL, Ko J, Patterson PH. Intrabody gene therapy ameliorates motor, cognitive, and neuropathological symptoms in multiple mouse models of Huntington's disease. Journal of Neuroscience October 28, 2009;29(43):13589−602.
[15] Emadi S, et al. Inhibiting aggregation of alpha-synuclein with human single chain antibody fragments. Biochemistry March 16, 2004;43(10):2871−8.
[16] Zhang H, et al. Several rAAV vectors efficiently cross the blood-brain barrier and transduce neurons and astrocytes in the neonatal mouse central nervous system. Molecular Therapy August 2011;19(8):1440−8.
[17] Olanow CW. Deprenyl in the treatment of Parkinson's disease: clinical effects and speculations on mechanism of action. Journal of Neural Transmission Supplementa 1996;48:75−84.
[18] Rascol O, et al. A double-blind, delayed-start trial of rasagiline in Parkinson's disease (the ADAGIO study): prespecified and post-hoc analyses of the need for additional

therapies, changes in UPDRS scores, and non-motor outcomes. Lancet Neurology May 2011;10(5):415423.
[19] Parkinson's UK. Blood pressure drug fails in phase 3 Parkinson's trial. May 2019. https://www.parkinsons.org.uk/news/blood-pressure-drug-fails-phase-3-parkinsons-trial.
[20] Hoffman M. Isradipine fails to slow early Parkinson disease progression in phase 3 study. May 2019. https://www.neurologylive.com/conferences/aan-2019/isradipine-fails-slow-early-parkinson-disease-progression-phase-3-study.
[21] Maiti B, Perlmetter JS. A clinical trial of isradipine: what went wrong? Annals of Internal Medicine May 5, 2020;172:9.
[22] Schneider JS, et al. GM1 ganglioside in Parkinson's disease: results of a five year open study. Journal of Neurological Science May 15, 2010;292(1−2):45−51.
[23] Fazzini E, et al. GM1 gangliosides alter acute MPTP-induced behavioral and neurochemical toxicity in mice. Journal of Neurological Science October 1990;99(1):59−68.
[24] Wu G, et al. Deficiency of ganglioside GM1 correlates with Parkinson's disease in mice and humans. Journal of Neuroscience Research October 2012;90(10):1997−2008.
[25] Sehgal SN. Sirolimus: its discovery, biological properties, and mechanism of action. Transplantation Proceedings May 2003;35(3 Suppl. l):7s−14s.
[26] Harrison DE, et al. Rapamycin fed late in life extends lifespan in genetically heterogeneous mice. Nature July 16, 2009;460(7253):392−5.
[27] Shang L, et al. Nutrient starvation elicits an acute autophagic response mediated by Ulk1 dephosphorylation and its subsequent dissociation from AMPK. Proceedings of the National Academy of Sciences of the United States of America March 22, 2011;108(12):4788−93.
[28] Jung CH, et al. mTOR regulation of autophagy. FEBS Letters April 2, 2010;584(7):1287−95.
[29] Inacio P. Phase 1/2 trial to test investigational therapy RTB101 in Parkinson's disease. April 5, 2019. https://parkinsonsnewstoday.com/2019/04/05/phase-1-2-trial-testrtb101-parkinsons/.
[30] Khalifeh M, Barreto GE, Sahebkar A. Trehalose as a promising therapeutic candidate for the treatment of Parkinson's disease. British Journal of Pharmacology and Chemotherapy May 2019;176(9):1173−89.
[31] Stecher B. The GBA-PD trials: expert survey. August 27, 2018. https://www.journalofparkinsonsdisease.com/blog/stecher/gba-pd-trials-expert-survey.
[32] Sawkar AR, et al. Chemical chaperones increase the cellular activity of N370S β-glucosidase: a therapeutic strategy for Gaucher disease. Proceedings of the National Academy of Sciences of the United States of America 2002;99:15428−33.
[33] Malcolm, E. Isofagomine. Available from: https://gaucherdiseasenews.com/chaperone-therapy/isofagomine/.
[34] Goldin E, et al. High throughput screening for small molecule therapy for Gaucher disease using patient tissue as the source of mutant glucocerebrosidase. PLoS One January 17, 2012;7(1):e29861.
[35] Aflaki E, et al. A new glucocerebrosidase chaperone reduces α-synuclein and glycolipid levels in iPSC-derived dopaminergic neurons from patients with Gaucher disease and parkinsonism. Journal of Neuroscience July 13, 2016;36(28):7441−52.
[36] Silveira CRA, et al. Ambroxol as a novel disease-modifying treatment for Parkinson's disease dementia: protocol for a single-centre, randomized, double-blind, placebo-controlled trial. BMC Neurology 2019;19:20.
[37] Gendelman HE, et al. Evaluation of the safety and immunomodulatory effects of sargramostim in a randomized, double-blind phase 1 clinical Parkinson's disease trial. NPJ Parkinson's Disease 2017;3:10.

[38] Staines R. FDA fast tracks Prevail's Parkinson's gene therapy. July 11, 2019. https://pharmaphorum.com/news/fda-fast-tracks-prevails-parkinsons-gene-therapy/.
[39] Sardi SP, et al. Glucosylceramide synthase inhibition alleviates aberrations in synucleinopathy models. Proceedings of the National Academy of Sciences of the United States of America March 7, 2017;114(10):2699−704.
[40] Zhao Y, Dzamko N. Recent developments in LRRK2-targeted therapy for Parkinson's disease. Drugs July 2019;79(10):1037−51.
[41] Gao X, et al. Use of ibuprofen and risk of Parkinson disease. Neurology March 8, 2011;76(10):863−9.
[42] Dehay B, et al. Targeting α-synuclein for treatment of Parkinson's disease: mechanistic and therapeutic considerations. Lancet Neurology August 2015;14(8):855−66.

CHAPTER 15

Journey's End – A recap of the chapters, the final conclusions, and new directions to explore

The light outside fades.
Familiar currents beckon.
Time to leave the cave.

It has been a long journey since we first met our coelacanth in Chapter 1. There, I naively suggested that the protein sequences of alpha- and beta-synuclein might have been highly conserved because of their importance to tetrapod locomotion. Having reached the end of our journey, it is clear that the actual reason for their conservation is more nuanced and complex. Their role in maintaining the ability to learn new motor tasks past early adulthood, as seen in the knockout mice, almost certainly contributed to their conservation, but I suspect the protective function of the synucleins in the innate immune system has had a greater role. This is just my opinion, of course, but no one can dispute that the perpetual struggle between pathogen and host is one of the greatest forces in evolution.

What else have we learned on the way? In "Bad News, Good News," we learned about the demographics of Parkinson's disease, how in the United States, males of European descent are most likely to get the disease, with 50% more males than females. We also learned that about one-third of cases appear to be due to genetic risk factors, another third to

Synuclein and the Coelacanth
ISBN 978-0-323-85707-9
https://doi.org/10.1016/B978-0-323-85707-9.00014-9

environmental factors, and the remaining third to some subtle combination of the two or just bad luck. So far, genome-wide association studies (GWAS) have identified about half of the putative risk factors implied by genome-wide complex trait analyses, and polymorphisms outside the coding regions of the *SNCA* gene are the largest risk factor, followed by coding region mutations in the *GBA1* and *LRRK2* genes. Of the environmental factors that have been identified, brain trauma involving loss of consciousness leads the list, followed by exposure to agricultural chemicals, although the greatest environmental risk factor, implied by the increased rates of Parkinson's disease in the Rust Belt states, has yet to be identified.

In "The Family Tremor" we were introduced to my father's family and to essential tremor, the most common movement-related neurodegenerative disease. More benign than Parkinson's disease, essential tremor involves action tremor in the hands, vocal cords, or neck. Having essential tremor increases the risk for developing Parkinson's disease, and some fraction of essential tremor cases might actually be due to synuclein-related pathology, although this is disputed, and more experimental evidence, from brain autopsies in particular, is needed to resolve the issue.

Next, we met "The Knockout." Despite the high sequence conservation of alpha- and beta-synuclein, we saw that much to my astonishment, the proteins are dispensable (at least in mice). Knocking out alpha-synuclein made no difference at all in younger mice, although in adulthood the mice started to display the movement-related learning deficits mentioned earlier. Only by knocking out all three synucleins were there serious neurological consequences, which suggests considerable functional redundancy for the proteins. Overexpressing alpha-synuclein in mice, whether the mouse version, the human version, or their mutants, is neurotoxic, although there are differences in how and where the synuclein pathology is manifested compared with humans. Overexpression of gamma-synuclein was also neurotoxic, but not beta-synuclein, which was actually neuroprotective against alpha-synuclein overexpression.

In "Shapeshifter," we glimpsed the mercurial protein structure of alpha-synuclein in its normal, nonpathological form. In solution its structure is random and dynamic, but bound to lipid membranes it adopts a curved, alpha-helical structure, probably important for stabilizing, shepherding, and outfitting presynaptic vesicles. In the following chapter, the pathological structure of alpha-synuclein was described in the form of beta-sheet rich amyloid fibrils found in the Lewy bodies and neurites that define Parkinson's pathology inside neurons. Then, two chapters later, we encountered alpha-synuclein in its most neurotoxic, oligomeric form, with a structure intermediate between its nonpathological and amyloid forms.

In "A New Kind of Pathogen," we learned about prion diseases such as mad cow disease. In prion diseases, the toxic, misfolded form of the protein induces the corresponding normally folded protein to assume the toxic fold, spreading the disease and slowly destroying the brains of the infected host. Lewy pathology in Parkinson's disease appears to spread in a similar manner, typically following the six Braak stages. In stage 1, Lewy pathology appears in the brain stem part of the vagus nerve and in the olfactory bulb above the nasal cavity; then, it spreads to the basal ganglia, reaching the substantia nigra in stage 3. There, the pathology slowly kills dopaminergic neurons, eventually causing the movement-related symptoms of Parkinson's disease. In stages 4–6, the pathology spreads through the limbic system and into the neocortex, eventually leading to dementia. Although the pathology of most Parkinson's disease patients appears to follow the Braak staging scheme, in many familial cases and in dementia with Lewy bodies, the spread of pathology can differ from the standard scheme.

Through the tragic misadventures described in "Addicts," the parkinsonism-inducing neurotoxin MPTP (1-methyl-4-phenyl-1,2,3,6-tetrahydropyridine) was discovered. It has been one of the most important tools in generating animal models to study Parkinson's disease. Studies of how it kills dopaminergic neurons have revealed their Achilles' heel, mitochondria, the power plants of cells. Mitochondria provide cells with adenosine triphosphate, the chemical energy molecule on which the cells' machinery runs. Inside the brain, MPTP is converted into its toxic form, MPP+ (1-methyl-4-phenylpyridinium), which then interferes with mitochondrial energy production, increasing the release of toxic free radicals and other chemicals. These chemicals trigger the release of inflammatory signals, spreading the inflammation outside the cell, stimulating the brain's immune cells to attack not only the stricken neuron, but neighboring neurons as well, amplifying the pathology.

In "Attack of The Oligomers," we saw how protein oligomers are the primary neurotoxic agents not just in Parkinson's disease, but also in Alzheimer's (where they were first discovered) and other amyloid-related diseases. The oligomers wreak havoc throughout the cell, disrupting membranes, throwing protein and ion homeostasis out of balance. However, the most critical damage in Parkinson's disease seems to be to mitochondria. Evidence for this is manifold, from studies of animal models and neural cell cultures, from the MPTP studies mentioned earlier, from the fact that paraquat and rotenone, toxins that target mitochondria, are linked to Parkinson's disease, but most importantly, from genetics. Two of the Mendelian mutations that lead to recessive familial Parkinson's disease occur in proteins, PINK1 and parkin, which are involved in recycling damaged mitochondria, strongly implicating a role for mitochondria in disease development. However, whereas alpha-synuclein oligomers cause the most damage, the disease does not seem to spread without the amyloid form also being present, because injecting just oligomers without amyloid fibrils into transgenic mice overexpressing alpha-synuclein does not trigger spreading pathology. Thus, alpha-synuclein oligomers and amyloid fibrils appear to be partners in crime; both are required for Parkinson's disease.

In "An Evolutionary Affair," we examined the inhibitory interaction between alpha-synuclein and glucocerebrosidase, the enzyme that causes Gaucher disease when two mutant forms are inherited. Mutations of glucocerebrosidase also increase the risk for developing Parkinson's disease, even when just one mutant form is inherited. Analysis of correlated changes of amino acids in the two proteins that have occurred in vertebrate species suggests the proteins coevolved. This coevolution implies that their interaction has been evolutionarily selected for, providing some benefit in vertebrates. However, because only humans develop Parkinson's disease, it is unlikely that the supposed benefit has anything to do with preventing the disease. Indeed, the interaction could promote the pathology. Instead, the evolutionary benefit might have something to do with gangliosides, complex lipids plentiful on the surface of neurons, because glucocerebrosidase is an enzyme involved in their degradation. The mutant version of the enzyme seems to perturb the balance of gangliosides and their lipid building blocks in a way that reduces lysosomal function, leading to less degradation of alpha-synuclein, increasing its levels, and thus increasing the risk for developing Parkinson's disease.

Why do only humans develop Parkinson's disease? In "Monkey Brains," we looked for clues comparing humans with our closest relatives, chimpanzees, and with other primates and nonprimate mammals. The most obvious difference is brain size; the human brain is 3.5 times larger than that of chimps, with 3.1 times as many neurons. Although on average chimps and humans seem to consume the same energy per neuron, this might not be true for all neurons. In particular, neurons with long-projecting nonmyelinated axons, such as the dopaminergic neurons of the substantia nigra, have to project about 50% farther owing to the larger size of human brains. Presumably, these longer axons, which must maintain the ion gradients across the axonal membrane required for neuron firing over longer distances, must use more energy proportional to their greater length. Evidence supporting increased metabolism comes from neuromelanin, a metabolic by-product found in catecholamine-producing neurons, including dopaminergic neurons. Humans have more neuromelanin in their substantia nigra neurons than any other animal. Thus, the answer might be that our bigger, gas-guzzler brains have pushed the dopaminergic neurons of the substantia nigra to near the point where the toxic by-products of metabolism can overwhelm the cell. Add to this the mitochondrial dysfunction caused by alpha-synuclein oligomers, and the neurons die off at an accelerated rate.

In "A Troublesome Subject," we assessed the possible connection between intelligence and Parkinson's disease. First, we reviewed the evidence suggesting that mutations in glucocerebrosidase and other lysosomal enzymes might have been selected for in Ashkenazi Jewish populations because they boosted IQ, which presumably would have been an advantage because many were moneylenders in the Middle Ages. On the other hand, neuronopathic forms of Gaucher disease and other lysosomal storage diseases often cause significant mental deficits, and so far, no study has proven a connection between IQ and mutations in glucocerebrosidase. Next, we examined the apparent correlation between Parkinson's disease and economic status, especially for those whose occupations require complex social interactions. Because IQ and economic status are clearly correlated, it is plausible that there might be some sort of relationship between IQ and Parkinson's disease, but definitive proof is lacking.

In "Appendix," we explored contradictory claims for a connection between appendectomies and the development of Parkinson's disease. High levels of aggregated alpha-synuclein in a nonamyloid, C-terminal truncated form have been found in the appendix, and it has long been speculated, first by Braak and colleagues, that the enteric nerves of the gut might be one of the starting points of Lewy pathology. The protein LRRK2 also seems to be involved in immune reactions in the gut, with mutant LRRK2 affecting vesicle transport and autophagy in ways that might impact alpha-synuclein levels. We also discovered that alpha-synuclein knockout mice are more susceptible to a variety of pathogens, especially viruses, which suggests that alpha-synuclein has an important role in innate immunity, as mentioned at the start of the chapter. In addition to the appendix, the olfactory epithelium, with connections to the olfactory bulb, is rich in alpha-synuclein and happens to be the other starting point in the Braak staging scheme. Amyloid-beta and the prion protein might also have protective roles in innate immunity, but whether the propensity to form amyloid is the common mechanistic thread tying their respective roles together in combatting microbial invaders is not yet known.

In "Circuitry," we followed the neurosurgical pioneers who not only sought the biological causes of Parkinson's disease but laid the foundation for its surgical treatment. We learned about the direct and indirect pathways in the basal ganglia and how the death of dopaminergic neurons of the substantia nigra interferes with them, causing the movement-related symptoms of Parkinson's disease. The discovery of levodopa treatment and growing public distrust of neurosurgery nearly brought an end to surgical treatment of Parkinson's disease. However, neurosurgeons persevered, and with the advent of deep brain stimulation, drug-based and surgical treatments now work hand in hand to improve and extend symptom reduction in patients.

In "Seeking a Magic Bullet," we visited the various ongoing efforts to develop Parkinson's treatments seeking to slow or halt progression of the disease. These efforts include antibodies engineered to cross the blood—brain barrier and virally delivered genes generating intracellular miniantibodies to counteract aggregated alpha-synuclein inside neurons. Other approaches include chemical chaperones to increase glucocerebrosidase activity and kinase inhibitors to reduce LRRK2 activity, both of which aim to reduce alpha-synuclein levels. The ganglioside GM1 has produced encouraging initial results, and autophagy-stimulating compounds such as rapamycin have also shown promise. Which of these or other approaches might be the magic bullet? Any of them, singly or in combination, could be the key to halting Parkinson's pathology.

In addition to disease-halting therapies, techniques to detect and measure the extent of Parkinson's disease pathology are also needed. First, simple biomarkers, such as disease-specific metabolites or characteristic changes in lipid molecules in the bloodstream, to name just a few possibilities, are needed for early detection to help catch the disease and treat it before serious damage is done. To date, there are no reliable biomarkers for detecting the early stages of Parkinson's disease, but there is reason for hope. In England, a woman, Joy Milne, can smell Parkinson's disease in men! She can even detect the disease before the main movement-related symptoms appear. She was tested using undershirts worn by men, either those with Parkinson's disease or controls without the disease. In one case, she smelled Parkinson's disease in the undershirt of one of the controls, who soon after developed symptoms of the disease. Inspired by Milne's olfactory ability, one study has already identified the potential biomarkers involved [1]. Maybe, in future editions of this book (I hope), I will be writing about skin swab tests, like those to detect traces of explosives at airports, for the early detection of Parkinson's disease.

Second, a technique to image Parkinson's pathology in the central nervous system is needed, to make sure therapies to slow or halt the disease are actually working. For Alzheimer's disease, the first imaging compound, created by a University of Pittsburgh team led by William E. Klunk and Chester A. Mathis, was successfully tested on a patient in 2002 [2]. Their compound, named Pittsburgh Compound B (PiB), was designed using a radioactively labeled derivative of thioflavin T, a compound commonly used in laboratories to bind and detect amyloid structure, not just amyloid fibrils of amyloid-beta, the Alzheimer's protein, but also alpha-synuclein and other proteins in their amyloid form. Despite its promiscuous amyloid binding ability, PiB does not work for imaging Lewy pathology, perhaps because not enough of the compound enters neurons. Recall that the amyloid plaques of Alzheimer's disease are extracellular whereas Lewy pathology is intracellular. In the decades since the creation of PiB, the search continues for compounds to image Parkinson's disease pathology.

Here are the molecular and evolutionary causes of Parkinson's disease, as I see it. Genetic mutations and risk factors seem to fall into two classes: those that make mitochondria more vulnerable, such as parkin and PINK1 mutations, and those that result in higher alpha-synuclein levels, such as LKKR2, glucocerebrosidase, and alpha-synuclein mutations and polymorphisms. An increase in alpha-synuclein concentration of as little as 8% could explain the increased risk for mutant *GBA1* carriers.[1]

[1] People with a gene duplication of *SNCA* have a 44% chance of developing Parkinson's disease in their lifetime [8]. Gene duplication implies that the concentration of alpha-synuclein could be as much as 50% higher than normal, although possibly the cell compensates to some degree and the percentage might be lower. Assuming that Parkinson's disease is initiated by the chance formation of aggregated synuclein, which depends on the concentration squared, how much of an increase over normal concentrations would one need to explain the increased risk for mutant *GBA1* carriers? Expressed mathematically, the chance for *SNCA* duplication carriers developing the disease is $a([1.5]^2 - 1) = (0.44 - 0.015)$, where a is a constant, 1.5 is the synuclein concentration (relative to the normal concentration, 1), and $0.44 - 0.015$ is the chance of duplication carriers developing the disease minus the average chance anyone gets the disease. This expression gives $a = 0.34$. For mutant *GBA1* carriers, the expression is $0.34(c^2 - 1) = (0.075 - 0.015)$, where c is the concentration and 0.075 is the chance the person gets the disease, which gives $c = 1.08$, that is, only an 8% increase in concentration.

To understand why, recall that the initial formation of aggregated synuclein is a chance, second-order process proportional to the square of the concentration. Imagine a crowded room with everyone moving randomly. If you double the number of people, the frequency of people bumping into each other will go up by a factor of four, and so it is with alpha-synuclein that even small increases in concentration will have an amplified effect.

As for mitochondria, evolution has increased the metabolism in our gas-guzzler, oversized brains to the point where mitochondria are perhaps under greater stress than in any other mammal, in particular in the dopaminergic neurons of the substantia nigra with their massive axonal arbors. Adding to that the extra stress caused by alpha-synuclein oligomers gumming up their works, the mitochondria likely accumulate damage at an accelerated rate, releasing reactive oxygen compounds and triggering the inflammation that ultimately leads to neuronal death.

Environmental risk factors for Parkinson's disease also implicate mitochondria, especially the toxins paraquat and rotenone. One risk factor that does not fit this simple picture is brain trauma with loss of consciousness. This risk factor appears to involve pathology of the tau protein, which is also responsible for chronic traumatic encephalopathy and other neurodegenerative diseases. Tau pathology manifests as neurofibrillary tangles of misfolded, hyperphosphorylated tau. These tangles are also a hallmark feature of Alzheimer's disease. Given its connection with the world's most common neurodegenerative disease, as well as with head trauma in some of the world's highest paid athletes, I imagine tau will get a book of its own someday.

Recall that polymorphisms of tau have also been identified as Parkinson's disease risk factors in GWAS, with the most common risk-associated polymorphism actually conferring a protective effect against the disease. Tau is involved in the structure of axons, interacting with microtubules, long tubelike structures that traverse the length of cells, including the axon, acting as highways for the transport of various cellular cargo including vesicles and small organelles such as mitochondria. My friends at work picture tau as a kind of speed bump helping to moderate traffic on the microtubule highway. Perhaps tau pathology increases the risk for developing Parkinson's owing to disruption of mitochondrial trafficking, but curiously, the neurofibrillary tangles associated with tau pathology accumulate not in axons where tau is normally most abundant, but in dendrites and in the neuron cell body [9]. Clearly, understanding the role of tau in neurodegeneration is going to require more study.

There is one more gaping hole regarding environmental risk factors for Parkinson's disease: to wit, the increased disease prevalence in the Rust Belt states [3]. Although pollutants such as industrial solvents and toxic metals have been proposed as possible causes of the increased prevalence, this is far from proven. Please allow me one final spate of speculation. I wonder whether there might be a connection between the Rust Belt effect and increased Parkinson's rates seen for those with higher socioeconomic status. I realize this must seem paradoxical, because wealthier people tend to live in less polluted areas, but I suspect the risk of developing the disease might be influenced by an allergic immune response.

Chronic pollution alters immune function, exacerbating and perhaps even causing asthma [4]. Similarly, being raised in an environment with less exposure to natural allergens can also alter immune function, leading to an increased prevalence of allergic rhinitis, or hay fever [5]. In the US Medicare study of Parkinson's disease prevalence, nonfarmers who lived in rural areas had a lower prevalence of the disease than did urban dwellers. In fact, studies have seen an increased risk for Parkinson's disease in asthmatics and people experiencing allergic rhinitis [6,7]. This is pure conjecture, but perhaps environmentally induced immune alterations involve alpha-synuclein. In particular, these alterations might ramp up the (still unproven) role of alpha-synuclein in the innate immune response, increasing its levels in the gut and nasal cavity, thus increasing the chance of triggering Lewy pathology.

And with that final flourish of hypothesizing, our journey through the molecular and evolutionary origins of Parkinson's disease ends. We have hacked through jungles of molecular biology, navigated tangled river systems of neuroanatomy, and traversed mountain ranges of genetics. You have indulged wild flights of theorizing and speculation, and I thank you. No doubt, the future will add more chapters to the saga of the synucleins, with new and unexpected pathways to explore. And now, let us sink deep beneath the nighttime waves of the Indian Ocean for a final visit to our coelacanth.

Still.

Drifting.

Scents floating in the current.

A presence approaching. Lunge, snap!

Got it! An eel.

Mmm.

References

[1] Trivedi DK, et al. Discovery of volatile biomarkers of Parkinson's disease from sebum. ACS Central Science April 24, 2019;5(4):599–606.
[2] Klunk WE, et al. Imaging brain amyloid in Alzheimer's disease with Pittsburgh Compound-B. Annals of Neurology March 2004;55(3):306–19.
[3] Wright Willis A, et al. Geographic and ethnic variation in Parkinson disease: a population-based study of US Medicare beneficiaries. Neuroepidemiology April 2010;34(3):143–51.
[4] Guarnieri M, Balmes JR. Outdoor air pollution and asthma. Lancet May 3, 2014;383(9928):1581–92.
[5] Nicolaou N, Siddique N, Custovic A. Allergic disease in urban and rural populations: increasing prevalence with increasing urbanization. Allergy November 2005;60(11):1357–60.
[6] Bower JH, et al. Immunologic diseases, anti-inflammatory drugs, and Parkinson disease: a case-control study. Neurology August 8, 2006;67(3):494–6.
[7] Cheng CM, et al. Risk of developing Parkinson's disease among patients with asthma: a nationwide longitudinal study. Allergy December 2015;70(12):1605–12.
[8] Devine MJ, et al. Parkinson's disease and α-synuclein expression. Movement Disorders October 2011;26(12):2160–8.
[9] Hoover BR, et al. Tau mislocalization to dendritic spines mediates synaptic dysfunction independently of neurodegeneration. Neuron 2010;68(6):1067–81.

Glossary

AAV **A**deno**a**ssociated **V**irus. A virus that can be engineered for use in **gene therapy**.

ablation In surgery, ablation means destruction of the target tissue. This may be done using heat, extreme cold, or toxic chemicals.

action tremor Rhythmic, involuntary shaking of a body part when it is involved in an action, including when the action involves moving the body part or when the body part is held still in a nonrelaxed position.

active immune therapy Vaccination.

adaptive immune system One of two branches of the immune system. Specific cells in this branch, of either the **B-** or **T-cell** type, respond to specific pathogens, in contrast to the **innate immune system**.

ADDLs **A**myloid beta—**D**erived **D**iffusible **L**igands. Another name for **amyloid-beta oligomers**.

aggresome A compartment containing aggregated, misfolded **proteins** that forms near the cell nucleus when the cell's **protein**-degradation system is overwhelmed, also called a **JUNQ**.

allele (uh-LEEL) The version of a **gene**. Different versions result from **mutations** and/or **polymorphisms**. Because individuals have two copies of each **gene**, they have two alleles, and the pair of alleles is called the **genotype**. The word was originally "allelomorph," from Greek, *allel-* "one another" and *morphe* "form."

alpha-helix One of the main **secondary structural** motifs in **protein** structure, consisting of the **peptide backbone** adopting a spiral structure stabilized by **hydrogen bonds**.

alpha-synuclein (see **synucleins**)

ALS **A**myotrophic **L**ateral **S**clerosis. A fatal neurodegenerative disease, also known as Lou Gehrig's disease.

Alzheimer's disease A fatal neurodegenerative disease with pathology caused by **amyloid** plaques made of the **peptide amyloid-beta** and **neurofibrillary tangles** made of **fibrils** of **tau protein**.

ambroxol An expectorant drug for treating coughs, being tested as a possible treatment for **Parkinson's disease**.

amino acid A molecular building block, consisting of a **peptide backbone** and a **side chain** that comes in 20 forms. The **peptide backbone** can be chemically bonded into a chain, forming a sequence of amino acids called a **protein**.

amygdala Latin for "almond." Part of the brain's **limbic system**, involved in emotional response.

amyloid **Protein** that has aggregated into a **polymeric** form, made of **fibrils**, long fibers consisting of the **protein** in a **cross-beta structure**.

amyloid-beta A **peptide** that forms **amyloid** and is found in dense, extracellular deposits called plaques in the brains of those with **Alzheimer's disease**.

ansa lenticularis Part of the nerve tract between the **thalamus** and the **globus pallidus**. Latin for "lentil-like loop."

antiparallel beta-sheet A **beta-sheet** in which the strands run in opposite directions.

antibody A **protein** made and secreted by **plasma cells**. Different antibodies recognize different **epitopes** of invasive pathogens, helping to neutralize them and causing the immune system to attack and/or engulf them.

apoptosis (ay-po-TOH-sis, or ay-pahp-TOH-sis) From Greek "falling from." Programmed cell death, whereupon receiving particular intracellular and/or extracellular signals, the cell releases **enzymes** that degrade and break up the cell such that its components are easily cleared by immune system cells, such as **microglia** and **macrophages**.

astrocyte A **glial** cell that helps to maintain neurons and other neural tissues such as the **blood–brain barrier** (**BBB**).

ATP **A**denosine **T**ri**p**hosphate. The primary chemical energy molecule used by the cell's molecular machinery. ATP is also incorporated into **DNA** as the base adenine.

autophagosome A double-**membraned organelle** involved in **macroautophagy** that engulfs large portions of the cell's interior and then fuses with a **lysosome** so that the engulfed contents can be broken down and recycled.

autophagy Several types of degradation mechanisms in the cell that involve **lysosomes**, including **macroautophagy**, **microautophagy**, **mitophagy**, and **chaperone-mediated autophagy**. From Greek *autos* "self" and *phagein* "to eat."

autosomal mutation A mutation in a gene on an autosomal **chromosome**. All **chromosomes** are autosomal except the sex **chromosomes**, X and Y.

auxilin A **protein**, encoded by the *DNAJC6* **gene**, which, once a new **vesicle** has been formed by **endocytosis**, helps to remove the scaffolding of **clathrin proteins** around the new **vesicle**.

axon The main conduit of a **synapse** that, upon receiving enough stimulatory signals, causes the neuron to fire; that is, the signals trigger a voltage wave that travels down the axon to the **axonal arbor**.

axonal arbor The branching structure at the end of the **axon** of a neuron, containing **axonal varicosities** (**boutons** and **en passant synapses**) that, when the neuron fires, release neurotransmitters into the **synapses** with the attached, postsynaptic neurons.

axonal varicosity A swelling, which in the context of the **axonal arbor**, includes **en passant synapses** and **presynaptic boutons**, because they are wider than the axonal branches to which they are connected.

B cells A class of **adaptive immune system** cells that respond to particular pathogens by making **antibodies** that recognize and bind to them.

balancing selection The situation in which inheriting one version of a **mutated gene** offers some sort of advantage, whereas inheriting two **mutated** versions is harmful.

basal ganglia Structures above the middle of the brain, including the **substantia nigra**, **striatum**, and **globus pallidus**, and other structures involved in motion, social interactions, learning, and other processes.

BBB **B**lood–**B**rain **B**arrier. A barrier formed by **endothelial cells** and **astrocytes** that prevents most biomolecules, especially charged ones, from passing from the blood into the central nervous system.

beta strand (see **beta-sheet**)

beta waves **Brain waves** in the 13- to 30-Hz (beats per second) range.

beta-carbolines A family of naturally occurring chemicals, often fluorescent, thought to increase risk for **Parkinson's disease**.

beta-sheet A **protein** secondary structural element consisting of extended segments of the **protein** chain side by side with other extended segments, with their **peptide backbones hydrogen-bonded** together to form a sheet-like structure. The extended segments are called **beta strands**, and their **amino acid** sequences can run in the same or opposite directions.

beta-synuclein (see **synucleins**)

bilayer (see **lipid membrane**)

bioinformatics A field of study that uses mathematics and computer programs to analyze large biological data sets.

biomarker A chemical, chemical imbalance, or other easily measured biological variable that indicates an underlying disease or condition.

body diagonal In geometry, the line running from one corner of a cube to the most opposite corner: that is, the corner most distant from it. The angle this line makes with the faces of the cube, 54.7 degrees, is the "**magic angle.**"

boutons (see **presynaptic boutons**)

Braak stage 1 First stage of the Braak staging scheme of Parkinson's pathology, with **Lewy pathology** in the **vagus nerve** and **olfactory bulb**.

Braak stage 2 **Lewy pathology** spreads into regions at the top of the brain stem, including the **locus coeruleus**.

Braak stage 3 **Lewy pathology** spreads beyond the brain stem into the **basal ganglia**, including the **substantia nigra pars compacta (SNc)**.

Braak stage 4 Substantial neuron loss in the **SNc,** with **Lewy pathology** spreading further throughout the **limbic system,** including the **amygdala**.

Braak stage 5 **Lewy pathology** spreads into the **neocortex**, starting with the temporal, parietal, and frontal lobes.

Braak stage 6 **Lewy pathology** reaches its peak, reaching sensory and motor areas of the **neocortex**.

bradykinesia (BRADD-ee-kin-EE-sia) Slowness of movement, from Greek *bradys* "slow" and *kinēsis* "movement."

brain waves The oscillating electric field fluctuations of the brain arising from large networks of neurons firing in a coordinated manner.

brain stem The region of the brain at the base of the brain atop the spinal cord, a relay center between the brain and nerves spreading throughout the body, with regions important for arousal and attention.

BSE **B**ovine **S**pongiform **E**ncephalopathy. A neurodegenerative disease in cows, more commonly known as "mad cow disease," caused by **PrPsc**.

C-terminus The last **amino acid** of a **protein** chain. More specifically, the C-terminus is the final carbon of the last amino acid which, instead of being bonded to an oxygen and a nitrogen, is in its carboxylic acid form (attached to two oxygen atoms).

calmodulin A calcium-binding protein involved in calcium signaling in the cell, with dozens of interaction partners, making it an important **hub protein**.

CAMKII Ca^{2+}/calmodulin(**CAM**)-dependent protein **K**inase **II**. A **protein kinase** whose activity is modulated by binding to **calmodulin**.

CAPRI Critical **A**ssessment of **Pr**ediction of **I**nteractions. A competition in which research laboratories can submit predictions of newly solved **protein** complexes, with the individual **protein** structures typically already known, to judge which prediction methods have the best success rate.

carbidopa A drug for treating the movement-related symptoms of **Parkinson's disease**. Taken with **L-DOPA**, carbidopa inhibits conversion of **L-DOPA** outside the brain, allowing more L-DOPA to reach the brain.

cardiolipin A lipid, first isolated from beef hearts, consisting of two phosphoryl groups and four hydrocarbon chains. Cardiolipin is abundant in **mitochondrial membranes**.

carrier In genetics, one who carries a copy of a mutated gene associated with a trait or a disease, but who does not have the associated trait or disease, because the mutation is either **recessive** or as the result of reduced **penetrance**.

CASP Critical Assessment of protein Structure Prediction. A competition in which research laboratories can submit predictions of newly solved **protein** structures, to see which **protein** structure prediction methods have the best success rate.

catecholamines A class of **neurotransmitters** that includes **dopamine** and **noradrenaline**.

caudate From Latin for "tail." Part of the dorsal **striatum**.

CD spectroscopy Circular Dichroism spectroscopy. A technique that measures the difference in interaction of clockwise versus counterclockwise circularly polarized light with a sample. See "Twisting light."

cecum A pouch at the beginning of the large intestine, important in herbivores for the digestion of leaves. Latin for "blind."

cell body For a neuron, the main part of the cell containing the nucleus from which **dendrites** and the **axon** branch off. Also called the **soma**.

cell membrane The **lipid membrane** surrounding a cell, also called the **plasma membrane**.

centrifugation A purification technique in which a sample is spun at high speeds, separating the sample components (i.e. different **proteins**) based on their specific densities, size, and other qualities.

ceramide A special **lipid**, not a **phospholipid**, but another kind called a **sphingolipid**, involved in signaling inside the cell; that is, its presence can alter particular cellular processes.

cerebellum Latin for "little brain," diminutive form of cerebrum. Located at the lower rear of the brain, important for coordinated muscular movement, speech, and balance.

cerebral edema (see **vasogenic edema**)

chaperone In molecular biology, a **protein** or **protein** complex that helps newly formed or misfolded **proteins** fold correctly.

chaperone-mediated autophagy A process in which particular proteins are transported through the **lysosomal membrane** by **protein** complexes to be degraded inside the **lysosome**. One of the main mechanisms by which **alpha-synuclein** is degraded.

chemical chaperone A small molecule that helps stabilize the structure of its target **protein**.

chimeric Adjective for an organism whose cells are a mixture of cells from two or more **genomes**.

CHO cells Cells derived from an ovary of the Chinese hamster, *Cricetulus griseus*.

chromosome From Greek *chroma* "color" and *soma* "body." A linear **DNA** structure made of double-stranded **DNA** helix and **proteins**. The genomes of **eukaryotes** are organized into chromosomes. Most **eukaryotes** are diploid; that is, their chromosomes are organized in pairs, one from each parent.

cilia (sing. cilium) Short hair-like structures extending from a cell. Latin for "eyelashes."

clathrin A **protein** that, interacting with other copies of itself, can form a stabilizing scaffold around a newly forming **vesicle** budding from a **membrane**.

CMA Correlated Mutation Analysis. An analysis that looks for similar patterns of **mutations** in **proteins** across species. Pairs of **protein** sequence positions that show correlations are often in contact in the **protein** or **protein** complex structure. See "Are we correlated?"

codon (See **genetic code**)

coelacanth (SEE-luh-canth) An ocean fish with two species, which, along with the freshwater lungfishes, make up the clade (class) of lobe-finned fishes, the closest relatives of the land vertebrates, or tetrapods, which include all amphibians, reptiles and birds, and mammals.

coevolution The correlated **mutation** process that occurs over time, across species, for interacting **proteins**. Interacting **proteins** with correlated mutations are said to have coevolved.

coimmunoprecipitation A **pull-down assay** in which an **antibody** to the **protein** of interest is used to bind the **protein,** along with any **proteins** interacting with it, which are said to coimmunoprecipitate with the **protein** of interest.

complementary strand The second strand of the **DNA** double helix, running in the opposite direction, whose sequence has guanine matched up with cytosine and vice versa, and adenine with thymine and vice versa.

complex I A large complex of 44 **proteins** located at the inner membrane of **mitochondria** involved in the transport of electrons and **protons**. Part of the **respiratory chain** of complexes that shuttle electrons and **protons** in **oxidative phosphorylation**.

COMT inhibitor Catechol-O-methyltransferase **inhibitor**. A drug that inhibits COMT, an **enzyme** that adds a methyl group to **dopamine** and **L-DOPA**, making more **dopamine** available in the brain, and making **L-DOPA** therapy more effective.

corpus callosum The **white matter** connecting the two hemispheres of the **neocortex**.

covariance Comparing two data sets, covariance is a measure of how similar their patterns of variation are with no emphasis on whether the variations are simple or complex.

CRISPR/Cas9 A **protein–RNA** complex that can be designed to edit a target **gene**.

cristae From Latin "plume, crest." The deep, regularly spaced folds of the inner **membrane** of **mitochondria**.

Crohn's disease An inflammatory bowel disease, perhaps caused by an excessive immune reaction to microbes in the gut, influenced by genetic risk factors.

cross-beta structure A (usually) **parallel beta-sheet** structure with intermolecular **hydrogen bonds** between the chains of the **protein** molecules forming an **amyloid fibril**. The strands of the **beta-sheet** run perpendicular to the axis of the **fibril**, and the **hydrogen bonds** are parallel to the axis.

cryo-EM **cryo-**Electron Microscopy on frozen samples. See "Ghost in the static."

cytochrome C oxidase The **mitochondrial protein** that uses energy derived from food molecules as well as oxygen to generate the **proton** gradient across the mitochondrial inner membrane necessary to produce the chemical energy molecule **ATP**, used to fuel the machinery of the cell.

cytosol The fluid inside the cell that surrounds all of the **organelles**.

DAT Dopamine Active Transporter. A **protein** in the **cell membrane** that transfers **dopamine** into neurons.

DCA **D**irect **C**oupling **A**nalysis. A **CMA** that uses an alternate method, rather than **Z-value analysis**, to correct for indirect correlations between data sets.

deep brain stimulation The application of an oscillating electric field to an electrode positioned at a target region deep in brain. Depending on the frequency and strength of the electric field, the functioning of the target region can be enhanced or inhibited.

dendrite Branching structures, termed "processes" in neurology jargon, extending from the neuron cell body. Dendrites form **synapses**, primarily with axonal **varicosities**, forming the postsynaptic part of the **synapse**, receiving chemical signals from attached neurons.

deprenyl An **MAO B** inhibitor drug used to treat **Parkinson's disease**.

diglyceride A **lipid** composed of two hydrocarbon chains attached to a glycerol (a small sugar, also called glycerine). The glycerol can have additional **hydrophilic** groups attached, and together with glycerol, they are called the **head group**.

dimer (see **multimer**)

direct pathway A neural circuit through the **motor cortex** and **basal ganglia** that results in a voluntary movement.

DJ-1 (see ***PARK7***)

DLB **D**ementia with **L**ewy **B**odies. A **synucleinopathy** related to **Parkinson's disease** in which the onset of dementia and movement-related symptoms occurs within a year of each other.

DNA **D**eoxyribo**n**ucleic **A**cid. A molecule consisting of a phosphate attached to a ribose sugar ring connected to one of four bases, **a**denine, **c**ytosine, **g**uanine, or **t**hymine. The phosphate can chemically bond to another **DNA** ribose, and so on, forming a chain called a strand, also called a **DNA** sequence, denoted using the first letters of the bases in the sequence.

dominant mutation A **mutation** in which inheriting at least one mutated **gene** from a parent leads to (or possibly leads to, in the case of reduced **penetrance** or **risk factors**) the associated trait or disease.

DOPA decarboxylase An **enzyme** that converts **L-DOPA** into **dopamine**.

DOPAL 3,4-**D**ihydr**o**xy**p**henylacet**al**dehyde. A toxic metabolite of **dopamine** that usually gets quickly degraded by cellular **enzymes**.

dopamine A **neurotransmitter** involved the initiation of movement, the brain's reward circuitry, and other processes in the brain and other parts of the body.

dopamine agonist A drug that binds the same (or a subset of the same) neuroreceptor **proteins** as **dopamine**, causing them to respond in a similar manner.

dopaminergic neuron A neuron that synthesizes and releases the **neurotransmitter dopamine**.

dyskinesia Involuntary writhing, bobbing, and fidgeting movements associated with several movement disorders. In **Parkinson's disease**, dyskinesia is often a side effect of **L-DOPA** treatment.

E. coli *Escherichia* ***coli*****.** A common bacterium in the intestines.

electroporation A procedure that causes temporary pores in the **membrane** of a cell by subjecting it to a strong electric field, allowing fluid to flow through the **membrane**. In this way, engineered genetic material added to the cell medium can flow into the cell.

en passant synapse *En passant* is French for "in passing." A **synapse** that occurs at an enlarged region along an **axonal** branch rather than its end.

endocytosis From Greek *endon* "within" and *kutos* "jar, container." The generation of a new **vesicle** from a larger **lipid membrane** by a budding process, often aided by the **protein clathrin**, which forms a scaffold stabilizing the nascent **vesicle**.

endoplasmic reticulum A network of **lipid membrane** throughout the cell where **lipids** and **proteins** inserted in the **membrane** are made.

endosome A **vesicle** inside the cell formed by **endocytosis** at a **lipid membrane**.

endothelial cell A cell that lines the interior of blood and lymph vessels.

enteroendocrine cell A specialized cell in the **intestinal epithelium** that can respond to contents and conditions inside the intestines by releasing hormones.

enzyme A **protein** that catalyzes a chemical reaction.

epitope The foreign **peptide** sequence or pathogen surface feature that a particular **antibody** recognizes.

essential tremor A neurological condition involving **action tremor**, typically in the muscles of the hands, arms, and/or neck, including muscles controlling the vocal cords.

ETC **E**lectron **T**ransport **C**hain (see **respiratory chain**)

eukaryote An organism whose cell or cells have a nucleus where DNA **chromosomes** are located.

exocytosis From Greek *ex* "out of" and *kutos* "jar, container." The process of a **vesicle** fusing its **lipid membrane** to a larger one, such as the **cell membrane** or the **membrane** of an **organelle,** releasing the contents of the **vesicle** to the other side of the large **membrane**.

exosome A **vesicle** released outside a cell.

fibril (FYE-bril) adj. fibrillar (FIB-bril-lar) (see **amyloid**)

fission (see **mitochondrial fission**)

fold (see **protein tertiary structure**).

founder effect The founder effect occurs when a new population is established by a small number of individuals from a larger population. **Polymorphisms** and **mutations** carried by these individuals, perhaps rare in the population from which they came, become common in their descendants.

fragmentation (see **mitochondrial fragmentation**)

frangibility The ease with which something is breakable.

FRET **F**örster **R**esonance **E**nergy **T**ransfer. An experimental technique for measuring how close one molecule, or molecular region, labeled with a particular fluorescent molecule, is to another molecule, or another region of the same molecule, labeled with a different fluorescent molecule.

frontal lobe The front part of the **neocortex**, responsible for thought and action.

frontal lobotomy Brain surgery in which the white matter between the **frontal lobe** and the **thalamus** is cut.

fusion (see **mitochondrial fusion**)

gain-of-function mutation (see **toxic gain-of-function mutation**)

gamma waves **Brain waves** in the 30- to 150-Hz (beats per second) range.

gamma-synuclein (see **synucleins**)

ganglioside a kind of **glycolipid** found in **cell membranes**, especially abundant in neurons. All gangliosides have **glucosylceramide** as their core structure, to which additional sugar molecules are attached. There are many different gangliosides, such as **GM1** with four additional sugar molecules, **GM2** with three, and **GM3** with two.

Gaucher cell In a **Gaucher disease** patient, a **macrophage** cell whose **lysosomes** become swollen with **glucosylceramide**, filling most of the cell.

Gaucher disease A **lysosomal storage disease** caused by inheriting two defective ***GBA1* genes**, which code for **glucocerebrosidase**.

gavage (ga-VAZH) Forced delivery of substances to the stomach through a tube down the esophagus.

GBA1 The **gene** for the protein **glucocerebrosidase**. Mutated *GBA1* is a **risk factor** for developing **Parkinson's disease**.

GCTA **G**enome-wide **C**omplex **T**rait **A**nalysis. An analysis comparing how many total **SNPs** patients share on average versus how much controls share on average. The analysis allows one to estimate how much genetic **risk factors** contribute to a particular disease (or trait).

GDP (see **GTP**)

gene A sequence of **DNA** bases that, once transcribed into **RNA**, typically (but not always) encodes a protein. **RNA** that encodes **protein** is called messenger **RNA** (**mRNA**).

gene therapy A treatment in which a beneficial or corrected version of a **gene** is delivered to cells.

genetic code The **DNA** triplets (sequences of three bases), called **codons**, which encode for 20 **amino acids**.

genome The sequence of **DNA** bases, which along with the **complementary strand**, forms a double helix structure that contains all of the genetic information (i.e. the **genes**) of the organism. In higher organisms, the genome is organized into **chromosomes**.

genotype The versions, or **alleles**, of the two copies of a **gene** inherited from one's parents, which result in a trait, called a **phenotype**.

glial cells From Greek *glia* "glue." Nonneural cells of the brain that support the function of neurons.

globus pallidus Latin for "pale globe." Part of the **basal ganglia** important for relaying signals from the **striatum** to other parts of the **basal ganglia** and other brain regions, such as the **thalamus**. The globus pallidus is divided into two regions, internal (the half facing the center of the brain) and external.

glucocerebrosidase A **lysosomal** enzyme that breaks down **glucosylceramide** into glucose and **ceramide**. Encoded by the gene ***GBA1***.

glucosylceramide (also known as **glucocerebroside**) A **glycolipid** composed of **ceramide**, a **sphingolipid**, which is composed of two hydrocarbon chains connected via a peptide bond (as opposed to the phosphoester bonds of **phospholipids**) with a glucose sugar attached, to which additional sugar molecules are attached to form different **gangliosides**.

glutamate An **amino acid** whose **side chain** has a negative charge.

glycolipid A **lipid** with one or more sugar molecules attached to the **head group**.

GM-CSF **G**ranulocyte **M**acrophage **C**olony **S**timulating **F**actor. A **protein** drug for the treatment of leukemia, being studied as a possible treatment for **Parkinson's disease**.

GM1 GM2 GM3 (see **ganglioside**).

Golgi apparatus Specialized extended **vesicles** in the cell in which **proteins** and **lipids** are further modified after synthesis, often by attaching chains of sugar molecules.

gray matter In the brain, regions consisting primarily of neuron cell bodies, **synapses**, and supporting **glial** cells.

Greek key motif A protein fold, consisting of **beta-sheet**, named for the interlocking geometric patterns that were commonly used to decorate ancient Greek pottery.

GTP **G**uanosine **Tri**phosphate. A molecule similar to **ATP**, which can be used as a signaling molecule in the cell, where cleavage of a phosphate yields **GDP** (**G**uanosine **Di**phosphate). GTP is also incorporated into **DNA** as the base guanine.

GTPase An enzyme that binds and cleaves **GTP** to **GDP**.

gut biome All of the microbes living in the intestines.

GWAS **G**enome-**W**ide **A**ssociation **S**tudy. A study comparing the **SNP** frequencies in patients with healthy controls (or comparing people with a particular trait with those without it). The **SNP** frequencies that are most different in patients versus controls map to **chromosomal** regions called loci that contain genetic **risk factors**. Usually, **GWAS** requires thousands of patients and controls to generate statistically significant results.

haploinsufficiency The situation in which one of two **gene** copies does not produce enough **protein**.

head group (see **diglyceride**)

hemoglobin The **protein** in red blood cells that transports oxygen to the body's tissues and carbon dioxide away from them. In adults, the **protein** consists of two alpha and two beta chains (other chains can be present in fetuses and infants).

hippocampus Greek for "seahorse." A brain region in the **limbic system** involved in memory.

homologous recombination A process in which, during a transient dissociation of strands at a point along the **DNA** double helix of a **chromosome**, another **DNA** strand, with a highly homologous sequence, comes in to form a double helix with the **complementary strand**, displacing the original strand. **DNA** repair enzymes can then incorporate the new **DNA** sequence into the **chromosome**.

homunculus In neuroanatomy, a cartoon diagram mapping the different parts of the body to the regions of the brain that control them or to which they provide sensory information. Latin for "little person."

hub protein A **protein** that has many important interaction partners, including other **proteins**, **DNA**, **RNA**, or other molecules in the cell.

hydrogen bond A bond formed by electric attraction between a hydrogen atom in a molecule and another atom (oxygen for instance) in the same or a different molecule. The hydrogen has a net positive charge, and the atom to which it is hydrogen-bonded has a net negative charge. Hydrogen bonds are weaker and longer than chemical bonds, such as the bond between carbon atoms.

hydrophilic Adjective describing a molecule, or a part of a molecule, that likes to interact with water.

hydrophobic Adjective describing a molecule, or a part of a molecule, that prefers to avoid water.

hyperdirect pathway A neural circuit through the **motor cortex** and **basal ganglia**, but bypassing the **striatum**, that suppresses movement.

hyperkinesia A class of disorders involving repetitive, involuntary muscle contractions.

hypophonia Soft speech, a common Parkinson's disease symptom. From Greek for "low sound."

idiopathic Adjective to describe a disease or condition whose cause in unknown, although in the context of **Parkinson's disease**, idiopathic means the disease is not obviously genetic or familial in origin.

iminosugar A molecule resembling a sugar molecule, but with one of the carbon atoms replaced by a nitrogen atom.

in silico Calculated, performed, or simulated on a computer. Fake Latin, alluding to the silicon chips of computers.

in vitro Latin for "in glass." Describes experiments done in an artificial setting, such as in a test tube or cell culture dish.

in vivo Latin for "in a living thing." Describes experiments or operations performed in a living organism.

in-register For a parallel **beta-sheet**, the situation in which each **beta strand** has the same sequence.

indirect pathway A neural circuit through the **motor cortex** and **basal ganglia** that results in suppression of undesired movements, at least in theory, during a voluntary movement.

innate immune system One of two branches of the immune system. Responds in a general way to pathogens and microbial invaders, in contrast to the **adaptive immune system**.

intestinal epithelium The cells lining the inside the intestines.

intrabodies Antibodies engineered to remain inside cells.

ion dyshomeostasis Ion imbalance owing to disruption of **membrane** integrity and ion transport mechanisms in cells.

isofagomine An **iminosugar** drug that both inhibits **glucocerebrosidase** and acts as a **chemical chaperone**.

isometric muscle contractions Flexing of particular muscles that keep a target body region unmoving.

isoquinolines A family of chemicals with a structure similar to **MPTP**, some of which have been shown to be neurotoxic.

isradipine A blood pressure drug that has been tested as a possible treatment for **Parkinson's disease**.

JUNQ Juxta **N**uclear **Q**uality control compartment. (see **aggresome**)

kinase A **protein** enzyme that attaches a phosphoryl group (see **phosphorylation**) to a biomolecule.

knockin mouse A mouse that has been genetically engineered to have a particular **mutation** in a target gene. The technique can also be used to correct a mutant gene. Using an additional technique called "floxing" (from "flanked by loxP sites"), any additional added DNA, such as the ***neo*** gene, can be cut out, leaving just the original sequence with the desired **mutation**.

knockout mouse A mouse that has been genetically engineered to have a particular gene inactivated, or "knocked out."

kuru A fatal neurodegenerative disease, also known as laughing sickness, that can be contracted by eating human flesh containing the **prion PrPsc**.

L-DOPA L-3,4-**d**ihydr**o**xy**p**henyl**a**lanine. **Levadopa**. A drug for treating the movement-related symptoms of **Parkinson's disease**. Unlike **dopamine**, L-DOPA can cross the **blood–brain barrier** (**BBB**), where it is converted to **dopamine** by the **enzyme DOPA decarboxylase**.

LATE Limbic-predominant **A**ge-related **T**DP-43 **E**ncephalopathy. A neurodegenerative disease involving the protein **TDP-43** that causes the same symptoms as **Alzheimer's disease**.

leaflet (see **lipid membrane**)

leucotomy The original term for **frontal lobotomy**.

levodopa (see **L-DOPA**)

Lewy body A large, spherical deposit inside a neuron cell body that contains aggregated **alpha-synuclein**. Lewy is pronounced "LOO-wee" or "LYOO-wee," although the original German pronunciation was probably closer to "LEH-vee."

Lewy neurite A small deposit consisting primarily of aggregated **alpha-synuclein,** typically found in the branches of (presumably) the **axonal arbors** of neurons.

Lewy pathology The presence of **Lewy bodies** and/or **Lewy neurites**.

LIGA20 A molecule similar to **GM1**, but with one **hydrophobic** chain instead of two, which is studied as a possible treatment for **Parkinson's disease**.

limbic system A group of structures in the middle part of the brain, including the **hippocampus** and **amygdala**, involved in long-term memory and emotional response and other aspects of behavior.

linkage disequilibrium The situation when two gene **alleles**, or **polymorphisms**, or **mutations** are found to occur together more often than would be predicted by chance.

lipid A fat molecule that consists of one or more fatty acid chains, usually attached to glycerol. With just one fatty acid chain attached, the lipid is a monoglyceride, with two chains, a **diglyceride**, and with three, a **triglyceride**.

lipid membrane A membrane made from two layers, a bilayer, made primarily of diglyceride lipids. The **hydrophobic** hydrocarbon tails of the lipids pack together in the middle of the membrane, whereas the **hydrophilic head groups**, most of which contain a phosphoryl group, are situated exposed to water at the two membrane surfaces. The two layers are called the outer and inner **leaflets**.

locus coeruleus (LO-cus seh-ROO-lee-us) Latin for "blue place." Brain region above the brain stem, important for attention and arousal.

loop A part of the **protein** sequence that connects **protein secondary structural** elements, like **alpha-helices** and **beta-sheets**. Unlike **alpha-helices** and **beta-sheets**, the **peptide backbone** of the loop does not necessarily make **hydrogen-bonds** or adopt any particular backbone structure.

loss-of-function mutation A **mutation** that makes the mutated **protein** nonfunctional.

LRRK2 The **gene** for the protein **L**eucine **R**ich **R**epeat domain **K**inase **2** (LRRK2), also known as dardarin. Mutated *LRRK2* is a **dominant Mendelian mutation** with reduced **penetrance** for developing **Parkinson's disease**.

LTP **L**ong-**T**erm **P**otentiation. The strengthening of **synaptic** connections, important in forming memories.

lymphoblastoid cell A cell created by infecting an immune system B cell (see **antibody**) with Epstein Barr virus (EBV) and treated with the immunosuppressant cyclosporin A, capable of growing and dividing in cell culture.

lysine An **amino acid** whose long **side chain** is **hydrophobic** except for the end, which has a positive charge, making it **hydrophilic**.

lysosomal storage diseases **Recessive** genetic diseases caused by defective **lysosomal enzymes** involved in degrading particular biomolecules such as **gangliosides** or in **proteins** involved in **lysosomal** function or transport of the target biomolecules or their breakdown products.

lysosome An acidic **organelle** that degrades **proteins**, **lipids**, and carbohydrates for reuse in the cell.

macroautophagy An **autophagy** process in which a large volume of the inside of a cell is enveloped by a double-**membrane** structure called an **autophagosome** and then fused with a **lysosome** to degrade the contents.

macrophage A white blood cell that clears cellular debris and pathogens from blood and tissue by enveloping them using **endocytosis** at the **plasma membrane**, sending the resulting **endosomes** to **lysosomes** to be degraded.

magic angle (see **body diagonal**)

MAO-B **M**onoamine **O**xidase-B. An **enzyme** that metabolizes a class of compounds, usually by removing a nitrogen-containing chemical group called an amine. MAO-B converts **dopamine** and **MPTP** into **DOPAL** and **MPP+**, respectively, both of which are neurotoxic.

MAPT The **gene** for the protein **tau**. Mutated *MAPT* is a **risk factor** for **Parkinson's disease**.

matrix (see **mitochondrial matrix**)

membrane (see **lipid membrane**)

Mendelian mutation A **mutation** in which inheriting one copy for a **dominant mutation** or two copies for a **recessive mutation** causes the associated trait or disease. If the disease or trait does not always occur, even when the right number of mutated **genes** has been inherited, the Mendelian mutation has reduced **penetrance**.

microautophagy An **autophagy** process in which small volumes of the **cytosol** are **endocytosed** into **multivesicular bodies** being delivered to the **lysosome** for degradation.

microglial cell A **glial** cell involved in immune response and cleanup of extracellular debris in the central nervous system.

micrographia Writing with abnormally small letters, or with letters that become smaller the longer the writer writes, a common **Parkinson's disease** symptom.

microinjection Injecting fluid, containing engineered genetic material, for instance, into a cell by piercing the **cell membrane** with a tiny, thin, hollow glass tube.

microtubules Part of the internal **protein** scaffolding of the cell, which is important structurally and is used as a route for the transport of cellular cargo. Microtubules are made of **polymers** of the protein **tubulin**.

miglustat An **iminosugar** drug that inhibits the **enzyme** that makes **glucosylceramide**, used to treat **Gaucher disease**.

mitochondrial fission The process in which a **mitochondrion** splits into two daughter cells. The process can be asymmetrical, with one daughter inheriting more damaged components, including **DNA**, and thus more likely to be targeted for **mitophagy**.

mitochondrial fragmentation The situation that occurs when **mitochondrial fission** is favored over **fusion**.

mitochondrial fusion. The process in which two **mitochondria** merge. In the merged **mitochondrion**, there is a chance that **mutations** in the **DNA** will be repaired via **homologous recombination**, generating a healthier **mitochondrion**.

mitochondrial matrix The volume inside the **mitochondrial** inner **membrane**, where its **DNA** resides and where food molecules are processed to provide the **respiratory chain** with electrons and **protons**.

mitochondrion (pl. mitochondria, adj. mitochondrial) From Greek *mitos* "thread," and *khondrion* "little granule" (diminutive of *khondros* "coarse, granular"). Compartments inside cells, enclosed in a double **membrane**, which produce the chemical energy molecule **ATP** that fuel the cell's molecular machinery.

mitophagy A special form of **autophagy** that breaks down damaged **mitochondria**.

mixed beta-sheet A **beta-sheet** in which some of the strands run in the same direction and some in the opposite direction.

molecular modeling Using prediction methods and computer graphics to generate and visualize biomolecular structures and interactions.

monoclonal antibodies **Antibodies** with the same **amino acid** sequence, and thus the same **epitope** specificity.

motor cortex The region at the rear of the **frontal lobe** of the **neocortex** that controls voluntary motion.

MPP+ 1-**M**ethyl-4-**p**henyl**p**yridinium. Neurotoxin specific to **dopaminergic** neurons of the **substantia nigra**. Product of the enzyme **MAO-B** acting on **MPTP**.

MPPP 1-**M**ethyl-4-**p**henyl-4-**p**ropionoxy**p**iperidine. Also called desmethylprodine. An opioid similar to Demerol.

MPTP 1-**M**ethyl-4-**p**henyl-1,2,3,6-**t**etrahydro**p**yridine. Precursor of the neurotoxin.

MPP+ A molecule that can be an accidental by-product of **MPPP** synthesis.

mRNA **m**essenger **RNA** (see **gene**)

multivesicular body (MVB) A large **vesicle** with many smaller **vesicles** within formed by **endocytosis**.

mTOR **M**echanistic **T**arget **of R**apamycin. A **protein** that is inhibited by **rapamycin** and part of the complexes **mTORC1** and **mTORC2**.

mTORC1 **mTOR** **C**omplex **1**. A **protein** complex that, along with the **protein RHEB**, is important for stimulating **macroautophagy**.

mTORC2 **mTOR** **C**omplex **2**. A **protein** complex involved in metabolism that includes **mTOR**.

MTS **M**itochondrial **T**argeting **S**equence. A sequence of 10–70 **amino acids** at the **N-terminus** of a **protein** that targets the **protein** to **mitochondria**, recognized by the TOM complex. The **N-terminal amino acids** of **alpha-synuclein** resemble an MTS.

mucolipidosis type IV A **lysosomal storage disease** caused by inheriting two defective *MCOLN1* **genes**, which codes for an ion channel **protein** that helps acidify **endosomes** being delivered to **lysosomes**.

multimer A structure composed of many identical subunits: proteins, for example. A structure with just one subunit is called a **monomer**; with two, it is a **dimer**; with three, a **trimer**; with four, a **tetramer**, and so on. A multimer with an undetermined but relatively small number of subunits is called an **oligomer**, whereas a multimer with the potential to have an unlimited number of subunits is a **polymer**. The words are formed from the roots *mono* "alone," *dis* "twice," *treis* "three," *tetteres* "four," *oligos* "few, little," *polys* "much," and *meros* "part," which are from Greek, and *multus* "many" from Latin.

muscle rigidity A condition in which involuntary, isometric muscle contractions keep body parts held in nonrelaxed positions, such as an extended arm. The condition is distinct from dystonia (a kind of muscle cramp), although dystonia is common in **Parkinson's disease**.

mutation In nature, the inadvertent changing, deletion, or addition of a base or bases in a **DNA** sequence. When this occurs in the coding region of a **gene**, the **mutation** can change an **amino acid** in the corresponding **protein**.

mutual information A measure of how similar the patterns of variation in two data sets are to each other, with an emphasis on more complex variations. See "Are we correlated?"

myelin The **lipid** sheath around **axons** composed primarily of the **sphingolipid sphingomyelin**. An **axon** with a myelin sheath is said to be myelinated.

myosin A **protein** that interacts with filaments of the **protein** actin, important for the internal scaffolding of the cell.

N-terminus The first **amino acid** of a **protein** chain. More specifically, the N-terminus is the **peptide backbone** nitrogen of the first amino acid which, instead of being bonded to a carbon atom of the preceding amino acid, is attached to three hydrogen atoms.

NAC **N**onamyloid-beta **C**omponent. The middle part of the **alpha-synuclein protein** sequence, first found as a component of an **amyloid-beta** plaque.

nanobodies A smaller, simpler **antibody** engineered from **antibodies** discovered in camelids and sharks.

necrosis From Greek for "dead." Cell death in a cell too disabled to carry out **apoptosis**.

neo The gene for resistance to neomycin and related antibiotics. Used in the generation of **knockout** (or **knockin**) mice. The gene codes for the enzyme amino 3′-glycosyl phosphotransferase.

neocortex The main part of the cerebrum, involved in higher-order brain functions such as sensory perception, cognition, generation of motor commands, and so forth.

neurofibrillary tangle Fibrillar, hyperphosphorylated, intraneuronal clumps of the **protein tau**, seen in several neurodegenerative diseases such as **Alzheimer's** and **Parkinson's diseases**.

neuromelanin A pigment molecule found in neurons that make **catecholamine neurotransmitters**

neurotransmitter Chemicals released by neurons into synapses that affect whether the attached neurons are stimulated to or suppressed from firing, depending on the particular neurotransmitter.

Niemann-Pick disease type A A **lysosomal storage disease** caused by inheriting two defective *SMPD1* **genes**, which codes for an **enzyme** that carries out a step in **sphingomyelin** degradation.

NMR **N**uclear **M**agnetic **R**esonance spectroscopy. A technique that measures the magnetic spin frequencies of atomic nuclei. It can be used to determine the structure of complex molecules, including **proteins**, and detect interactions between them. See "The nuclei sing."

noradrenaline A **neurotransmitter** made by particular neurons, including those of the **locus coeruleus**. Also called norepinephrine.

off-pathway Adjective describing an **oligomer** incapable of forming **amyloid,** as opposed to **on-pathway**.

olfactory bulb The part of the brain connected to the odor-sensing neurons of the nose.

olfactory ensheathing glia **Glial** cells that form a sheath around the **axons** of scent-detecting neurons.

olfactory epithelium The layer of cells at the top of the nasal cavity embedded with scent-detecting neurons.

oligomer (see **multimer**)

on-pathway Adjective describing an **oligomer** capable of forming **amyloid**, as opposed to **off-pathway**.

organelle Any specialized compartment inside a cell, often surrounded by a **lipid membrane**.

oxidative phosphorylation The process of using food molecules to fuel the synthesis of **ATP** via the **respiratory chain**.

parallel beta-sheet A **beta-sheet** in which the strands run in the same direction.

paraquat An herbicide with a chemical structure similar to **MPP+**. Paraquat exposure increases the risk for developing **Parkinson's disease**.

parietal lobe The top-rear part of the **neocortex**, behind the **frontal lobe** and above the temporal lobes to the side and occipital lobe to the rear. The parietal lobe receives and processes sensory information.

PARK7 The gene for the protein DJ-1. Mutated *PARK7* is a **recessive Mendelian mutation** for developing **Parkinson's disease**.

parkinsonism A condition in which one or more of the movement-related symptoms of **Parkinson's disease** is present but does not involve alpha-synuclein pathology.

Parkinson's disease A neurodegenerative disease of the brain in which at least two of the following movement-related symptoms are present: **resting tremor**, **bradykinesia**, **muscle rigidity**, and **postural instability**. The disease involves **alpha-synuclein** pathology at the cellular level.

pars cerebellaris loci coerulei Latin for the "cerebellar part of the blue place." A portion of the **cerebellum** adjacent to and with connections to the **locus coeruleus**.

passive immune therapy Treatment of a disease or condition by injection of **antibodies** produced outside the body.

penetrance The chance that a **dominant** or **recessive Mendelian mutation** causes the associated disease.

peptide A short **protein** or fragment of a **protein**, typically less than 50 **amino acids** long.

peptide backbone The part of the chemical structure of **amino acids** that is the same for all of them and in which each **amino acid** is chemically bonded to the preceding and following **amino acid**s to form the **protein** chain.

PH domain **P**leckstrin **H**omology domain. A common protein fold consisting of two stacked **beta-sheets** and a long **alpha-helix** packed together in a way that resembles a wedge, like a guitar pick; hence the name pleckstrin from the Greek word for "harp pick," *plektron*.

phasic firing When a neuron receives sufficient excitatory signals, it fires strongly, releasing more **neurotransmitters** than in **tonic firing**.

phenotype The trait caused by the two copies of a **gene**, which might differ owing to **mutations** or **polymorphisms**, which is called the **genotype**. From Greek *phainein* "to show" and *tupos* "impression, type."

phospholipid A **diglyceride** with a phosphate group attached to the glycerol.

phosphorylation The chemical process of adding a phosphoryl group (a phosphorus atom with three oxygen atoms attached) to a molecule. In **proteins**, the process is carried out by enzymes called **kinases**.

PiB **Pi**ttsburgh compound **B**. A radioactively labeled compound that binds to **amyloid** plaques, used to visualize Alzheimer's brain pathology in living patients.

PINK1 The gene for the protein **PTEN-In**duced putative **K**inase **1** (PINK1). Mutated *PINK1* is a **recessive Mendelian mutation** for developing Parkinson's disease.

plasma cell A transformed **B cell** that has recognized its **epitope**, multiplying and releasing **antibodies** recognizing the **epitope**.

plasma membrane (see **cell membrane**)

pluripotent stem cell A cell that can made to differentiate into several cell types.

point mutation A mutation in a **gene** that causes the change of an **amino acid** in the corresponding **protein** sequence.

polygenic trait A trait affected by many **genes**.

polymer (see **multimer**)

polymorphism A **mutation** that has little or no impact on the survival or reproduction of an organism.

postural instability Difficulty in making corrections to one's posture to regain balance after a perturbation, such as someone unexpectedly bumping into you.

presynaptic boutons *Bouton* is French for "button." The nerve terminal connections at the end of the branches of the **axonal arbor**, which release **neurotransmitters** when the neuron fires into the **synaptic cleft**.

prion (PREE-on) A misfolded, toxic **protein** that, when it encounters correctly folded copies of the same **protein**, causes them to adopt the misfolded, toxic form.

PRKN The gene for the **protein** parkin. Mutated *PRKN* is a **recessive Mendelian mutation** for developing **Parkinson's disease**.

proteasome A large **protein** complex that degrades **proteins** that have been recognized as misfolded or damaged by the cell's **protein** quality control machinery.

protein A complex molecule consisting of a chain of **amino acids** encoded by a **gene**. **Proteins** make up most of the important machinery necessary for cell functioning.

protein primary structure Lowest level in the hierarchical description of **protein** structure, consisting of the sequence of **amino acids** along with their individual chemical structures.

protein quaternary structure The three-dimensional structure of a complex of **proteins**.

protein secondary structure The structural elements, such as **alpha-helices**, **beta-sheets**, and **loops** that make up the structure, or **fold**, of a **protein**.

protein tertiary structure The three-dimensional structure, or **fold**, of a **protein**.

protein threading A technique for predicting the three-dimensional **fold**, or **tertiary structure**, of an unknown **protein** based on its similarities to proteins with known **folds**.

proton The nucleus of a hydrogen atom. Also a component of the nuclei of all other elements.

PrPsc Prion **P**rotein in its **Sc**rapie form. A prion that causes **scrapie** in sheep, **BSE** (bovine spongiform encephalopathy, more commonly known as "mad cow disease") in cows, and **vCJD** and **kuru** in humans. All of these diseases are **TSEs**.

PSP Progressive **S**upranuclear **P**alsy. A **tauopathy** that includes **parkinsonism**.

pull-down assay An experiment in which a **protein**, or protein fragment, is attached to a synthetic bead, and these beads exposed to cell extract. The beads are then isolated, and the cellular **proteins** that bind to the attached **protein** or protein fragment are identified.

putamen From Latin for "husk, nutshell." Part of the dorsal **striatum**.

Rab A family of **GTPase** proteins involved in vesicle trafficking, including Rab1 and RAB7L1.

rapamycin (see **sirolimus**)

rasagiline An **MAO B** inhibitor drug used to treat **Parkinson's disease**.

RBD **R**apid eye movement (REM) sleep **B**ehavior **D**isorder. A brain disorder in which patients move during REM sleep, acting out movements in their dreams.

reactive oxygen species Molecules whose oxygen atom(s) can easily form chemical bonds with other biomolecules, damaging them.

recessive mutation A **mutation** in which one must inherit two copies of the mutated **gene**, one from each parent, to get the associated trait or disease.

respiratory chain Also called the **electron transport chain** (**ETC**). The **protein** complexes I—IV at the inner **membrane** of **mitochondria** that shuttle electrons and **protons** during **oxidative phosphorylation**.

resting tremor Rhythmic shaking of a body part, typically fingers, hands, and/or arms in **Parkinson's disease**, when the body part is relaxed (i.e., at rest).

retrorubral area Also called retrorubral nucleus or field. Retrorubral is from Latin for "behind the red [area]." Part of the **basal ganglia** that supplies the **globus pallidus** with **dopamine**.

RHEB **R**as **H**omology **E**nriched in the **B**rain. A **GTPase protein** similar to the **Rab proteins** that helps recruit **mTORC1** to the **lysosomal** surface.

risk factor In epidemiology, an environmental or genetic factor that increases or decreases the chance of developing a disease.

RNA **R**ibo**n**ucleic **A**cid. Like **DNA** but with an extra oxygen atom in its ribose sugar ring, and with the same bases except instead of thymine, the closely related base uracil is used. Unlike **DNA**, RNA is typically single stranded, often forming complementary base pairs between different parts of the same strand.

ROC **R**as **o**f **C**omplex proteins**.** **GTPase** domain and part of the **LRRK2 protein**, involved in regulating the **LRKK2 kinase** domain.

rotenone A pesticide that inhibits **complex I**. Rotenone exposure increase the risk for developing **Parkinson's disease**.

saddle point The special place on a surface where along one direction, both paths lead down, but on the path crossing roughly perpendicular the same point, both paths lead up.

salt bridge A structural element where two oppositely charged chemical groups interact.

sargramostim (see **GM-CSF**)

sciatic nerve One of the main nerves to the leg, mostly innervating below the thigh.

sensory cortex The region at the front of the parietal lobe of the **neocortex** that receives sensory information from the body.

side chain The part of an **amino acid** that branches off from the **peptide backbone** and which comes in 20 forms, one for each type of **amino acid**.

sirolimus Also known as rapamycin. An immunosuppressive drug that is being studied as a possible treatment for **Parkinson's disease**.

size exclusion chromatography A purification technique in which a solution is flowed through a long tube, or column, filled with a gel. The speed at which the various components of the solution, different **proteins**, and so on, depends on molecule size; the bigger the molecule, the faster it flows through.

SNARE complex A complex of **proteins**, including **synaptobrevin-2 (VAMP-2)**, which, when triggered by calcium binding to the **protein synaptotagmin**, cause **vesicles** to fuse with **lipid membranes**, such as the **cell membrane**, releasing their fluid contents.

SNc **S**ubstantia **N**igra pars **c**ompacta. Latin for "black substance, compact part," so named owing to the high amount of the dark pigment **neuromelanin** in the densely packed neurons of this region. Part of the **basal ganglia**. The **axons** of neurons in SNc extend to the **striatum**, releasing **dopamine** there. The loss of SNc neurons causes most of the movement-related symptoms of **Parkinson's disease**.

SNCA The **gene** for alpha-synuclein. **Polymorphisms** outside the coding regions of *SNCA* are **risk factors** for **Parkinson's disease**.

SNP **S**ingle **N**ucleotide **P**olymorphism, pronounced "snip," corresponding to a position in the genome sequence where the DNA base can vary among individuals.

solid-state NMR **NMR spectroscopy** of samples in a solid or gel state. Technically much more challenging than solution-state NMR.

soma (see **cell body**)

sphingolipid A **lipid** with one or two fatty acid chains, in which the second fatty acid is connected via a peptide bond. "Sphingo" come from sphinx, the mythological creature, because when sphingolipids were first discovered, their biological role was enigmatic.

sphingomyelin A sphingolipid found in **myelin**.

sphingosine A **sphingolipid** that has just one fatty acid chain.

standard error The amount a mean value is expected to vary when calculated from a random subset of a particular size compared with the mean for the full data set.

stereotactic frame A frame fixed to the skull that enables neurosurgeons to direct surgical instruments more precisely to regions deep in the brain.

stereotactic surgery Brain surgery using a **stereotactic frame**.

STN **S**ub**t**halamic **N**ucleus. A region in the **basal ganglia** located beneath **thalamus** involved in the **indirect pathway**.

striatum Latin for "grooved." Part of the **basal ganglia** that receives signals from the **neocortex** involved in voluntary motion in its dorsal (in this case meaning "top") section, social interactions, and other behavior in its central and ventral (in this case meaning "bottom") sections. In primates, the dorsal striatum is divided into the **caudate** and the **putamen**.

superhelical twist The bending of the axis of a helix that occurs when forcing its helix structure to be wound tighter or looser than average.

synapse A connection between neurons in which, upon the firing of the presynaptic neuron, **neurotransmitters** are released into the **synaptic cleft**, to be bound by receptors on the postsynaptic neuron. The **neurotransmitters** then either stimulate or suppress firing of the postsynaptic neuron, depending on the particular **neurotransmitter**.

synaptic cleft The small, extracellular space between the presynaptic and postsynaptic regions of a **synapse**.

synaptic vesicle A small **vesicle** containing **neurotransmitters**.

synaptobrevin (VAMP-2) (see **SNARE complex**)

synaptotagmin A **protein** that, when bound to calcium, triggers the fusion of **synaptic vesicles** to the **cell membrane** of the **synapse**.

synucleinopathy Any neurodegenerative disease that involves synuclein-related pathology at the cellular level, such as **Parkinson's disease** and **DLB**.

synucleins (sih–NOO–clee-ins) A family of **proteins** involved in neural function, modulating **neurotransmitter** release. Three synucleins are found in humans: alpha, beta, and gamma. Alpha-synuclein is associated with **Parkinson's disease**.

T cells A class of **adaptive immune system** cells that respond to particular **peptides** from pathogens displayed on the surface of cells.

tau A neuronal **protein** that stabilizes and modulates cellular traffic on **microtubules** in **axons**. It can aggregate to form **neurofibrillary tangles**, pathological intracellular **protein** deposits present in the **tauopathy** diseases, **Alzheimer's disease**, and, to a lesser extent, **Parkinson's disease**.

tauopathy Any neurodegenerative disease in which the predominant cellular-level pathology involves the protein **tau**. Whereas **Alzheimer's** and **Parkinson's disease** can involve **tau**-related pathology, other pathology is predominant, so these diseases are not considered tauopathies.

Tay-Sachs disease A **lysosomal storage disease** caused by inheriting two defective *HEXA* **genes**, which codes for an **enzyme** that carries out a step in **ganglioside** degradation.

TDP-43 TAR DNA-binding Protein **43**. A **DNA**-binding **protein** involved **gene** regulation, also involved in the pathology of **ALS** and **LATE**.

tetramer (see **multimer**)

TFEB Transcription Factor **EB**. A **protein** that increases expression of **genes** that stimulate **macroautophagy**. When **phosphorylated** by **mTORC1**, TFEB stays in the **cytosol** and **macroautophagy** is not stimulated.

tk The **gene** for thymidine kinase, an **enzyme** that converts antiviral drugs such as ganciclovir into their toxic forms. Used in the generation of **knockout** (or **knockin**) mice.

TOM20 and TOM40 Proteins in the TOM (Translocase of the Outer Membrane) complex. The TOM complex transports **proteins** through the outer **membrane** of **mitochondria**. TOM20 helps to recognize the **N-terminal mitochondrial-targeting sequence** (MTS) of the protein being transported, and TOM40 forms a pore through the **membrane**.

tonic firing Steady firing of a neuron, releasing a moderate amount of **neurotransmitters**.

toxic gain-of-function mutation A **mutation** that makes the mutated **protein** toxic to cells.

transgenic Adjective for an organism with foreign **DNA** incorporated into its **genome**.

transorbital lobotomy A **frontal lobotomy** performed by piercing the bone at the rear of the eye socket with an ice pick to a particular depth, which was then swiveled to sever the **white matter**.

trehalose A sugar molecule that can stimulate **macroautophagy**.

triglyceride A **lipid** (fat molecule) composed of three hydrocarbon chains attached to a glycerol (a small sugar, also called glycerine).

trimer (see **multimer**)

TSE Transmissible Spongiform Encephalopathy. Any disease spread via **PrPsc**.

tubulin (see **microtubules**)

two-hybrid system A technique for detecting interactions among different **proteins**. In this technique, the DNA binding domain of the **protein** GAL4 is attached to one **protein**, the "bait," and the activating domain of GAL4 is attached to the other **protein**, the "prey." If the bait and prey interact in the yeast cell, the activator domain will be in near proximity to a reporter **gene** next to the GAL4 DNA binding site, causing it to be transcribed, ultimately leading the reporting **protein** to be made, which is then detected.

ubiquinone The molecule that shuttles electrons and **protons** between **complexes I** and III of the **respiratory chain**.

ubiquitin A **protein** found in nearly every cell type, which, when attached to **proteins** in particular way, directs them to the proper cellular machinery, such as the **proteasome**.

UPS **U**biquitin-**P**roteasome **S**ystem. One of the major pathways for **protein** degradation in the cell, in which **proteins**, typically damaged or misfolded, are attached to a string of **ubiquitin proteins**, which then direct the **protein** to the **proteasome**.

vagus nerve From Latin *vagus* "wandering." A nerve that originates in the **brain stem** and extends down, branching to the organs of the body down to the intestines.

vasogenic edema Also called cerebral edema. Excessive fluid inside and around cells of the brain owing to breakdown of the **blood–brain barrier**.

vCJD **V**ariant **C**reutzfeldt-**J**akob **D**isease. A fatal neurodegenerative disease that can be contracted by eating beef containing the **prion PrPsc**.

vesicle A microscopic, fluid-filled sac enclosed by a **lipid membrane**.

VTA **V**entral **T**egmental **A**rea. Also called the ventral tegmentum, in which tegmentum is Latin for "covering." Neurons in this area supply **dopamine** to the brain's pleasure and reward circuitry.

white matter In the brain, regions consisting primarily of **myelinated axons**.

wild type The unmutated, normal form of a **gene** and the corresponding **protein**.

X-ray crystallography A technique using x-ray diffraction to determine the structure of molecules making up a crystal. See "Crop rows."

Z-value analysis An analysis that measures how unique a particular pattern of variation in data set is compared with all of the other patterns of variation.

Index

Note: 'Page numbers followed by "*f*" indicate figures and "*t*" indicate tables'.

Q

R

S

Kolibri 360 GmbH
Ferdinand-Jühlke-Straße 7
99095 Erfurt, Deutschland
produktsicherheit@kolibri360.de